Building Physics, Structural and Safety Engineering

Building Physics, Structural and Safety Engineering

Editors

Xin Ren
Yaolin Lin
Teng Shao

MDPI

Basel • Beijing • Wuhan • Barcelona • Belgrade • Novi Sad • Cluj • Manchester

Editors

Xin Ren
Nanjing Tech University
Nanjing
China

Yaolin Lin
University of Shanghai for
Science and Technology
Shanghai
China

Teng Shao
Northwestern Polytechnical
University
Xi'an
China

Editorial Office
MDPI
St. Alban-Anlage 66
4052 Basel, Switzerland

This is a reprint of articles from the Special Issue published online in the open access journal *Buildings* (ISSN 2075-5309) (available at: https://www.mdpi.com/journal/buildings/special_issues/physics_struc_safe).

For citation purposes, cite each article independently as indicated on the article page online and as indicated below:

Lastname, A.A.; Lastname, B.B. Article Title. *Journal Name* **Year**, *Volume Number*, Page Range.

ISBN 978-3-7258-1079-6 (Hbk)
ISBN 978-3-7258-1080-2 (PDF)
doi.org/10.3390/books978-3-7258-1080-2

Contents

About the Editors

Xin Ren

Xin Ren is a Full Professor at the College of Civil Engineering, and the Chair of the Department of Mechanics and Engineering Materials, Nanjing Tech University. In both 2022 and 2023, he was included in the list of the Top 2% Highly Cited Researchers worldwide by Stanford University. He has published more than 70 high-level papers, including nearly 50 JCR-Q1 papers as the first/corresponding author and 15 "ESI Highly Cited Papers", and has a single Google Scholar academic highest citation number of more than 800. He has received more than 3500 citations, and his H-Index is 29. He serves as an editorial board member, youth editorial board member, or guest editor for several international journals such as *Engineering Structures, Thin-Walled Structures, Defence Technology*, and *Mechanics of Advanced Materials and Structures*.

Yaolin Lin

Dr. Yaolin Lin is an associate professor at Shanghai University of Science and Technology. Before that, he worked as an associate professor at Wuhan University of Technology, and as a principle engineer at a leading building energy efficiency consulting company in California, USA. He received his PhD degree in building engineering from Concordia University in Canada. He has over 7 years of industrial experience in building HVAC controls, building energy management, retro-commissioning building systems, and advanced controller development, and over 10 years of research experience on green and sustainable buildings, HVAC system control optimization, and building energy system modeling and simulation, leading over 50 industrial projects across five different countries, including the USA, Canada, France, Singapore, and China. He has published over 90 journal papers, book chapters, and peer-reviewed conference papers on building energy efficiency and simulation. Dr. Lin is an associate editor of the international journal of *Frontiers in Built Environment*, a topic editor for the international journal of *Energies*, a leading guest editor for the international journal of *Energy Exploration & Exploitation*, and a guest editor for the international journal of *Sustainability*. He serves as an editorial member for the SCI journals *Buildings* and *Advances in Meteorology* and for the EI journal *Energy Engineering*. He is a reviewer for over 40 renowned international journals, such as *Energy, Applied Energy, Energy and Buildings, Building and Environment*, and *Renewable Energy*.

Teng Shao

Teng Shao is an Associate Professor, Master's supervisor, and vice-head of the Architecture Department at Northwestern Polytechnical University. His research interests include green building and building energy efficiency, urban and building physical environment, building multi-objective performance optimization, and solar energy utilization in buildings. He has presided over or participated in fifteen national and provincial research projects. He has published more than forty academic papers, co-authored two national key books, and co-edited a national "13th Five-Year Plan" textbook. He has participated in drafting six association and local standards, authorized five patents, and obtained four computer software copyrights. His awards include the China Award for Science and Technology in Construction (first prize), the Shaanxi Province Award for Science and Technology (third prize), and the Heilongjiang Province Award for Science and Technology in Urban and Rural Construction (first prize).

buildings

MDPI

Article

An Approach to In Situ Evaluation of Timber Structures Based on Equalization of Non-Destructive and Mechanical Test Parameters

Andrej Josifovski [1,*], Nebojša Todorović [2], Jelena Milošević [1], Marko Veizović [2], Filip Pantelić [3], Marina Aškrabić [4], Miomir Vasov [5] and Aleksandar Rajčić [1]

1 Faculty of Architecture, University of Belgrade, 11000 Belgrade, Serbia; jelena@arh.bg.ac.rs (J.M.); rajcic@arh.bg.ac.rs (A.R.)
2 Faculty of Forestry, University of Belgrade, 11000 Belgrade, Serbia; nebojsa.todorovic@sfb.bg.ac.rs (N.T.); marko.veizovic@sfb.bg.ac.rs (M.V.)
3 School of Electrical and Computer Engineering, Academy of Technical and Art Applied Studies, 11000 Belgrade, Serbia; filipp@viser.edu.rs
4 Faculty of Civil Engineering, University of Belgrade, 11000 Belgrade, Serbia; amarina@imk.grf.bg.ac.rs
5 Faculty of Civil Engineering and Architecture, University of Niš, 18000 Niš, Serbia; miomir.vasov@gaf.ni.ac.rs
* Correspondence: andrej@arh.bg.ac.rs

Citation: Josifovski, A.; Todorović, N.; Milošević, J.; Veizović, M.; Pantelić, F.; Aškrabić, M.; Vasov, M.; Rajčić, A. An Approach to In Situ Evaluation of Timber Structures Based on Equalization of Non-Destructive and Mechanical Test Parameters. *Buildings* **2023**, *13*, 1405. https://doi.org/10.3390/buildings13061405

Academic Editors: Xin Ren, Yaolin Lin and Teng Shao

Received: 20 April 2023
Revised: 13 May 2023
Accepted: 19 May 2023
Published: 29 May 2023

Abstract: This paper addresses the challenges in evaluating the structural performance of built structures using non-destructive methods and in situ tests. Such an examination of structural properties, without their sampling, is a diagnostic improvement, especially for historical heritage buildings, where it is not allowed to violate their physical integrity. Therefore, the research proposes a non-destructive testing method based on the equalization of the mechanically and non-destructively determined parameters of the strength of built-in timber. The research included three phases: (1) a preliminary examination; (2) a calibration procedure of the non-destructive method, and (3) in situ application of the established non-destructive method. The preliminary examination involved testing specimens using X-rays and ultrasonic waves by directing them, analogous to mechanical testing, in the fibers' longitudinal, radial, and tangential directions. In the second phase, it was shown that equalizing the parameters of mechanical and non-destructive testing using ultrasound and X-rays of timber was feasible. Furthermore, mechanical calibration was conducted to establish an applicable non-destructive in situ method. Finally, in the third phase, an in situ assessment of timber architectural elements confirms the effectiveness of the suggested non-destructive approach in diagnosing architectural structures.

Keywords: building diagnostics; in situ testing; non-destructive testing (NDT); ultrasonic testing (UT); X-ray imaging; structural performances; timber structures; physical and mechanical properties

1. Introduction

Preserving existing structures and protecting architectural heritage has received significant attention as part of the worldwide goal of sustainable development. In situ diagnostics of heritage buildings require special care and consideration due to their cultural and historical significance. Consideration of diverse approaches to evaluating the performances of these buildings represents the focus of various papers [1–6], which provide insight into the different in situ diagnostics methods. When performing diagnostics on historical heritage buildings, it is essential to use non-destructive testing (NDT) methods [7] that do not cause damage to the building's original materials. It is also essential to work with experts in conservation and restoration to ensure that any necessary repairs are made with sensitivity to the building's original character.

Generally, in situ building diagnostics refer to the process of evaluating the condition and performance of a building using NDT methods. The diverse methods used in building inspection (including thermography, air tightness testing, acoustic testing, electrical testing, and water testing) can provide valuable information about the condition and performance of a building and can be used to identify areas for improvement and optimize efficiency.

In situ examination of the structural performances of timber architectural elements without their sampling, focused on by this paper, represents a diagnostic improvement, especially when it is not allowed to damage the physical integrity of buildings, such as in the case of historical heritage buildings [8]. Application of NDT for evaluating structural integrity and properties of timber buildings without causing damage includes diverse methods such as:

- Visual Inspection (VI)—a method that involves a thorough visual examination of the timber to identify any visible defects or damage. VI can include using magnifying lenses, borescopes, endoscopes, and other specialized equipment to access hard-to-reach areas.
- Impact Echo Tests (IET)—a method that involves striking timber with a hammer and measuring the resulting vibrations. The data can be used to evaluate the thickness and stiffness of the wood, as well as identify any internal decay or insect damage.
- Drill Resistance Tests (DRT) involve drilling a small-diameter hole into the timber and measuring the resistance to drilling. DRT can be used to evaluate the internal condition of the wood, including density and moisture content, as well as identify any decay or insect damage.
- Infrared Thermography (IRT)—a method that uses infrared cameras to detect differences in temperature on a building's surface. IRT can be used to identify areas of heat loss, air leakage, and moisture intrusion.
- Ultrasonic Tests (UT)—a method that uses high-frequency sound waves to detect cracks, voids, and other flaws in timber. UT can be used to evaluate the stiffness and density of the wood, as well as identify any internal decay or damage.
- X-ray Imaging—a method that uses X-rays to create an image of the internal structure of timber. X-ray imaging can be used to identify the presence of knots, cracks, and decay in the wood.

NDT methods, further discussed in [9], can effectively evaluate the condition of timber buildings without causing damage. However, it requires specialized training and equipment to be performed accurately.

The relevance of condition evaluation and ongoing monitoring of the existing timber structures has been discussed by various authors. For example, Cruz et al. [10] present guidelines for the on-site assessment of historic timber structures covering principles and possible approaches for the safety assessment of old timber structures of historical relevance that could be used as the basis for possible European Standards. On the other hand, Nowak et al. [11] present a survey of the state of the art of applying DRT methods as diagnostics techniques for testing timber structures and examples of their applications. Moreover, the paper by Gomes et al. [12] presents the assessment and diagnosis of two collar timber trusses using visual grading and NDT.

Furthermore, Stepinac et al. [13] present various methods for assessing the condition of the existing timber structures through the example of condition assessment of the Nikola Tesla Technical Museum in Zagreb. Conversely, De Matteis et al. [14] propose a design procedure for the structural assessment of ancient timber structures using resistographic NDT and apply this procedure to the case study of the Croce di Lucca Church in Naples. Significant structural flaws in the system are discovered when the roof structure's carrying capability is assessed using the suggested experimental methods, particularly regarding the bottom tie beam.

Moreover, Wang et al. [15] propose a comprehensive evaluation method for historical timber structures and apply it to assess a typical historical Chinese timber structural building named the Fujiu Zhou house. Furthermore, Cruz et al. [16] propose joint use of non-destructive techniques, including acoustic emission, the elastic weave technique, and finite element numerical modeling, to make multi-feature decisions about repairing or replacing elements. They evaluate the effectiveness of the proposed approach on an old beam floor from a historic building in the center of Granada, Spain.

On the other hand, Martinez and Martinez [17] describe the use of infrared thermography (IRT) for the NDT assessment of a historic timber-roofed building in Madrid, Spain. Moreover, Nasir et al. [18] critically review diverse NDT methods, including spectroscopy, stress waves, guided wave propagation, X-ray computed tomography, and thermography, and introduce the concept of acoustic emission (AE) and its experimentation and analyze its possible application. Finally, Xin et al. [19] present an approach for evaluating ancient timber members' density and mechanical properties that combines NDT, color parameters, and machine learning (ML).

A significant portion of historical architecture comprises timber floors and roofs, which must withstand vertical loads and adhere to safety regulations set forth by modern building codes. However, because it is biodegradable, wood is highly prone to harmful environmental factors, including xylophage attacks and persistently high humidity. Therefore, periodic assessments, preferably non- or minimally invasive, are required to protect old timber structures. Evaluating material properties is one of the critical steps in these assessments. Unfortunately, the previous task is not a simple one since (1) timber is a complex inhomogeneous material, (2) defects have a significant impact on a structural element's total bearing capacity, and (3) in most cases, it may not be possible to perform comprehensive destructive testing.

Starting from the premise that it is possible to equalize the mechanical and non-destructive parameters of wood testing, the paper investigates whether the mechanical potentials of load-bearing timber beams may be immediately evaluated in situ with the stress-sound test techniques and their stability determined. The research aims to propose a reliable, non-destructive method based on equalizing the mechanical parameters of static resistance to pressure [20] and its extrapolation of the derived resistance to bending [21], the most common and most pronounced in timber structures in general [22], with the parameters of non-destructive methods, on the same standard profiled test samples. An in situ assessment of timber architectural elements of the roof structure of the Technical Faculties Building in Belgrade, Serbia, confirmed the proposed method's effectiveness.

2. Methodologies

The research was planned and carried out in three phases (Figure 1), through a preliminary examination, calibration procedure of the non-destructive method and, finally, the immediate in situ application of the established non-destructive method.

2.1. Preliminary Examination

As wood is a heterotopic and orthotropic material [23], preliminary research was carried out using X-rays and ultrasonic waves. Analogous to mechanical testing, X-rays and ultrasonic waves were directed in the longitudinal, radial, and tangential directions of the mechanical fibers of the test samples.

The preliminary examinations showed that the achieved equalization of the mechanical parameters of static resistance to pressure and its extrapolation in the bending resistance of wood, with the parameters of ultrasound propagation speed and X-ray image brightness intensity, reflected the mechanical characteristics of a certain type of wood with identical regularity in relation to the standard anatomical directions of examination [24]. That the equalization of the parameters of mechanical and non-destructive testing of wood is a realistically feasible procedure is also confirmed by the fact that the results of research, obtained by approximating the mechanical parameters of the tested wood samples, by taking the highest mean table values, exhibit static resistance to pressure and bending, of samples of diffuse porous hardwoods, ring porous hardwoods and softwoods [25].

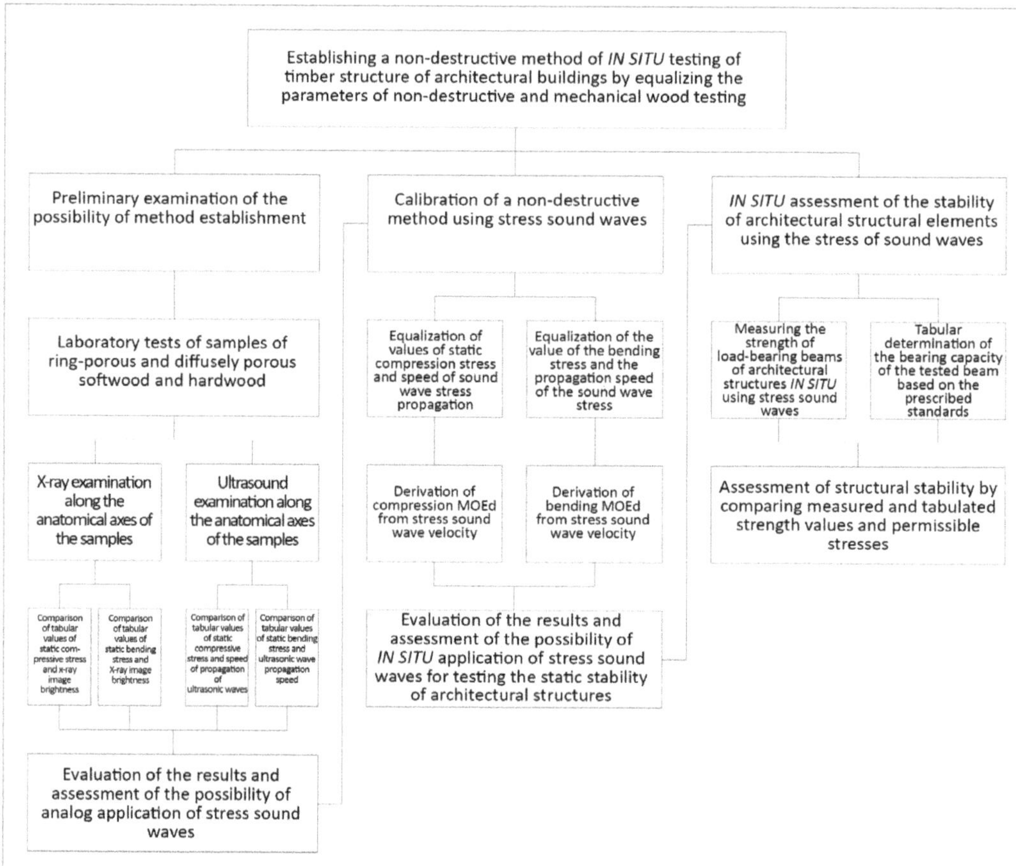

Figure 1. Flowchart of the in situ evaluation of timber structures.

The fastest propagation of ultrasonic waves and the highest brightness of the X-ray image, completely analogous to the results of the static pressure resistance test, were obtained by directing the waves in the longitudinal direction of the flow of the mechanical fibers of the test samples, with their gradual reduction in the radial and tangential directions [26]. A similar regularity was also manifested for certain tree species, where the velocity of propagation of ultrasonic waves and the brightness of the X-ray image, in proportion to the table values of static resistance to pressure [27], were the highest in diffusely porous hardwood, and proportionally lower in annular porous hardwood and softwood [28] (p. 32).

Preliminary studies were carried out on standardized laboratory samples for testing the static resistance of wood to pressure, which are small in size, which enabled the principal confirmation of the functioning of the method, but due to the physical properties of scattering and weakening of ultrasonic and X-ray waves, proportional to the length of the path, their application in the third phase of testing structural elements, such as supporting wooden beams, was practically unfeasible. Taking into account the physical properties of rectilinear oscillatory sound wave propagation [27], and its negligible attenuation along the traveled path, the analogy with the methodology of conducting preliminary research could be fully achieved, with the fact that for the establishment of a non-destructive in situ method and its application, it was necessary to carry out its proper mechanical calibration.

The process of calibrating the method of applying stress sound waves should therefore be performed by directing the sound waves at right angles, analogous to the direction of static pressure forces on standard laboratory test tubes, in the longitudinal, radial, and tangential direction of the flow of mechanical fibers and compare the bending resistance values obtained by calculation [29] and by applying mechanical compression force. In this phase of the test, it was necessary to make a certain modification of the calibration procedure, because the application of stress sound waves on small test tubes for mechanical tests, as was done with X-ray and ultrasound waves in preliminary research, was not technically feasible.

2.2. Calibration of Non-Destructive Method

Calibration of the stress-sound method was performed by mechanical testing of static bending and compression stress, on standard hardwood and softwood test samples, cut from boards which were previously tested with stress-sound waves. To additionally check the correctness of the procedure, the same set of test samples from the preliminary research was not used in this test. The test samples were cut exactly from those parts of the boards through which the stress-sound wave was previously directed, so that the mechanical calibration of the method was performed on the same structure through which the sound wave went.

To equalize the test conditions on the test samples, before measuring the compressive stress, the equilibrium moisture content (EMC) and density were determined [28] (p. 190). Before the sound testing, the moisture level of the boards, which had been in the room at room temperature until that moment, was checked with a hygrometer.

The procedure of stress-sound testing is in principle similar to the one performed in preliminary research using ultrasonic waves. The values of static compressive stress obtained by mechanical testing of test samples were close to the above table values for the given type of wood [29]. Based on these values, the values of the static bending stress were determined by the relationship established between the complex bending stress and the resultant compressive and tensile stresses (compressive stress: bending stress: tensile stress; 1:2:2.8—hardwood; 1:1.5:2.5—softwood) [24]. Later, direct examination of the bending stress yielded values similar to the calculated values, which was also a confirmation of the validity of the conclusions reached in the premillennial research. The results of the mechanical calibration of the stress-sound wave method are presented in the given tables.

2.2.1. Determination of Wood Density

In this test, the wood density was calculated according to the given formula [28] (p. 213) in relation to 12% EMC of previously dried test tubes in the chamber (relative humidity—65 ± 5%; temperature 21 ± 2 °C). The test samples were previously dried because they are in a relatively stable relationship within the limits of up to 20% EMC (Kollman) [30]. The mass (m) of the test sample of each type of wood was formally measured with an electronic scale and was expressed in grams (g), and its volume was calculated according to the mathematical formula for a prismatic geometric body. The previous procedure is important for in situ testing, because within these limits, the mechanical potential of the wood does not change significantly with a change in moisture content [28] (p. 190).

2.2.2. Determination of Wood Moisture Content (MC)

Although the test samples were previously dried in the chamber, the EMC was checked immediately before their calibration, so that the stress-sound test and the determination of the static pressure stress were performed under the same conditions. EMC was determined only in standard cut test tubes so that these data could also be used as a benchmark when testing the elements of wooden structures with stress-sound waves, in the third phase of the research.

The determination of the MC of test samples and elements of wooden constructions was carried out using an electric hygrometer (SRPS EN 13163-2) because in the standard static calculations of elements of wooden architectural structures, the degree of hygroscopic MC is taken in the range of 7–25%, where this device is very reliable (Table 1) [28] (p. 212).

The use of this hygrometer was also convenient because the decrease in the measuring electrical resistance of the device, by increasing the level of humidity in the range from absolutely dry to the point of fiber saturation point and, vice versa, its increase, do not require the introduction of a correction in relation to the type of wood and its density.

Table 1. Values of hygroscopic humidity and density of test samples determined in the calibration procedure of the stress sound wave method.

	Softwood					Hardwood					
						Ring Porous				Diffuse Porous	
	Spruce 1	**Fir 1**	**Fir 2**	**Pine 1**	**Pine 2**	**Oak 1**	**Oak 2**	**Black Locust 1**	**Black Locust 2**	**Beech 1**	**Beech 2**
Axial [mm]	30.19	28.94	28.81	29.24	28.66	28.94	29.32	29.18	28.08	29.42	29.35
Radial [mm]	19.77	19.82	19.75	20.05	19.99	19.93	19.92	19.74	20.12	19.95	19.75
Tangential [mm]	19.73	19.78	19.81	20	19.84	19.83	19.94	19.9	20.35	19.94	19.81
Mass [g]	5.64	4.72	4.82	7.84	6.45	7.73	9.55	8.55	8.67	9.1	8.02
Axial [mm]	30.08	28.92	28.81	29.18	28.58	28.94	29.24	29.14	28.03	29.35	29.28
Radial [mm]	19.33	19.39	19.36	19.61	19.77	19.65	19.68	19.37	19.72	19.66	19.49
Tangential [mm]	18.89	19.31	19.29	19.56	19.53	19.47	19.49	19.33	19.78	19.35	19.41
Mass [g]	4.92	4.33	4.42	7.19	5.91	7.08	8.81	7.86	8.02	8.4	7.41
MC [%]	10.9	9.0	9.0	9.0	9.1	9.2	8.4	8.8	8.1	8.3	8.2
Oven-dry density [g/cm^3]	0.448	0.400	0.411	0.642	0.536	0.639	0.786	0.720	0.734	0.752	0.669

2.2.3. Determination of Compression and Bending Stresses

For this test, all test samples were of the normalized parallelopiped shape and standard dimensions SRPS ISO 13061. The calibration samples were cut in such a way that their longest side was parallel to the grain so that the plane that cut the sample longitudinally, at a right angle, parallel to the lines of the drawing, corresponded to the axial plane. Sections made by imaginary planes cutting two longitudinal adjacent sides at right angles corresponded to the radial, i.e., tangential, plane of the test sample. The test samples were cut in such a way that a corresponding projection of part of the growth ring of the wood was painted on their sides, which showed the direction of the flow of mechanical fibers, i.e., the plane of their cutting.

Calibration of test sample by applying pressure force was performed, as in the preliminary research, by directing it parallel and perpendicular to the direction of the fiber in the radial and tangential direction, as for the directions of ultrasound waves and X-rays. For the mechanical test of compressive stress, in parallel with the extension of mechanical fibers, the surface of the presser, according to the standard method, covered the entire surface of the upper side of the vertically placed test sample. When testing the compressive stress perpendicular to the grain, the presser covered the middle third of the longitudinal side of the test sample in the radial and tangential direction. The increase in the intensity of the pressure force was set to last a maximum of 5 min, and the strength of the pressure force, just before the test sample broke, was registered, equivalent to the stress at the limit of proportionality. The expression of the intensity of the compressive force was standard in N/cm/min, and the compressive stress in MPa [28] (p. 100) (see Figure 2, Table 2).

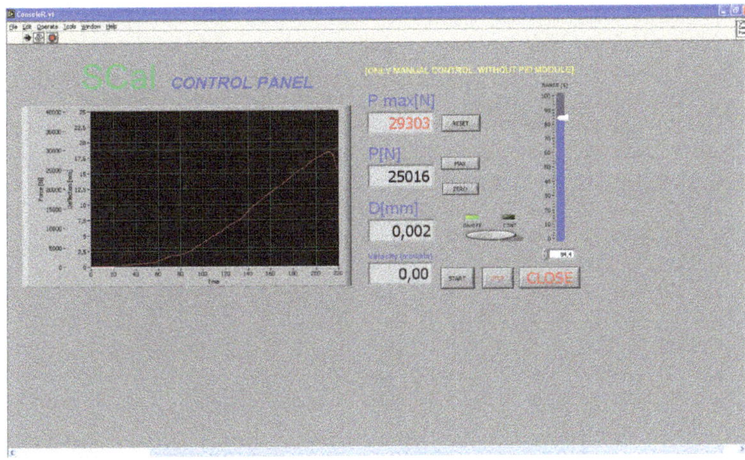

Figure 2. The graph of the ratio of applied force (N/cm^2) and stress (MPa), shown on the display of the control panel of the measuring instrument, during the determination of the compression stress, in the calibration procedure of the stress-sound wave application method.

Table 2. Values of compressive stresses determined on test sample in the calibration procedure of the stress-sound wave application method.

	Softwood					Hardwood					
						Ring Porous				Diffuse Porous	
	Spruce 1	**Fir 1**	**Fir 2**	**Pine 1**	**Pine 2**	**Oak 1**	**Oak 2**	**Black Locust1**	**Black Locust2**	**Beech 1**	**Beech 2**
Length (mm)	39.72	40.2	39.13	39.37	39.1	39.04	39.3	39.32	39.3	39.48	39.57
Width (mm)	20.18	20.16	19.96	20.12	19.93	19.92	19.86	19.79	20.27	19.94	19.8
Thickness	/	/	/	/	/	/	/	/	/	/	/
Compressive force—transversal (lateral)	3113	1883	1648	13,129	4979	4149	4621	5839	4736	4624	4760
Compressive force—axial (transverse)	/	/	/	/	/	/	/	/	/	/	/
Transversal stress	3.9	2.3	2.1	16.6	6.4	5.3	5.9	7.5	5.9	5.9	6.0
Axial stress	/	/	/	/	/	/	/	/	/	/	/
Length (mm)	/	/	/	/	/	/	/	/	/	/	/
Width (mm)	19.75	10.86	19.94	20.08	20.29	19.84	19.84	19.93	20.02	19.82	19.96
Thickness	19.86	19.85	19.89	20.09	20.08	19.77	19.79	19.9	20.12	19.19	19.9
Compressive force—transversal (lateral)	/	/	/	/	/	/	/	/	/	/	/
Compressive force—axial (transverse)	19,550	2116	21,212	34,214	27,739	24,223	27,742	39,241	37,842	31,563	29,053
Transversal stress	/	/	/	/	/	/	/	/	/	/	/
Axial stress	49.8	53.7	53.5	84.8	60.0	61.8	70.7	98.9	93.9	80.5	73.1

To assess the predictive potential of the stress-sound method, the dynamic flexural modulus was used (Figure 3, Table 3) as a significant indicator of the acoustic property of wood, which is calculated according to the formula based on the speed of propagation of the stress-sound wave [28] (p. 100) (Tables 4 and 5). By correlating it with the parameters of static compressive strength and static bending strength (Figures 4 and 5), correlation coefficients r-76 and r-77 were obtained, which, given that wood is an anisotropic material, can be considered very good. As the speed of sound, when it comes to wood, is not strongly dependent on the density:

$$c = \sqrt{E}/\rho, \tag{1}$$

then the dynamic modulus of elasticity (MOEdyn) can be considered a reliable predictor of the influence of bending and compression force on the wood.

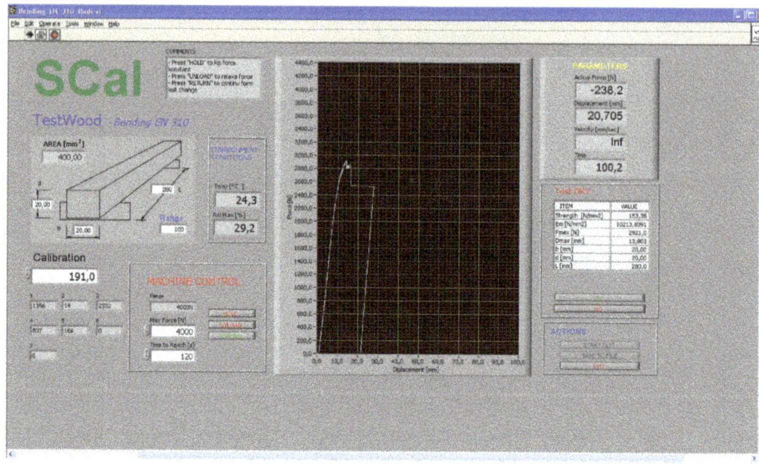

Figure 3. The graph of the ratio of applied force (N/cm^2) and stress (MPa), shown on the display of the control panel of the measuring instrument, during the determination of the bending stress, in the calibration procedure of the stress sound wave application method.

Table 3. Values of bending stress determined on test samples in the calibration procedure of the stress sound wave application method.

	Softwood				Hardwood					
							Ring Porous		Diffuse Porous	
	Spruce 1	Fir 1	Fir 2	Pine 1	Pine 2	Oak 1	Oak 2	Black Locust1	Black Locust Acacia 2	Beech 1	Beech 2
h [mm]	19.74	19.86	19.84	20.01	19.78	19.89	19.85	19.79	20.1	19.88	19.91
b [mm]	19.72	19.83	19.88	20.11	19.91	19.88	19.84	19.94	20.06	19.92	19.81
Bending strength MPa [N/mm^2]	88.9	79.9	71.6	132.9	102.8	91.09	124.5	86.4	170.3	169.8	128
Modulus of elasticity MPa [N/mm^2]	6108	8137	7905	11,567	7448	6792	8232	11,020	14,788	11,987	9145
Force max. [N]	1627	1488	1334	2572	1907	1706	2316	1605	3288	3182	2393
Deflection max. [mm]	13.85	11.78	10.28	9.39	17.95	16.58	15.04	10.27	14.5	12.64	16.67
h [mm]	19.74	19.83	19.89	19.98	20.03	19.85	19.94	19.9	10.07	19.9	19.8
b [mm]	19.78	19.79	19.86	19.92	20.14	19.83	19.85	19.92	20.31	19.84	19.8
Bending strength MPa [N/mm^2]	84.1	81.3	76.9	152.1	119.1	103.3	105.8	200.9	185.2	135.9	133.3
Modulus of elasticity MPa [N/mm^2]	8784	8262	7304	12,778	10,337	7722	6882	15,175	13,920	10,315	9722
Force max. [N]	1544	1506	1438	2880	2291	1921	1987	3773	3607	2542	2464
Deflection max. [mm]	12.58	11.26	11.06	18.21	12.6	17.36	20.05	13.16	14.74	11.61	12.59

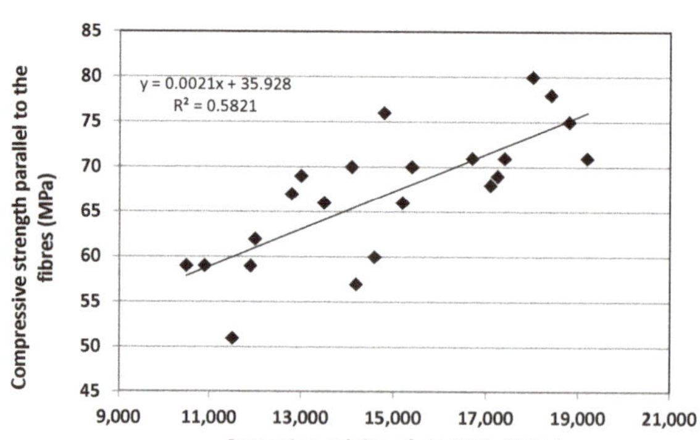

Figure 4. The relation between the dynamic modulus of elasticity and compressive strength parallel to the fibers.

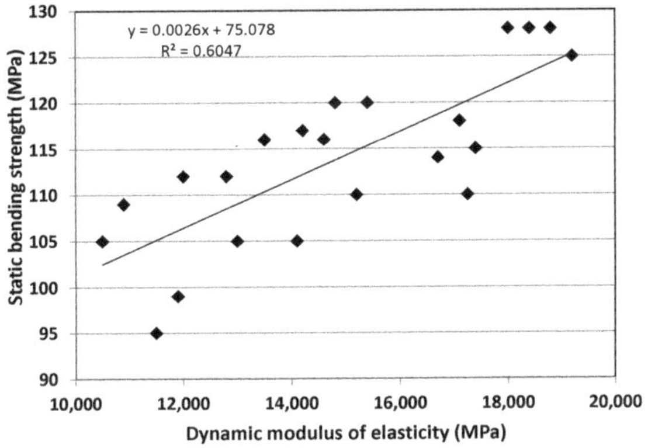

Figure 5. The relation between the dynamic modulus of elasticity and the bending strength.

2.2.4. Stress-Sound Measurement Scheme and Equipment

In this research, the speed of stress-sound waves was measured. In the first part of the research, the measurement was performed on test samples, parallel to the grain and perpendicular to the grain in the radial and tangential directions, in a similar way as it was performed in the preliminary research using ultrasonic waves. The waves were transmitted from one side to the opposite side of the test tube mechanically, via the emissive transducer, and were registered via the receiving transducer. At the same time, the transducer function was performed by microphones, connected via a tuner to a computer program for signal analysis.

2.2.5. Stress-Sound Measurement Implementation

Waves were produced by the impact of the emission transducer on the surface of the tested sample, on the opposite side from the one on which the receiving transducer was located. Considering the small dimensions of the test sample for determining the static compressive stress, the sound test was performed on larger test tubes to determine the modulus of elasticity under pressure. The strength of the input signal was adjusted via the tuner, and the propagation speed of the sound wave was calculated based on the elapsed time between the two measurement points. The speed was determined in relation to the time and length of the path, in the same way it was used to determine the speed of conducting ultrasound:

$$V = L/T, \tag{2}$$

as well as to correlate the ratio, wavelength, speed, and frequency of the ultrasound wave:

$$\Lambda = V/f, \tag{3}$$

as proposed in reference [30].

2.2.6. Spectrogram

Spectral analysis of the frequency of stress-sound waves in this research was applied both in the calibration process of the method and in the in situ examination of the elements of the wooden structure. Speed and attenuation were important parameters of the calculation when evaluating the results obtained by this method. The degree of attenuation of sound wave stress is expressed by the natural logarithm of the decrease in the amplitude of the sound wave stress per unit of distance traveled using the equation:

$$\alpha = 1/x \text{ Ln } A0/Ax, \tag{4}$$

(α attenuation factor, l-length of the wave propagation path, A0-initial wave amplitude, and Ax-wave amplitude at the end point of the distance), is directly correlated with the dynamic modulus of elasticity (MOEdyn), to consider the possibility of such behavior in situ examination phase.

Since wood is by its nature a heterotrophic and anisotropic material, the calculation of the dynamic modulus of elasticity required modifying the formula in order to obtain a true representative of wave propagation in a three-dimensional frame, which, in addition to wave speed and ρ—the wood density of the wood substance:

$$MOE_{dyn} = \rho \cdot V^2 \quad (5)$$

according to Newton's general equation [30]:

$$V = \sqrt{\frac{E}{\rho}} \left(\frac{m}{s}\right) \quad (6)$$

the speed of the propagating sound wave directly depends on the structural properties of the wood, i.e., from its specific density, and is directly proportional to the square root of the modulus of elasticity, and thus we had objective parameters for the immediate assessment of the strength, i.e., the load capacity of wooden structural elements based on the propagation speed of the stress sound wave.

The assessment was based on the fact that, due to decay or an excessive increase in humidity, wood loses a percentage of the content of the basic substance in its total mass, and thus its strength, so it was expected that the propagation of sound waves through damp or decayed wood (Table 5) would be significantly lower and exhibit irregular oscillation frequencies compared to the one recorded in healthy and dried wood (Table 4) [26]. These in situ measurements were made in three standard directions of the tested beam.

Table 4. The speed of propagation of stress sound waves in intact test samples.

	No. of Samples	Type of Wood	Length [m]	No. of Selections	Time [μs]	Speed [m/s]	Thickness [mm]	Width [cm]	Mass [kg]	MC [%]
Softwood	1	Spruce	2.055	131	682.292	3012	47.29	20.2	8.39	10.3
	1	Fir	1.9965	153	796.875	2505	46.44	23.9	8.96	9.6
	2	Fir	1.795	88	458.333	3916	47.29	20.1	7.99	9.6
	1	Pine	2.141	136	708.333	3022	49.44	23.9	14.38	7.9
	2	Pine	2.1225	161	838.542	2531	52.39	23.9	15.69	7.7
Hardwood	1	Oak	2.04	108	562.500	3627	27.61	17.25	6.67	10.4
	2	Oak	2.02	111	578.125	3494	26.81	21.25	9.03	9.8
	1	Black locust	1.543	103	536.458	2876	29.08	12.4	4.11	10
	2	Black locust	1.505	65	338.542	4445	28.42	10.6	3.49	9.3
	1	Beech	2.066	101	526.042	3927	51.03	15.6	12.46	7.3
	2	Beech	2.099	100	520.833	4030	50.09	16.1	11.96	6.4

Table 5. The speed of propagation of stress sound waves in perforated test samples.

Damage	No. of Samples	TYPE OF WOOD	Length [m]	No. of Selections	Time [μs]	Speed [m/s]	Thickness [mm]	Width [cm]	Mass [kg]	MC [%]
1	1	Fir	1.9965	141	734.375	2718	46.44	23.9	8.96	9.6
1	1	Fir	1.9965	90	468.750	4259	46.44	23.9	8.96	9.6
1	1	Fir	1.9965	94	489.583	4078	46.44	23.9	8.96	9.6
0	1	Oak	2.04	108	562.500	3627	27.61	17.25	6.67	10.4
1	1	Oak	2.04	115	598.958	3406	27.61	17.25	6.67	10.4
2	1	Oak	2.04	109	567.708	3593	27.61	17.25	6.67	10.4
1 + 1	1	Oak	2.02	110	572.917	3560	27.61	17.25	6.67	10.4
2	1	Oak	2.04	107	557.292	3660	27.61	17.25	6.67	10.4
3	1	Oak	2.04	108	562.500	3627	27.61	17.25	6.67	10.4
5	1	Oak	2.04	107	557.292	3660	27.61	17.25	6.67	10.4
5	1	Oak	2.04	117	609.375	3348	27.61	17.25	6.67	10.4
5 + 5	2	Oak	2.04	131	682.292	2990	27.61	17.25	6.67	10.4

2.2.7. Assessment of Timber Structural Elements

The approach used in this research for the in situ assessment of the structural performance of a timber structure is based on the standard allowable loads given according to the type and quality of wood (SRPS EN 1995-1-1) [31]. For modern truss wooden roof structures, the values defined by the Eurocode (EN 1991) [32] were used, calculated based on the weight of the loadbearing element and all the wooden elements of the supporting assembly.

Dimensioning of elements in the calculation of wooden structures is a concept that uses the values of their limit states, namely: limit state of elasticity—limit of failure; and usability limit state. The procedure itself is based on knowledge of technical mechanics, as well as on experimental validations. As, according to this adopted concept, the mechanically equalized parameter of the speed of sound wave progression, obtained on a wood sample of the first quality, was used for the assessment of the state of the supporting beam, the table value for the allowable compressive load, perpendicular to the wood fibers, given for the first-class wood, was taken for this in situ test.

The Eurocode (EN 1991) [32] prescribes normative values similar to these, so for trusses with a span of up to 20 m, a load of up to q = 0.25 kN/m^2 is predicted, for spans of 20–25 m, that value is q = 0.35 kN/m^2 and for spans over 25 m, the value q = 0.45 kN/m^2 is provided. The values of the allowable load of the supporting beams according to the (SRPS EN 1995-1-1) [31] standard is shown in (Appendix A).

The assessment of the quality of the supporting beams of timber structures in the third phase of the test was intended to be performed by comparing the directly obtained, mechanically equalized, stress propagation speed parameter of the sound wave, using the calculated dynamic modulus of elasticity (MOEdyn) and its correlation with the values obtained by the static compressive and bending stress (Figures 3 and 4), and the corresponding tabular values of the allowable load in N/cm^2 defined by the standard.

2.2.8. Comparative Analysis of Table and Measured Load Values

This part of the research considers the limited state of static equilibrium of timber structural systems of architectural buildings based on the given values of destabilizing and stabilizing loads. The values of these loads were indirectly determined (laboratory) during the static pressure test during the calibration procedure of the experimental test tubes. By comparison, it is checked whether the measured load (destabilizing load) of the supporting structural element is excessive, in accordance with the laboratory and standardized values (stabilizing load) for timber structures (SRPS EN 1995-1-1) [31].

Based on the results of previous research, X-ray and ultrasound equivalents of static compressive stress were obtained, and by transposing tabular values, tensile, and shear stresses within the complex bending stress were obtained. According to the same principle, the parameters of the speed of propagation of stress-sound waves were reduced, with the fact that, now in this procedure, the mechanical calibration of the experimental test tubes was preceded, which achieved the establishment of a method by realistically comparing mechanical and stress-sound parameters.

The basic principle used in the assessment of static equilibrium was that the case where the result was measured on the element with the largest cross-section was valid [31]. In accordance with the recommendation and considering that the in situ research was carried out on a building with a timber structure (over 70 years old), it was acceptable that except for basic loads, other combinations of actions, such as vertical and horizontal loads, could be abstracted. In accordance with that allowable load of the structure, it would be acceptable that it is 15% higher than defined for the examined type of wood [28] (p. 170).

In order to somewhat computationally compensate the influence of abstracted additional loads (friction on bearings, temperature changes, shrinkage and swelling) and special loads (seismic, fire load), the values of the allowable stresses of the supporting beams were considered in relation to those values obtained by testing the test tubes, made of the highest quality wood. Considering the fact that the standardization of permissible stresses was established in relation to the humidity of the wood and that the test was carried out in

Belgrade, it is in relation to the climatic conditions [28] in this work. A tabular model of the installed wood of the first exploitation class of 18% hygroscopic humidity was taken for calculation.

3. In Situ Measurement of the Strength of Timber Structural Elements

The results of in situ testing of the quality of the supporting beam of the timber roof structure (Figure 6) indicate that the speed of propagation of stress-sound waves, identically as in laboratory conditions, was the highest in the direction parallel to the axial axis of the supporting beam, and proportionally less along its radial and tangential direction. On the other hand, the directly measured moisture level of the beam was 10%, which meant that this circumstance, as well as the age of the beam (70 years) and the undoubted rheological influence [33,34], obtained values of the velocity of the sound wave stress through the beam, as a representative of its mechanical potential and from those points of view are considered and interpreted. Finally, due to the relatively small examined sample, it was impossible to take an absolute position in evaluating the obtained results. However, the primary goal was to present and confirm the new methodological approach to structural diagnostics.

The in situ test was carried out with an accelerometer (Bruel & Kjaer), which was used as an impulse hammer to provoke sound waves on one side of the beam, and with a measuring microphone (NTi), which registered the incoming stress-sound wave on the other side of the beam. The measuring microphone was positioned on the receiving side in a very near field, only 2 mm from beam. Both signals were recorded with the audio interface (Steinberg) with a sampling frequency of fs = 192 kHz. The minimum interval that can be registered with this device is $\Delta t = 1/fs = 5.2083$ μs. The number of samples (N) that the microphone signal receives after the provoking impulse is detected by comparing those signals. The time it takes for sound to pass through the beam was calculated by multiplying the number of samples N by the value Δt. The shortest recorded time was 2.6041 ms, and the longest was 3.1354 ms.

The maximum measured speed of the stress-sound wave along the central axial axis of the 14.9 m long beam, from the point of intersection of the diagonals of its cross-section (Figure 5), was 5721 m/s (2.6 ms), which corresponded to the time of sound passing through healthy wood (3000–5000 m/s)*, which was indicated by recording the propagation, and a gradual decrease in the wave amplitude (Table 4). By recording the wave propagation of the axial axis in 4/5* of the cross-section of the supporting beam, a gradual decrease in its amplitude was recorded, which indicated a healthy tree. That stress-sound wave velocity coefficient during axial propagation corresponded to MOEdyn of 14,744 MPa, which was obtained through the Formula (5), where ρ is the value of dense wood in absolute in a dry state V is the speed of sound passing through wood material, and ν is Poisson's coefficient. According to the scatter line of the given graph, this MOEdyn value corresponded to a static compressive stress value of 6800 N/cm^2 (68 MPa) and a static compressive stress value of 11,400 N/cm^2 (114 MPa). If we consider that for our types of wood the average maximum value of bending pressure is around 100 MPa and that the limit of proportionality is 60–70% of that value, then according to the standard (SRPS EN 1995-1-1) [31] these are permissible stresses for class II European conifers, from which, according to the construction documentation, the roof structure of the building of the Technical Faculty in Belgrade was built. According to the same standard, the permissible static compressive stress perpendicular to the flow of mechanical fibers for that category of construction is 250 N/cm^2 (2.5 MPa), which, according to the Eurocode standard (EN 1991) [32], even for lattice girders with a span of up to 20 m in this case, represents a certain coincidence and a kind of confirmation that the mechanical potential of the tested beam is within the limits of permissible loads.

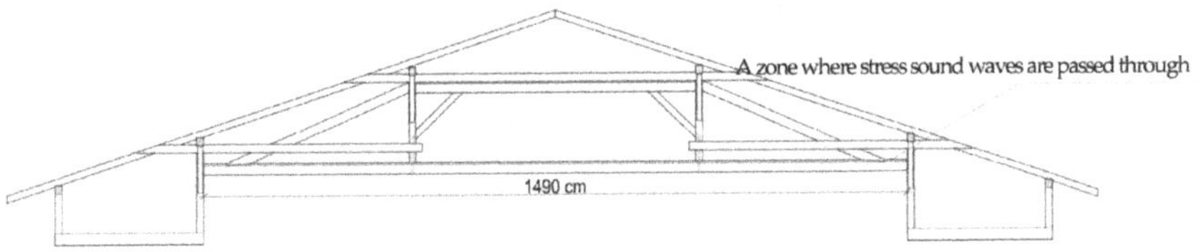

Figure 6. Roof girder, a segment of the roof structure through which the stress-sound wave was passed.

Propagation of sound in the outer part of the upper fifth of the cross-section, along the line of the axial direction, was recorded at a lower speed (3.31354 ms) compared to the maximum (2.6 ms), recorded in the central part of the section, which indicated a defect in the structure on that line wave propagation (Table 5). Upon inspection of the beam, 10 m from the place of propagation, a slightly slanted crack with a length of about thirty centimetres was observed at the upper surface of the beam. Technically, however, the sound mapping of the cross-section was not possible on that surface due to limited access. From the findings of the inspection, the drop in the speed of sound propagation in the range of normal values, as well as the other results of the sound tests, indicating a negligible defect and healthy wood, it could be concluded that the mechanical potential of the beam is within the limits of permissible loads. Certain deviations of directly obtained values, static compressive stress, and static bending stress by comparing the non-destructive method and given tabular values could be explained by the fact that tabular average values were obtained by testing many samples and that the values of our small sample could roughly, therefore, be classified as extremes. The preservation of a very old supporting element of the timber architectural structure, with above-average mechanical potential, can be explained in this case by the rheological influence in conditions of isolation from moisture and extreme temperatures, as is the case here with the roof structure of the Technical Faculties Building in Belgrade, Serbia.

4. Discussion and Conclusions

To establish the NDT method for assessment of the quality of timber structural elements of architectural heritage buildings, all the given conditions were met according to the postulates of scientific research. This claim is based on consistently cited scientific facts related to the chemical and anatomical structure of wood, as well as the physical and mechanical properties of wood—therefore also on the laws in the behavior of wood as a material, on which the plan was conceived, and this research was carried out.

From the aspect of the methodological approach, this in situ method of examination of timber architectural structures was fully set following the valid conventional postulates for architectural buildings of cultural and historical importance. Considering that this test was successfully carried out under such conditions on the roof structure of the Technical Faculties Building in Belgrade, we can conclude that such an approach to testing guarantees respect for the given principle of preserving the authenticity and integrity of the protected object, as well as that it is generally applicable.

The presentation of other conclusions refers to examining the mechanical properties of wood. It is based on preliminary research results, laboratory establishment of the in situ method, and its immediate application. The application of X-rays and ultrasonic waves, analogous to the standard direction of the pressure force along the flow of wood fibers or perpendicular to them, confirmed the real possibility of equalization, perceptual gradation of the brightness of the X-ray image, and parameters of the speed of propagation of ultrasonic waves through the timber structure, with average table values of static stress parameters on pressure, taken for given, non-destructively tested samples—test tubes, of certain types of wood.

Based on one such numerical simulation of the mechanical testing of wood, where it was confirmed that the wood with a tabularly higher strength gave a brighter X-ray image and passed ultrasonic waves faster [5], it could be said that the mechanical calibration of any non-destructive method that can be applied analogously to the standard direction of a force in the normal stress test is feasible.

The mechanical calibration of the method of applying stress-sound waves for testing the quality of wood confirmed that this method is related to the ultrasonic method and that the results obtained by its application can be interpreted according to the same principle as in the previous reports [2–4]. Previous results meant that in this case, too, according to the physical laws of wave propagation, wood with a higher density, i.e., a higher value of static compressive stress (strength) and a higher dynamic modulus of elasticity (MOEdyn), will transmit stress-sound waves faster.

The calibration of the stress-sound method was carried out in conditions of standard hygroscopic humidity (12%) of test samples—test tubes, similar to the calibration of the ultrasonic method based on tabular approximation—a qualitatively new approach to in situ testing of elements of timber architectural structures was then established. Based on this, it finally follows that in equal conditions of hygroscopic humidity, the tested wooden structures will transmit stress-sound waves proportionately to their strength, thus giving a realistic reflection of their mechanical potential.

The results of the in situ tests were entirely in line with the expectations that, based on the equalization of the stress-sound wave propagation speed parameters, derived (MOEdyn) and parameters of static pressure and bending stress, obtained from test tubes made of healthy wood, can comparatively examine the mechanical potential of supporting beams of timber architectural structures. At the same time, both methods of comparison proved to be reliable in relation to the speed of stress-sound waves in healthy wood and according to the standard permissible compressive stresses for supporting wooden beams. By confirming the initial hypothetical assumptions about the possibility of equalizing the mechanical and non-destructive parameters of wood testing, we arrive at the main answer sought through this research—that the mechanical potential of load-bearing timber beams can be directly tested in situ with the stress-sound test method, and thus their structural stability.

The establishment of such a diagnostic approach, based on the equalization of the dynamic parameter of the speed of sound propagation, i.e., MOEdyn, and the parameter of static pressure and bending stress, opens the possibility of further improvement of the method. The preliminary research of this paper also shows that the principle of equalization can be applied to other NDT methods, where the diagnostic procedure is carried out analogously to the direction of the pressure force during the static pressure test. The development perspective of X-ray diagnostics is certainly of particular interest due to the possibility of direct visualization of the wood structure.

Another aspect of improving the methodology of the in situ diagnostic approach to testing the quality of wooden structures refers to introducing the degree of hygroscopic humidity as an exponential factor in the function of wave propagation speed through the wooden material. This reduces the parameters of non-destructive and mechanical testing of timber to a common denominator, which increases the precision of measurements, and, thus, the reliability of assessing the quality of built-in timber elements.

The practical application of the mechanically calibrated method of stress-sound waves could improve the process of restoration and renovation of architectural buildings of cultural and historical importance, as well as the construction process itself, primarily because of the method of application and because of a simpler choice of the suitable timber material for such purposes.

Author Contributions: Conceptualization, A.J., N.T. and J.M.; methodology, A.J., N.T. and J.M.; software, N.T., M.V. (Marko Veizović) and F.P.; validation, A.J., N.T. and J.M.; formal analysis, A.J. and N.T.; investigation, A.J., N.T., M.V. (Marko Veizović), F.P. and M.A.; resources, A.J. and N.T.; data curation, A.J. and N.T.; writing—original draft preparation, A.J. and J.M.; writing—review and editing, A.J. and J.M.; visualization, A.J.; supervision, N.T., J.M., M.V. (Miomir Vasov) and A.R. All authors have read and agreed to the published version of the manuscript.

Funding: This research was funded by the Ministry of Education, Science and Technological Development of the Republic of Serbia, grant number 451-03-68/2020-14/200090. The research was carried out under the research lab of the University of Belgrade, Faculty of Architecture—Laboratory for Innovative Structures in Architecture (LISA).

Data Availability Statement: Not applicable.

Conflicts of Interest: The authors declare no conflict of interest. The funders had no role in the design of the study; in the collection, analyses, or interpretation of data; in the writing of the manuscript; or in the decision to publish the results.

Appendix A

Table A1. Basic allowable stresses for wood moisture content of 18 and 15% in daN/cm^2 for structures made of solid wood and glued laminated wood (SRPS EN 1995-1-1:2012) [31].

Permissible Stress		**Solid Wood w = 18%**					**Glued Laminated Wood**			
		Softwood			**Oak, Beach**		**Softwood**		**Oak, Beach**	
Stress Type	**Label**	**Class**			**Class**		**Class**		**Class**	
		I	II	III	I	II	I	II	I	II
Bending	σmd	1300	1000	700	1400	1200	1400	1100	1620	1370
Tension	σt∥d	1050	850	0	1150	1000	1050	850	1800	1080
Compression	σc∥d	1100	850	600	1200	1000	1100	850	1500	1200
Compression perpendicular to fibers	σc⊥d	200	200	200	300	300	200	200		
		250	250	250	400	400	250	250	490	430
Shearing	τ∥d	90	90	90	120	120	90	90	150	150
Shearing from T forces	τm∥d	90	90	90	120	120	120	120	130	110
Fiber cutting	τ⊥d	350	300	250	400	350	350	300	250	400

References

1. Avdelidis, N.P.; Moropoulou, A. Applications of Infrared Thermography for the Investigation of historic structures. *J. Cult. Herit.* **2004**, *5*, 119–127. [CrossRef]
2. Baar, J.; Tippner, J.; Rademacher, P. Prediction of mechanical properties-modulus of rupture and modulus of elasticity-of five tropical species by nondestructive methods. Maderas. *Cienc. Tecnol.* **2015**, *17*, 239–252. [CrossRef]
3. Bachtiar, E.V.; Sanabria, S.J.; Mittig, J.P.; Niemz, P. Moisture-dependent elastic characteristics of walnut and cherry wood by means of mechanical and ultrasonic test incorporating three different ultrasound data evaluation techniques. *Wood Sci. Technol.* **2017**, *51*, 47–67. [CrossRef]
4. Bucur, V.; Böhnke, I. Factors affecting ultrasonic measurements in solid wood. *Ultrasonics* **1994**, *32*, 385–390. [CrossRef]
5. Keunecke, D.; Sonderegger, W.; Pereteanu, K.; Lüthi, T.; Niemz, P. Determination of Young's and shear moduli of common yew and Norway spruce by means of ultrasonic waves. *Wood Sci. Technol.* **2007**, *41*, 309–327. [CrossRef]
6. Kandemir-Yucel, A.; Tavukcuoglu, A.; Caner-Saltik, E.N. In situ assessment of structural timber elements of a historic building by infrared thermography and ultrasonic velocity. *Infrared Phys. Technol.* **2007**, *49*, 243–248. [CrossRef]
7. Hussain, A.; Akhtar, S. Review of Non-Destructive Tests for Evaluation of Historic Masonry and Concrete Structures. *Arab. J. Sci. Eng.* **2017**, *42*, 925–940. [CrossRef]
8. Fais, S.; Casula, G.; Cuccuru, F.; Ligas, P.; Bianchi, M.G. An innovative methodology for the non-destructive diagnosis of architectural elements of ancient historical buildings. *Sci. Rep.* **2018**, *8*, 4334. [CrossRef]
9. Barnaure, M.; Cincu, M. Testing methods for the assessment of material properties in historical masonry structures: A review. *Proc. IOP Conf. Ser.: Mater. Sci. Eng.* **2019**, *662*, 032040. [CrossRef]
10. Tejedor, B.; Lucchi, E.; Bienvenido-Huertas, D.; Nardi, I. Non-destructive Techniques (NDT) for the Diagnosis of Heritage Buildings: Traditional Procedures and Future Perspectives. *Energy Build.* **2022**, *236*, 112029. [CrossRef]
11. Paipetis, A.S.; Matikas, T.E.; Aggelis, D.G.; Van Hemelrijck, D. *Emerging Technologies in Non-Destructive Testing V*, 1st ed.; CRC Press: Boca Raton, FL, USA, 2012.
12. Jokileto, J. Aspekti autentičnosti. *GDKS* **2002**, *26*, 11–16.
13. Klapálek, P.; Melzerová, L. Methods of Non-Destructive Assessment of Timber. *Appl. Mech. Mater.* **2015**, *732*, 369–372. [CrossRef]
14. Cruz, H.; Yeomans, D.; Tsakanika, E.; Macchioni, N.; Jorissen, A.; Touza, M.; Mannucci, M.; Lourenço, P.B. Guidelines for On-Site Assessment of Historic Timber Structures. *Int. J. Archit. Herit.* **2014**, *9*, 277–289. [CrossRef]
15. Nowak, T.P.; Jasieńko, J.; Hamrol-Bielecka, K. In situ assessment of structural timber using the resistance drilling method—Evaluation of usefulness. *Constr. Build. Mater.* **2016**, *102*, 403–415. [CrossRef]

16. Gomes, I.D.; Kondis, F.; Sousa, H.S.; Branco, J.M.; Lourenço, P.B. Assessment and Diagnosis of Two Collar Timber Trusses by Means of Visual Grading and Non-destructive Tests. In *Historical Earthquake-Resistant Timber Framing in the Mediterranean Area. Lecture Notes in Civil Engineering*; Cruz, H., Saporiti Machado, J., Campos Costa, A., Xavier Candeias, P., Ruggieri, N., Manuel Catarino, J., Eds.; Springer: Berlin/Heidelberg, Germany, 2016; Volume 1, pp. 311–320. [CrossRef]
17. Stepinc, M.; Rajčić, V.; Barbalić, J. Inspection and Condition Assessment of Existing Timber Structures. *Građevinar* **2017**, *69*, 869–873. [CrossRef]
18. De Matteis, G.; Ricci, L.; Chisari, C.; Mandara, A.; Panico, S. Structural Assessment of Historical Timber Roof Trusses: The Case of the Croce di Lucca Church in Naples. In *Protection of Historical Constructions. PROHITECH 2021. Lecture Notes in Civil Engineering*; Vayas, I., Mazzolani, F.M., Eds.; Springer: Berlin/Heidelberg, Germany, 2022; Volume 209, pp. 1111–1125.
19. Wang, P.; Li, S.; Macchioni, N.; Palanti, S.; Milani, G. Comprehensive Evaluation Method of Historical Timber Structural Building Taking Fujiu Zhou House as an Example. *Forests* **2021**, *12*, 1172. [CrossRef]
20. Cruz, C.; Gaju, M.; Gallego, A.; Rescalvo, F.; Suarez, E. Non-Destructive Multi-Feature Analysis of a Historic Wooden Floor. *Buildings* **2022**, *12*, 2193. [CrossRef]
21. Martinez, I.; Martinez, E. Qualitative Timber Structure Assessment with Passive IR Thermography. Case Study of Source of Common Errors. *Case Stud. Constr. Mater.* **2022**, *16*, e00789. [CrossRef]
22. Nasir, V.; Ayanleye, S.; Kazemirad, S.; Sassan, F.; Adamopoulos, S. Acoustic Emission Monitoring of Wood Materials and Timber Structures: A Critical Review. *Build. Mater.* **2022**, *350*, 12887. [CrossRef]
23. Xin, Z.; Ke, D.; Zhang, H.; Yu, Y.; Liu, F. Non-destructive Evaluating the Density and Mechanical Properties of Ancient Timber Members Based on Machine Learning Approach. *Build. Mater.* **2022**, *341*, 127855. [CrossRef]
24. Galligan, W.J.; Courteau, R.W. Measurement of Elasticity of Lumber with Longitudinal Stress Waves and the Piezoelectric Effect of Wood. In Proceedings of the 2nd International Symposium of Nondestructive Testing of Wood, National Science Foundation, Washington State University, Pullman, WA, USA, April 1965; pp. 223–224.
25. Gerhards, C.C. Bending Creep and Load Duration of Douglas-Fir 2 by 4s under Constant Load. *Wood. Fiber. Sci.* **1991**, *23*, 384–409.
26. Gojković, M.; Stojić, D. *Drvene Konstrukcije*; Grosknjiga, Građevinski Fakultet Univerziteta: Belgrade, Serbia, 1996.
27. Sisojević, D. *Anatomija Drveta*; Šumarski fakultet, Institut za Preradu Drveta: Belgrade, Serbia, 1987.
28. Šoškić, B.; Popović, Z. *Svojstva Drveta*; Šumarski Fakultet Univerziteta: Belgrade, Serbia, 2002.
29. Josifovski, A.; Savanović, D. Non-destructive testing of wood structures by X-rays. *SAJ* **2017**, *1*, 51–64.
30. Armstrong, L.D. Effects of Moisture Changes of Crep in Wood. *Nature* **1965**, *185*, 862–863. [CrossRef]
31. *SRPS EN 1995-1-1:2012*; Eurocode 5: Design of Timber Structures—Part 1-1: General—Common Rules and Rules for Buildings. Institute for Standardization of Serbia: Belgrade, Serbia, 2016. Available online: https://iss.rs/en/ (accessed on 10 March 2023).
32. *EN 1991-1-1:2002*; Eurocode 1: Actions on Structures—Part 1-1: General Actions—Densities, Self-Weight, Imposed Loads for Buildings. European Committee for Standardization: Brussels, Belgium, 2002. Available online: https://eurocodes.jrc.ec.europa.eu (accessed on 10 March 2023).
33. Popović, Z. Uticaj Vlažnosti i Temperature na Modul Elastičnosti i Savitljivost Bukovog Drveta. Master's Thesis, Šumarski Fakultet Univerziteta, Beograd, Serbia, 1990.
34. Pentoney, R.E.; Davidson, R.W. Rheology and the study of wood. *Forest. Pro. J.* **1962**, *12*, 243–248.

buildings

MDPI

Article

Hygrothermal Performance of Thick PCM Mortar behind PV Panels in Energy-Activated ETICS Facades

Martin Talvik [1,*], Simo Ilomets [1], Paul Klõšeiko [1], Targo Kalamees [1], Mattias Põldaru [1] and Dariusz Heim [2]

[1] Department of Civil Engineering and Architecture, Tallinn University of Technology, 19086 Tallinn, Estonia
[2] Department of Environmental Engineering, Lodz University of Technology, 90924 Lodz, Poland
* Correspondence: martin.talvik1@taltech.ee

Abstract: The concept of integrating PV panels into traditional ETICS facades has been developing for several years. Problems concerning the options for passively controlling the temperatures of PV panels with PCM and directing excess moisture out of the wall via diffusion channels have been previously studied theoretically. During this study, real wall-scale experiments were conducted to test the thermal and hygrothermal performance of the wall system in an extreme climatic environment, as well as in a real outdoor environment in Tallinn, Estonia. Finally, a simulation model was calibrated according to the measured data. It was found that in case of test walls with diffusion channels, it was possible to keep the moisture content of PCM mortar under 0.11 m^3/m^3. Excess water drained out via channels leading to the external environment. Without diffusion channels, the moisture content rose as high as 0.18 m^3/m^3. Both the experiments and hygrothermal modelling showed that the high moisture content of PCM mortar, caused by water leakage, dropped to 0.08 m^3/m^3 over 10 solar cycles as moisture escaped via the diffusion channels. PCM mortar with a moisture content of 0.08 m^3/m^3 endured extreme rain and freeze-thaw cycles without visual damage, and PV panels retained their electrical production capabilities.

Keywords: ETICS; building-integrated PV (BIPV); PCM; En-ActivETICS; hygrothermal performance; hygrothermal modelling

Citation: Talvik, M.; Ilomets, S.; Klõšeiko, P.; Kalamees, T.; Põldaru, M.; Heim, D. Hygrothermal Performance of Thick PCM Mortar behind PV Panels in Energy-Activated ETICS Facades. *Buildings* **2023**, *13*, 1572. https://doi.org/10.3390/buildings13061572

Academic Editor: Danny Hin Wa Li

Received: 20 April 2023
Revised: 19 May 2023
Accepted: 15 June 2023
Published: 20 June 2023

1. Introduction

Building-integrated photovoltaics (BIPV) are becoming increasingly widespread to reduce the CO_2 emissions of buildings, which account for 36% of emissions [1], and to produce green electricity locally. There is a large demand for visually appealing BIPV systems [2], but unfortunately, the majority of the BIPV systems still depend on fixed-tilt PV module systems [3].

For ventilated-façade-integrated photovoltaic (PV) panels, there are many solutions to choose from. Example procedures are available to clarify how to develop PV systems on ventilated facades [4]. The majority of BIPV facades are constructed as ventilated facade technologies using PV panels as a rain screen cladding system, which adds an additional protection layer against hygrothermal problems. There is some recent research on developing a system to apply PV panels onto existing and new ETICS (external thermal insulation composite system) facades, but there is still a ventilation gap planned between PV and ETICS [5].

The traditional ETICS concept consists of insulation material, glued directly to the structural layer with adhesive mortar and the plaster system, consisting of base coat and finishing plaster, which is applied directly to the insulation layer. This means that there is no air cavity, and the outer layer of the wall system has relatively high resistance to water vapour diffusion.

If PV panels were applied directly on thermal insulation, there could be some undesirable effects, e.g., overheating or moisture condensation. To address the aspect of overheating, there has been some research about using latent heat storage of phase change materials (PCM) for temperature stabilisation of the PV panels, both passively [6] and actively [7]. There have also been studies conducted with double-skin facades with integrated PCM layers and PV glazing [8], but in this case there was an air chamber between the PV and PCM, with the possibility to open and close the ventilation gaps. Combination of PV and PCM has also shown potential in reducing overall energy consumption of buildings in cold climates [9]. Although PCM applications are a relatively new technology in the construction sector, there are some data from indoor applications stating that after a decade, PCM was still working as a passive heat sink, despite showing a significant decrease in the PCM capacity [10].

However, the PCM and PV solution has not been combined with traditional ETICS facade systems before. The application of diffusion gaps in the PCM layer to avoid moisture condensation and accumulation has not been studied in the past.

The concept of an En-ActivETICS facade integrates traditional ETICS facade technology with PV panels. There has been theoretical work conducted towards the development of the facade concept [11–13]. Measurements with real test walls regarding the electrical and thermal performance have also been conducted [14], as well as comparisons between different PV-ETICS integration solutions and free standing PV panels. However, the hygrothermal performance of this wall concept has not been studied by experimental methods before.

Earlier studies have proven that applying a thick layer of PCM mortar is effective in avoiding the overheating of PV panels [14]. However, applying a vapour-tight layer on the outermost side of the facade creates a situation in colder climates where moisture is not able to get out of the wall via diffusion in the same way it does in the case of traditional ETICS. This problem was identified and analysed numerically in [15]. A solution utilising diffusion gaps connected to the outside environment between a pair of steel sheets at the lower part of the PV has been proposed and developed [16]. However, the overall hygrothermal performance of this solution has not been studied before, except for the initial drying-out process of PCM mortar just after the application [16].

In the results of several empirical studies, it has been found that one common mechanism for ETICS facades to deteriorate prematurely is water leaks through the cracks [17–19]. During the lifespan of ETICS facades, micro-cracks tend to appear after several years. As water seeps into the micro-cracks, they become wider, leading to even more water penetrating the facade plaster, which starts the vicious cycle of degradation. Integrating a dark-coloured PV panel into the light-coloured facade significantly increases the theoretical risk of cracks appearing, due to the number of joint details between different materials. Unfortunately, ETICS systems are known to be not very robust, and minor defects could lead to deterioration [20]. Therefore, if more complexity is added to the ETICS system via the integration of PV, diffusion channels should be recommended as a countermeasure to increase the robustness of the system.

Therefore, it is necessary to thoroughly test this novel facade system, which takes into account the expectation that there will be cracks in the facade and that water will eventually leak into the wall. This is a new approach that has not yet been tried on a wall that combines PV panels, PCM mortar, and an ETICS facade. If a hygrothermally well-performing facade solution can be developed, then this novel En-activETICS facade could have a huge market potential to alleviate the energy crisis. There has been a great demand for BIPV for several years, and there are also many different BIPV solutions on the market, but there are many concerns among architects about the BIPV facade solutions available today, as they are difficult to blend in as a part of the architecture [21]. In contrast to ventilated BIPV solutions, En-ActivETICS would be on the same surface plane as facade plaster and would make it possible to create interesting patterns of PV and facade plaster areas.

It was found in [22] that despite ETICS facades being often preferred for building renovation as being a cost-optimal solution, they tend to have a higher carbon footprint than pre-fabricated timber elements. If PV panels are integrated into otherwise affordable ETICS facades, they could offset some of the carbon footprint and become closer to also being CO_2e optimal. Therefore, there could be a large market potential for En-ActivETICS for both new and renovated buildings, but the hygrothermal performance of the novel wall concept must be thoroughly studied first.

The aim of the study was to test the constructability and study the hygrothermal performance of a novel facade system that combines PV panels, PCM mortar and ETICS.

2. Methods

To test the concept of hygrothermal performance, two real-scale test walls were built. One wall was built in a climate chamber to test the system in a controlled and critical environment. The test package consisted of water injection, solar radiation, wind-driven rain (WDR) and freezing tests. Another wall was built into the existing facade of a test building to be studied under real outdoor climate conditions. All measurements were compared with data from calculations in order to calibrate the model.

The cross-section of an En-ActivETICS element (Figure 1b) consists of a structural layer, 160 mm of insulation (EPS or mineral wool), 30 mm thick PCM mortar (cement-based mortar, mixed with PCM granules) and a 2 mm flexible PV (FPV) panel. Thin FPV was chosen because of its relatively light weight and its ability to be glued directly onto the surface of PCM mortar [11]. The 100 W FPV panel [23] used on the wall has monocrystalline silicon wafers in the functional layer, TPT (Tedlar Polyester Tedlar) as the back sheet and flexible transparent plastic as the outermost layer for weather protection. The FPV does not have any frame for structural integrity and therefore needs to be glued to the surface.

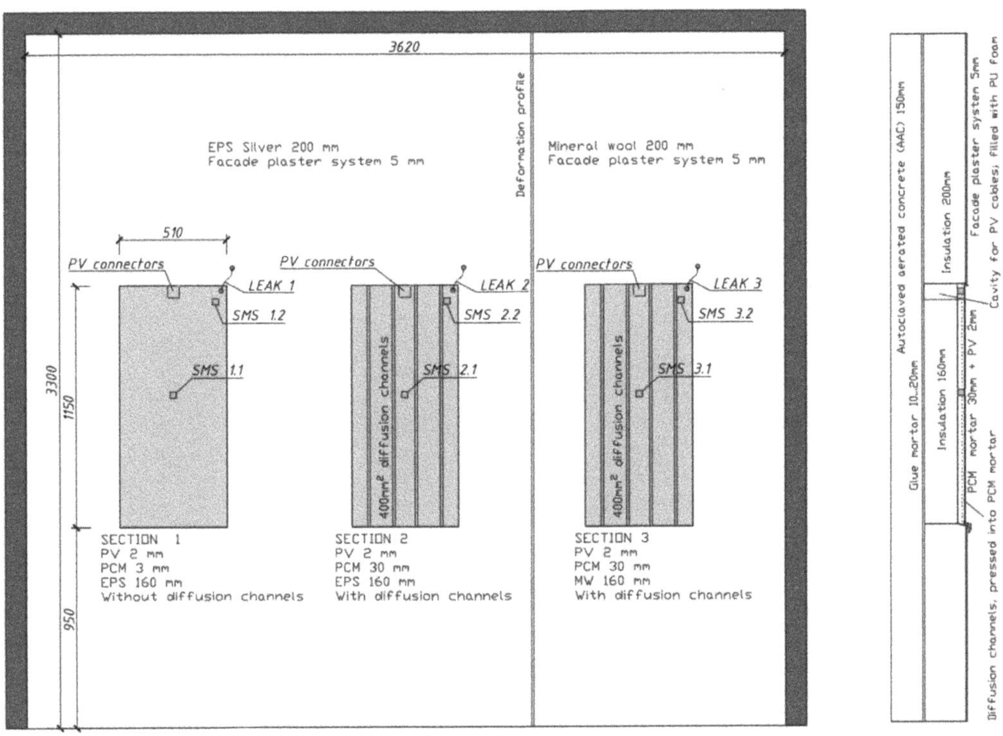

(a) (b)

Figure 1. Layout (**a**) and cross section (**b**) of a test wall in the climate chamber. Locations of soil moisture sensors are marked as "SMS" and points where water leaked in are marked as "LEAK".

Four diffusion channels for each panel were carved into the PCM mortar before hardening. Cable connectors were placed inside a cavity in the insulation layer, on the upper part of the PV and during installation, and the cavity was filled with PU foam. As waterproof MC4 connectors were used to make the connections, no special junction box was needed. The exact building process and detailing of the solution has been described before, in [16].

Between and all around the PV panels, a traditional ETICS with 200 mm of insulation was installed. The load-bearing structure was built from 150 mm of AAC (autoclaved aerated concrete) without internal plaster, to maximise the indoor moisture load corresponding to the worst-case scenario. The reasons for choosing AAC as the critical structural material are high levels of built-in moisture and relatively low water vapour diffusion resistance factor ($\mu = 7$).

2.1. Test Wall in Climate Chamber

The climate chamber can simulate both indoor and outdoor temperature (−25 °C to +70 °C) and relative humidity (RH = 5 ... 98%). In the outdoor chamber, WDR was simulated with additional water sprinkler stands and short-wave thermal radiation was simulated with halogen stands.

A 3.6 m wide and 3.3 m high test wall was built in the climate chamber (Figure 2a) with three different sections: (1) Reference section with PCM mortar without diffusion channels; (2) PCM mortar on expanded polystyrene (EPS) with diffusion channels behind PV; and (3) PCM mortar on mineral wool (MW) insulation with diffusion channels behind PV (Figure 1). EPS and mineral wool areas were separated with PU foam and the facade plaster was separated with a deformation profile on top of the PU foam.

(**a**) (**b**)

Figure 2. Climate chamber used for building the test wall (**a**) and the test wall inside the climate chamber (**b**).

2.2. Test Wall under Real Outdoor Climate Conditions

Besides the mock-up developed for investigations under a critical and controlled environment (climate chamber), it was decided to build another test wall to enable tests on the actual facade under real outdoor climate conditions in Tallinn, Estonia. Tallinn (59.44° N, 24.75° E) is located in a humid continental climate with warm to hot (and often humid) summers and freezing cold (Dfb according to the Koppen–Geiger climate classification system) [24]. The facade faces directly south (180° azimuth) and is located on the first storey, therefore not receiving the most extreme WDR loads (Figure 3). The wall itself was exactly the same as the one built in the climate chamber as section no. 2—PCM mortar on EPS with diffusion channels (Figure 1a), in order for the results to be comparable. The analysed period of measurements was from June to August 2022.

(a) (b)

Figure 3. Test wall in Tallinn under real climate conditions (**a**) and overview of the test house (**b**). Pyranometer, used for measuring radiation, is visible above the roof line of test house.

2.3. Sensors Used for Measurement

Temperature, relative humidity (RH), volumetric moisture content and heat flux were measured inside the test walls (climate chamber and real facade) and on the surface. Solar radiation on the wall, wind velocity, outside air temperature and RH were also measured. Two or three alternative sensors were used to measure each parameter independently to optimise for data availability, cost efficiency and reliability of the measured data. To measure the temperature inside the walls, the following temperature sensors were used: Onset Hobo TMC15-HD with data logger UX120-006M and Littelfuse KS103J2 NTC thermistor with data logger Grant Squirrel SQ2020 1F8. RH and temperature were measured with an Onset Hobo UX100-023; a Rotronic Hygroclip HC2-C05 with A/D converter Siemens SM331 6ES7331 and a Sensirion SHT35 & tailor-made logger based on Arduino Mega.

Moreover, volumetric moisture content inside the PCM mortar was measured with EC-5 soil moisture sensors [25]. The sensors were calibrated specifically for measuring this PCM mortar, as the result of several small-scale experiments and formulas for calculating volumetric moisture content with temperature corrections [16].

2.4. Injecting Water into the Wall

As the climate chamber test is an accelerated version of real-life processes, real cracks were unlikely to develop during the test. Moreover, the overall facade plaster area was much smaller, and the construction process was carried out in a favourable environment. But it was still necessary to imitate water leakages through the cracks in the system as they tend to appear on real ETICS facades.

To imitate the critical leakages, water hoses were installed into the wall. It is most likely that the cracks on a real facade would appear on the corners, where PV panels and facade plaster are joined together. Therefore, the ends of the water hoses were installed to the upper corners, behind the PV panel and into the PCM mortar layer. Tubes were designed to make it possible to inject water into the wall using a common medical syringe (Figure 4b). PCM mortar moisture content was measured in the area immediately exposed to the increased moisture content caused by the leakage (Figure 4a). Another sensor was further away. During the drying-out phase and WDR experiments, the injection tube ends were sealed with vapour-tight tape.

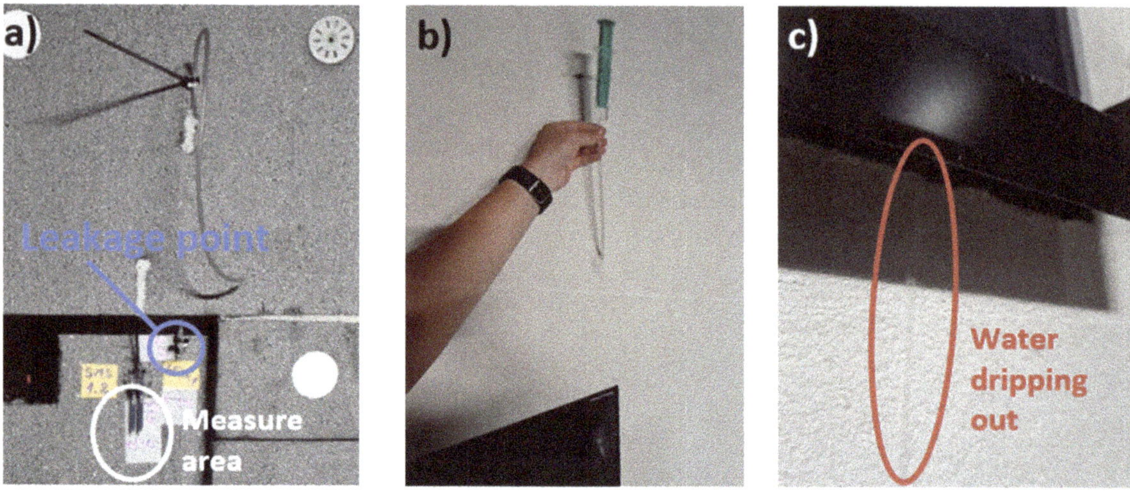

Figure 4. Water tubes for injecting water into the wall were installed before the application of PCM mortar (**a**); after finishing the facade system, water was injected behind the PV panel with a syringe (**b**); from sections with diffusion channels, water dripped out between bottom steel sheets (**c**).

The amount of water chosen to inject into the wall was chosen to be on the extreme side. It has been found out experimentally that through a 0.4 × 4 mm crack in the cement fibre board, it is possible for 18 mL/min of water to enter the wall when rain is forced into the wall with 150 Pa of air pressure difference and 12 mL/min in the case of no air pressure difference [26]. It takes into account the WDR entering directly into the crack, and also the water flowing down the surface of the facade and entering the crack on the way [26].

The water was injected into all three wall sections simultaneously, with quantities of 40 mL for one 5-min session. All three injection pipes were filled with 10 mL of water, as pipe volume was limited. After a minute, water was absorbed by the wall and the tube was filled again. One injection session consisted of 4 injections, 10 mL each, for each test wall section. 40 mL of water represents heavy rain for approximately 3 min through a typical 0.4 × 4 mm crack [26]. The next injection session was conducted after the moisture content sensor readings had been stabilised. The total amount of water injected into all of the wall sections was 220 mL over a period of 5 days.

2.5. Radiation Cycles with Solar Simulators

To simulate solar radiation in the climate chamber, three radiation modules were built, each containing eight tungsten halogen light bulbs, approximately 230 W each (see Figure 5). Halogen lights were fixed on the aluminium sheet to help to distribute the radiation more evenly on the wall surface and to remove excess heat from the bulbs. The distance to the wall was appropriately selected to achieve a radiation intensity of 500 W/m^2, which corresponds to the peak solar radiation on a vertical surface during the summer in Estonia. The regularity of the radiation intensity was measured with a pyranometer at different points inside the illuminated area. In total, an area of 1.2 m height and 2.5 m width was illuminated with radiation of 500 W/m^2 with a deviation of less than 10%. For 3 modules, 24 halogen bulbs were used with a total nominal power of 5520 W.

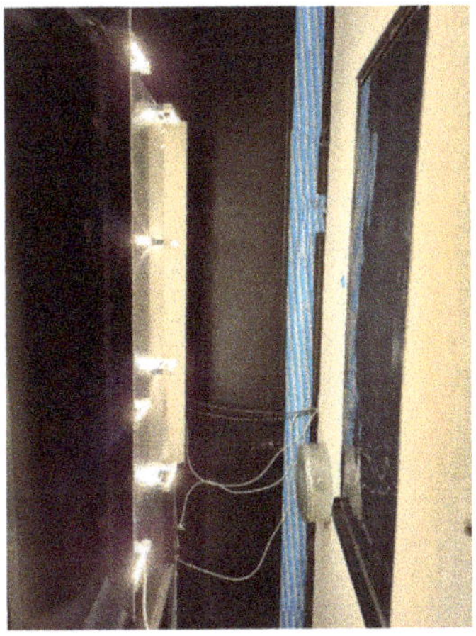

(a)

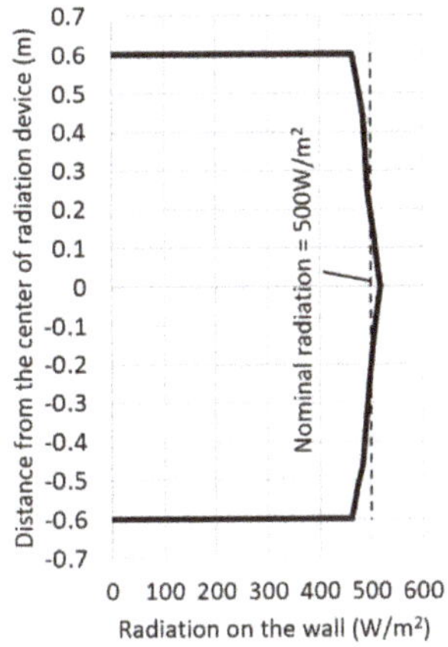

(b)

Figure 5. Solar radiation emulator, made of 8 Tungsten halogen lamps and an aluminum sheet (**a**). Three of these radiation stands were used, one for each wall section tested. Radiation intensity graph (**b**) for the radiation device at distance of 60 cm (right). Zipp Zonen CMP11 Pyranometer was used to calibrate the radiation devices.

Tungsten halogen bulbs were chosen as the radiation source because they give out high light intensity with relatively low cost and have near-sunlight spectral distribution [27]. Compared to sunlight, halogen radiation is a little weaker in the ultraviolet and stronger in the infrared range [28]. As the electrical output of PV panels was not the focus of the study, it can be stated that the thermal performances obtained under artificial and real environmental conditions were similar.

In total, 10 solar radiation cycles were performed to test the drying out of the leaked-in moisture. As an example, climate chamber temperature and RH for 2 of the cycles are shown in Figure 6. The temperature of the climate chamber was set to +20 °C, but the ventilator speed was limited to keep the air flow between the wall and the solar radiators within reasonable limits. There was stable vertical air flow at the wall of approximately 1.0 m/s.

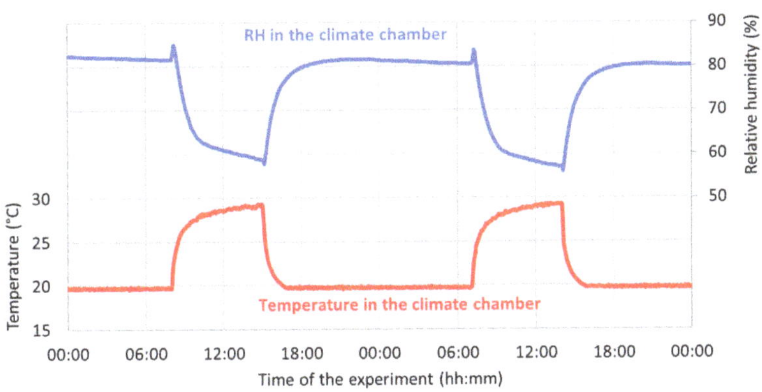

Figure 6. Temperature and RH inside the climate chamber during drying-out experiments. The increase in air temperature was due to the solar radiation simulators used to test the facade during 7 h cycles.

During the irradiation period, the air temperature close to the wall was measured at 29 °C, which simulates the air temperature on hot summer days in the Nordic climate. After approximately 7 h of stable radiation, the halogens were turned off and air in the chamber was mechanically cooled down to 20 °C, over approximately 1.5 h. Then it was kept stable during the following night for 17 h. Air flow was kept at the same level of 1.0 m/s during the whole experiment.

2.6. Wind-Driven Rain Experiment in Climate Chamber

For generating WDR load on the facade surface, water sprinkler stands were used. A method for calibrating exactly the same stands has been previously defined by Kotkas [29]. Three identical sprinkler stands were used, with 18 sprinklers in total. Each sprinkler was calibrated to spray water at 0.4 L/min. The distance between sprinklers and the wall was 55 cm and the amount of water reaching the wall was approximately 1.0 L/(min·m^2).

The character of WDR-freeze cycles is described in Figure 7. The wall was sprinkled for 1 min and then there was a 9-min break for the water to absorb into the facade plaster. This cycle was repeated 36 times, resulting in 6 h of WDR. The external air temperature was stabilised at a level of 5 °C; however, as the tap water being used was warmer than 5 °C, it caused some temperature fluctuations in the climate chamber. During the experiment, the indoor climate was kept at +21 °C and 55% RH. External RH was not controlled, but rain cycles did raise it around 94% on average at +5 °C.

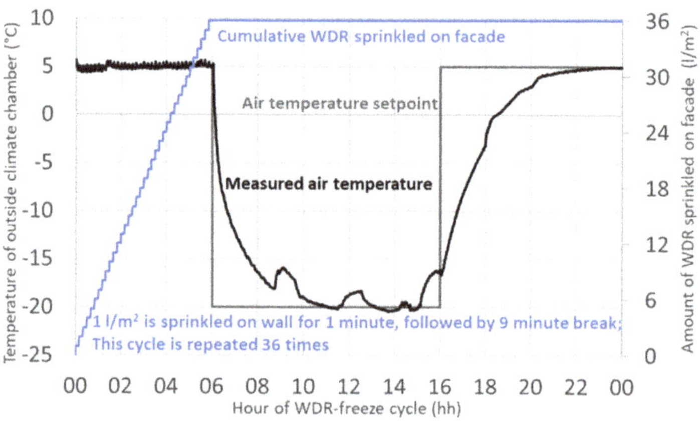

Figure 7. Characters of one WDR-freeze cycle. 6 h of WDR is followed by 10 h of extreme cold and 8 h of melting. This 24 h cycle was repeated 10 times.

After the 6 h rain period, a period of extreme cooling down to −20 °C followed (Figure 7). Cooling down the chamber was conducted as quickly as technically possible just after the sprinkler tubes were emptied. It resulted in a thin layer of ice emerging on the facade surface (Figure 8). After 10 h, the chamber was heated up to 5 °C. After 24 h, a rain period started again. This cycle was carried out 10 times.

(a) (b) (c)

Figure 8. Sprinklers for imitating WDR in climate chamber (**a**). Frozen facade after rain period (**b**,**c**).

2.7. Hygrothermal Simulations

All the results from the climate chamber experiments and test building experiments were compared to results from hygrothermal modelling. Delphin 6 software was used to develop a model of an En-ActivETICS wall. Delphin 6 uses the finite volume method to simulate the heat and moisture flow inside construction materials. Mathematical equations used in hygrothermal simulations are given in detail in [30,31]. More information about Delphin 6 software and about numerical validation can be found in [32].

For modelling the WDR and freeze-thaw cycles, two different approaches were used: the 'Kirchhoff potential for liquid flux calculation' and the 'equilibrium ice model'. The 'equilibrium ice model' takes into account the phase change effect of water, when the liquid water content inside the PCM mortar and facade plaster freezes. The 'Kirchhoff potential for liquid flux calculation' uses a function generated from the Kl function to model the liquid water transport potential. Therefore, when a wet material freezes, calculations with Kirchoff potential do not account for additional energy that is absorbed by the freezing of the water inside material pores, which leads to underestimation of the time needed for the freezing and melting processes. On the other hand, this effect is taken into account when the 'equilibrium ice model' is used to conduct calculations. Both calculation methods take into account the phase change effect of PCM itself, as this effect is written as a u(T) function into the material file.

Material properties used in the modelling are given in Table 1. For autoclaved aerated concrete (AAC), adhesive mortar, EPS and mineral wool, materials from the Delphin database were used, with some minor changes according to local material manufacturers' data sheets. The hygrothermal properties of the particular facade plaster system used in the experiment were previously measured at the Tallinn University of Technology [33]. Properties of the PCM composite mortar (made from PCM granules and cement filling) were measured both at Tallinn University of Technology and Lodz University of Technology [16].

Figure 9 shows a horizontal cross section of the 2D calculation model. The PV panel is modelled not as a material layer, but as a vapour-tight boundary condition. Diffusion channels behind the PV panel are modelled as 20 mm wide sections of boundary condition with equivalent vapour diffusion thickness Sd = 0.5 m, which considers the physical length of the channel as well as the micro-convection effect inside the channel.

Table 1. Material properties used in hygrothermal modelling.

	AAC	Adhesive Mortar	EPS Silver	Mineral Wool	PCM Mortar	Facade Plaster System	PV Panel
Absorption coefficient α, -	-	-	-	-	-	0.3	0.8
Bulk density ρ, kg/m^3	390	707	35	134	1150	1365	- [1]
Porosity θ, m^3/m^3	0.87	0.733	0.935	0.92	0.5	0.42	-
Heat capacity c, J/(kg·K)	1081	945	1500	840	According to u(T) function [2]	1000	-
Thermal conductivity λ_{dry}, W/(m·K)	0.095	0.21	0.032	0.038	0.63	0.7	-
Water vapour diffusion resistance factor μ, -	7	25	30	1	154	100	-
Equivalent vapour diffusion thickness Sd, m	-	-	-	-	-	-	10,000
Water uptake coefficient A_w, kg/(m^2·s$^{0.5}$)	0.043	0.0376	1×10^{-5}	0	0.045	0.00345	-
Liquid water conductivityK_l, kg/(m·s·Pa)	6.026×10^{-10}	3.182×10^{-9}	0	0	4.1×10^{-11}	4.18×10^{-13}	-

[1] Most of the material properties of the PV panel are not defined, because the PV was modelled as a watertight boundary condition with high water vapour resistance. [2] PCM heat capacity is described as a function for every temperature; however, outside the phase change region, the heat capacity for dry material is 1500 J/kgK. PCM mortar is a mixture of RT25HC (6% by mass; nominal melting temperature, 25 °C), RT28HC (6%; 28 °C), RT35HC (6%; 35 °C) [34], aluminium powder (24%) and cement based mortar (58%). More information about the thermal properties of the PCM mortar can be found in [14].

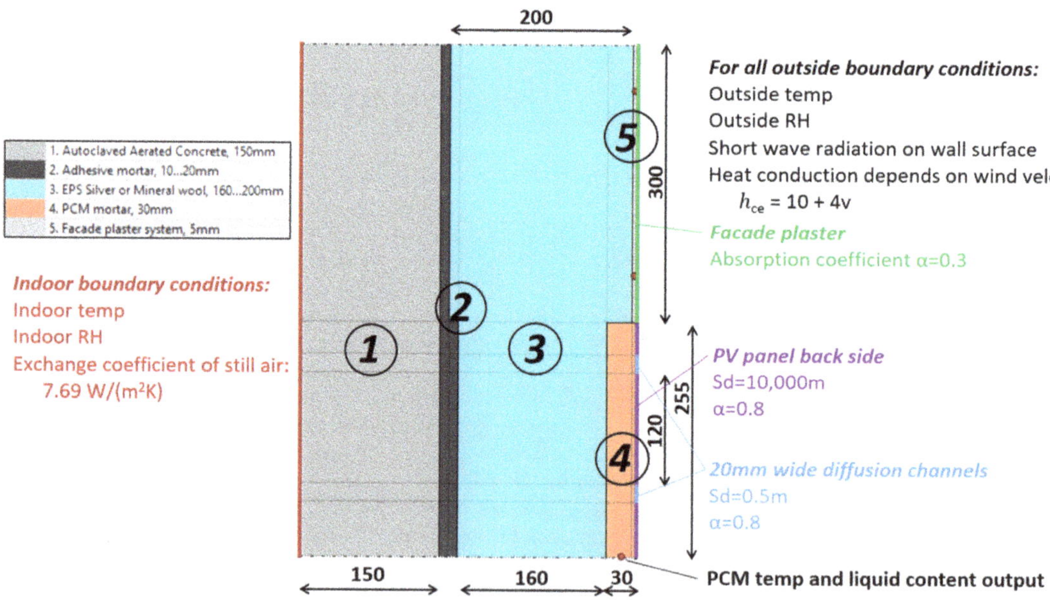

Figure 9. Two-dimensional calculation model in Delphin 6. Horizontal cross-section representing the wall section with diffusion channels. Numbers in the figure reference to the modelled materials, shown in the table left from the main figure. In case of calculation without the diffusion channels, corresponding boundary condition (blue) was replaced by PV panel boundary condition (purple).

Boundary conditions used for indoor and outdoor climates are also described in Figure 8. During all experiments, the indoor and outdoor temperature and RH were measured with sensors. Solar radiation was measured with a pyranometer (total radiation for wall surface in the case of the climate chamber and only diffuse radiation for the real facade, because the wall was behind scaffolding coverings). Before WDR experiments, the amount of water reaching the facade surface in one minute was measured according to guidelines by [29] and the data were used as a boundary condition in the calculation model.

Preliminary model validation, mainly concerning the PCM mortar moisture content, has been conducted previously and is described in detail in [16]. All the experiments described in the current study were simulated, using measured indoor and outdoor conditions as boundary condition inputs, to further validate the outputs of the calculation model.

3. Results

3.1. *Impact of Diffusion Channels to Work as Drainage Channels*

Despite injecting equal amounts of water into the PCM mortar layer for each wall section (Figure 4), it turned out to be impossible to increase the moisture content of PCM mortar above 0.11 m^3/m^3 in the example with diffusion channels (Figure 10). The reason for this was that, after the second injection, water was starting to drip out from the PCM mortar between two steel sheets that were installed below the PCM mortar (Figure 4c). The first drops of water were visible below the PV panels after approximately 40 mL of water had been injected in the EPS example and 60 mL in the mineral wool example.

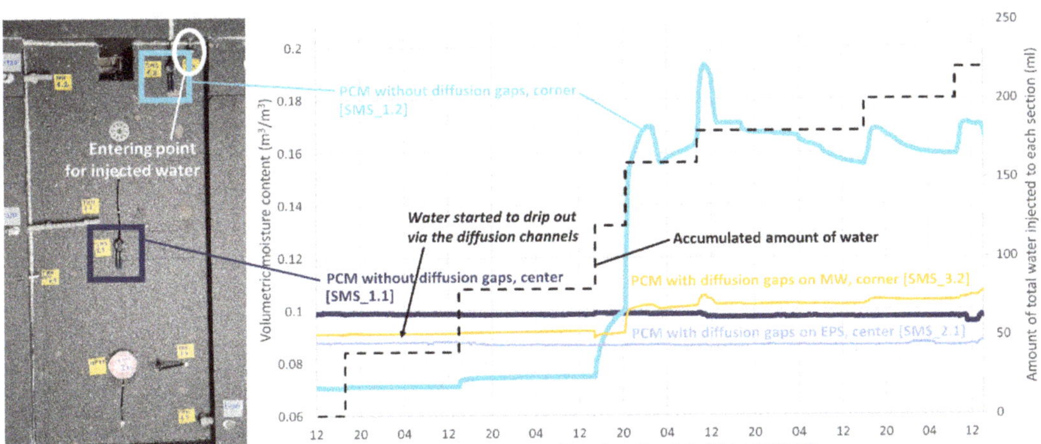

Figure 10. Moisture content of PCM mortar during water leakage simulation into the wall by injection.

In the case of the wall section without diffusion channels, all the water stayed inside the wall, and it was possible to increase the moisture content of the PCM mortar, short-term and locally, up to 0.195 m^3/m^3, but over the long term, it stabilised around the 0.16 m^3/m^3 level, as localised high moisture content spread around the leak source. However, even after 2 weeks, the moisture content registered by the soil moisture sensor in centre of the wall without diffusion channels did not change significantly, signifying that most of the 220 mL of water was absorbed locally into the PCM mortar and it takes time for the moisture levels to stabilise.

In the case of the two specimens with diffusion channels, most of the water injected into the wall found its way out via the channels. Therefore, they do act as a drainage channels as well. However, based on just two leakage tube locations, it cannot be stated that it is always 100% guaranteed that water will find its way to the diffusion channel after penetrating into the wall.

3.2. *Drying out of PCM Mortar with Controlled Radiation Cycles in Climate Chamber*

Twenty-four hours after the water injection experiment, a controlled solar radiation experiment (Figure 5) was started in the climate chamber.

It can be seen from Figure 11 that the drying out of moisture in PCM mortar without the diffusion channels was significantly slower. This is also backed up by the calculation model results. The worst-case scenario (without diffusion channels, and high initial moisture content from the water leakage) showed very small changes of dampness, and the high moisture level even short-circuited the soil moisture sensor used. Overall, the modelled results show that 10 cycles do not reduce the moisture level below 0.148 m^3/m^3.

Besides modelling the measured cases, one hypothetical and more extreme case was modelled and analysed. This simulated the possibility that larger quantities of water could be injected into the wall section with diffusion channels and that the water would not drain out of the wall through the diffusion channels. Results (Figure 11, dashed black line) show that diffusion channels help the excess moisture to dry out during the summer period faster than without diffusion channels.

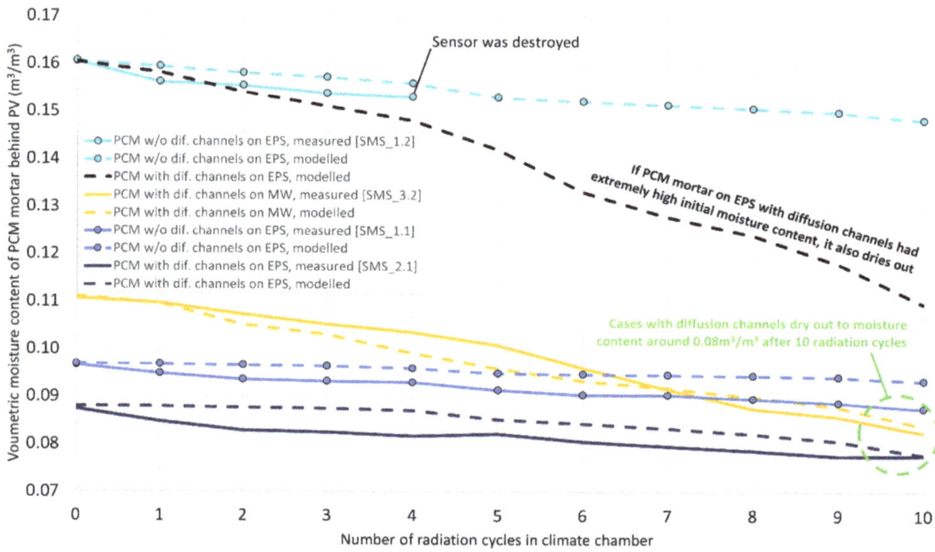

Figure 11. Drying out process of PCM mortar with solar radiation cycles in climate chamber. Measured moisture content is compared to modelled moisture contents of the PCM mortar with and without (*w/o*) diffusion channels behind vapour-tight PV panel.

Overall, diffusion channels with a cross section of 400 mm^2 spaced at every 130 mm do add additional moisture safety against water accumulation, and if a certain amount of water is still absorbed by the PCM mortar, it dries out much more easily during the summer period than without diffusion channels. Therefore, it can be assured that the moisture content will not build up over subsequent years.

Without diffusion channels, in cases where multiple PV panels are applied on the facade side-by-side, the situation becomes worse as the ratio between PV perimeter length and PV area becomes smaller. If diffusion channels are used, hygrothermal performance does not deteriorate if multiple PV panels are applied side-by-side as long as the height of the PV panel (and the height of the diffusion channel) does not change.

3.3. *Initial Moisture Content and Drying out of PCM Mortar on Test Building Facade*

Besides the climate chamber test, one test wall was built on the southern facade of the test facility and tested under real outdoor climate conditions (see Figure 3). AAC masonry was built in late April and insulation was applied within early May 2022. PCM mortar was applied in the middle of May and the facade was left to dry out under white scaffolding coverings, protected from direct sunlight and rain. Two weeks of drying out time has been proven to be enough, according to previous research [16]. During the dry-out period, the real walls tended to dry out faster than model predicts (Figure 12). This could be related to some inaccuracies in the liquid water conductivity function of the novel PCM mortar material. However, in the hygroscopic region, which is more relevant during the service life for the facade, the modelling accuracy is higher. After the dry-out period, a PV panel was glued to the PCM mortar, and facade plaster was applied to the rest of the test facade.

It can be seen from Figure 12 that despite the warm and sunny summer months of June and July, after covering the PCM mortar with the PV panel, the volumetric moisture content decreases slowly for one month. It can be seen that the drying process of PCM mortar, after covering with PV, is faster in the immediate region of the diffusion channel (corner point), than between two channels (middle point).

There is an unexpected increase in moisture content after application of the PV panel on the corner sensor (solid light blue line in Figure 12), which is significantly higher than the modelled moisture content representing the same measuring spot (dashed blue line). This could be explained by the non-homogeneous moisture load from the PV panel glue and by the micro-convection processes inside the diffusion channels. As there was warm weather after the application of the PV panel, moisture from glue mortar could have been travelling upwards via the diffusion channel with warm air. Because of its complexity, micro-convection inside the diffusion channels was not modelled.

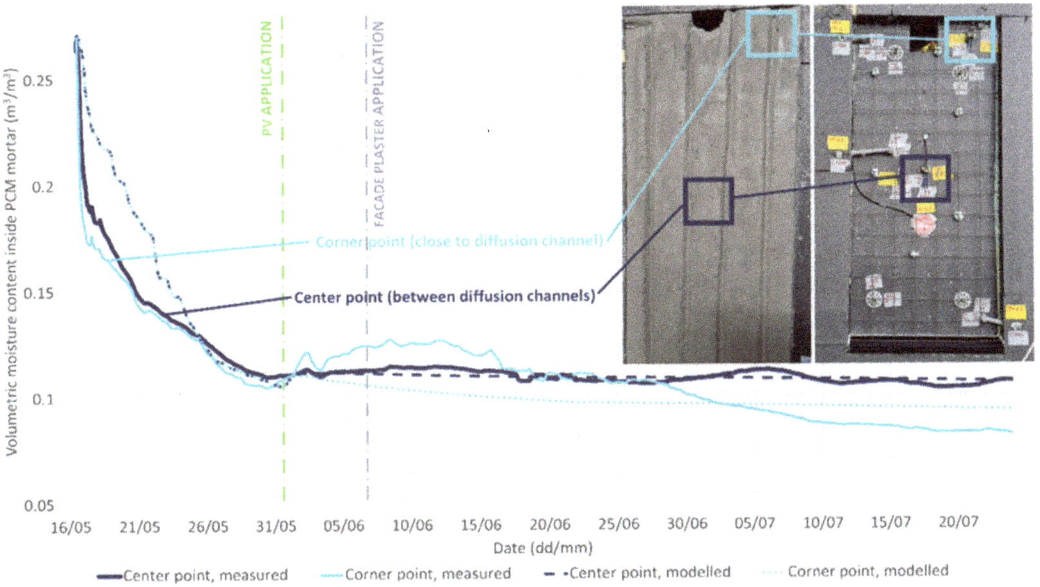

Figure 12. Drying out of PCM mortar under real conditions on southern facade in Tallinn, Estonia during summer. Locations of soil moisture sensors are highlighted on the photos before and after application of PCM mortar with diffusion channels. Comparison with calculation model results is shown with dashed lines.

3.4. Solar Radiation Experiment in Climate Chamber and Test Building

Solar radiation cycles in the case of the real wall were imitated in the climate chamber, but in a more controlled manner. In the climate chamber, there was constant irradiation for 7 h. Therefore, it made the hygrothermal performance of the wall easier to model. Overall, the temperatures of the PV panel, PCM mortar and facade plaster in the climate chamber and under real conditions during some warmer days were relatively similar (Figure 13). However, because of the controlled environment, the examples in the climate chamber are much more stable and give a higher correlation when compared to modelled results. It should be noted that peak temperatures are dependent on wind velocity—in the climate chamber this was a constant 1.0 m/s, but under real-world conditions, it varied.

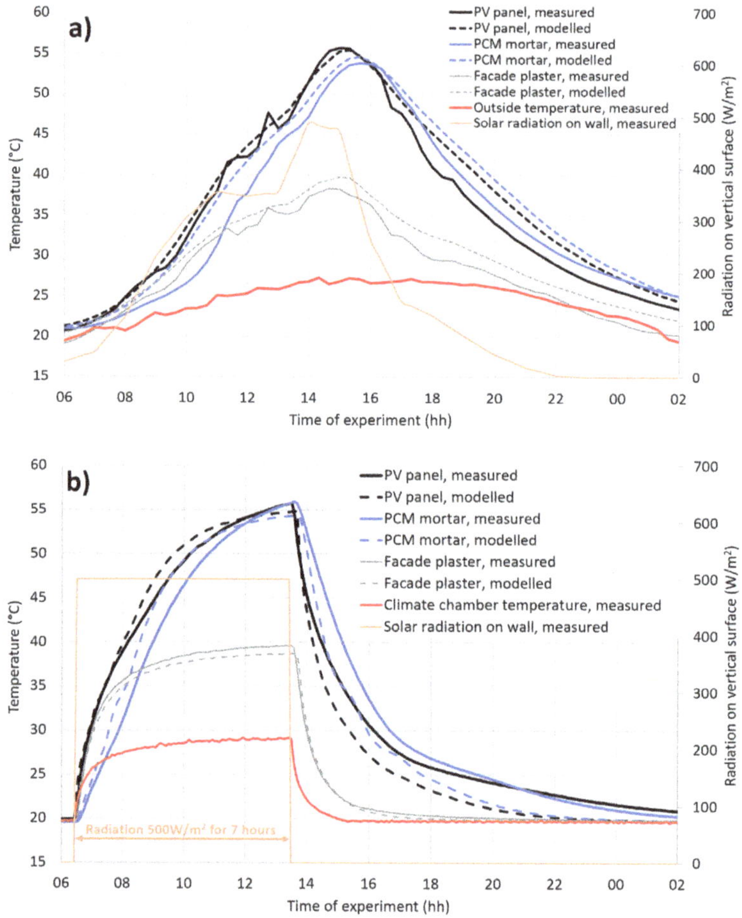

Figure 13. Temperature dynamics of energy activated facade, measured under real conditions on 30 June 2022 (**a**) and in climate chamber under simulated conditions (**b**).

There are noticeable differences between the modelled and measured temperatures in Figure 13. Although the maximum values are similar, there are differences in the heating up and cooling down speeds, especially under real climate conditions, which had much more unstable outside boundary conditions. However, if the results from the climate chamber are compared with those from the calculation model, it is visible that the real PCM mortar is heating up and cooling down slower than the modelled material. This could be most likely because of some inaccuracies of modelling heat capacity of relatively wet PCM mortar, but also because of some inhomogeneity in the mixture of PCM granules and mortar. Some differences could also be due to limitation of the Delphin model, being able to only use one average latent heat (u(T)) function for both the solidifying and melting processes, despite these processes being slightly different in the case of real material.

Differences in the facade plaster system temperatures are less significant, and, in the case of real wall, they are probably caused by some inaccuracies in modelling the wind effect and calculating the heat conduction of the surface. Differences in PV temperatures are probably a combination of differences in PCM mortar temperatures and inaccurate boundary conditions.

However, considering the prolonged nature of hygrothermal processes behind PV panels, the moisture content outputs are not so dependent on short-term inaccuracies of thermal modelling. Having a calculation model slightly underestimating the PCM effect could compensate the potential PCM capacity decrease over time [10].

3.5. Extreme Wind-Driven Rain and Freeze-Thaw Cycles in Climate Chamber

As expected, the temperature of the facade plaster changed much faster during cooling cycles than the temperature of the PCM mortar (Figure 14). It appeared that, during freezing cycles, there was a difference whether the 'Kirchhoff potential for liquid flux calculation' or the 'equilibrium ice model' was used in Delphin. The ice model does take into account the phase change effect when liquid water content inside the PCM mortar and facade plaster freezes. Using the ice model gives better correlation with measured results in the case of PCM mortar. However, as there is very little liquid water inside facade plaster, plaster temperatures are more accurate with the Kirchoff model.

It should be noted that the difference between measured facade plaster temperatures and modelled ones during the WDR phase is caused by the fact that the tap water used in the experiment was slightly warmer than the air temperature (5 °C) in the chamber.

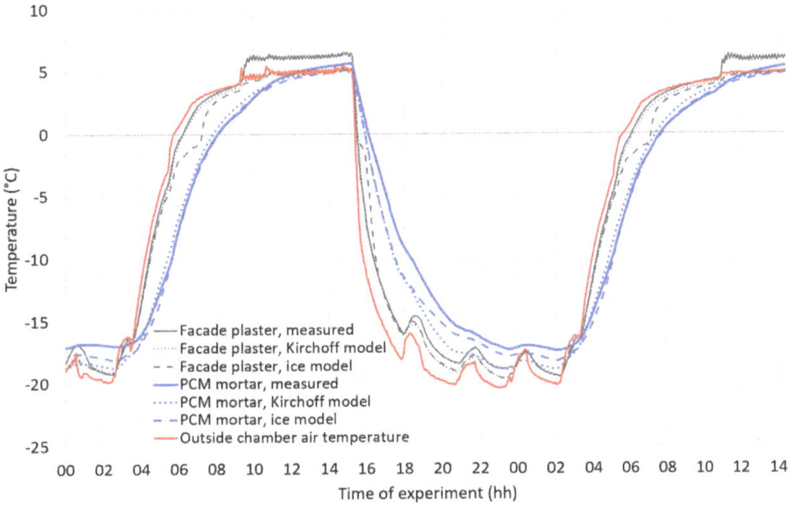

Figure 14. Facade plaster and PCM mortar temperatures measured in experiments and modelled with two different methods.

Besides temperature, the moisture content of the PCM mortar was measured with a soil moisture sensor during the WDR-freeze experiment. Overall, the moisture content of PCM mortar does not change significantly after 10 WDR-freezing cycles. At the beginning of the experiment, it was 0.078 m^3/m^3 for EPS and 0.075 m^3/m^3 for MW. After 10 cycles, volumetric moisture content was 0.076 m^3/m^3 for EPS and 0.080 m^3/m^3 for MW. Both test walls had diffusion channels.

Near-constant volumetric moisture content of PCM mortar during WDR-freeze cycles was also supported by modelled data. The main reason is that there were no cracks in the test facade and rainwater did not leak inside. Indoor excess moisture was not able to increase the moisture content of the PCM mortar significantly because there were also diffusion channels that decreased the moisture content of the PCM mortar at the same time.

After the experiment, the facade was observed for cracks visually from a distance of 30 cm with light intensity of 13,000 lux perpendicular to the wall. No visible cracks appeared after 10 extreme WDR and freezing cycles. Also, none of the PV panels were damaged during the experiment and they were able to produce the same amount of electricity.

4. Discussion

Overall, the facade system survived the series of tests without showing any signs of deterioration. The method for measuring liquid moisture content with specifically calibrated soil moisture sensors proved to be appropriate. Although, only one sensor survived all the experiments in the harsh, humid environment, so the correlation with modelled results could have deviations.

High WDR loads did not increase the moisture content behind the PV panel. This signifies that in the case of a relatively small surface area of the wall, under correct application conditions, and with good workmanship, the probability of cracks appearing is relatively low. Without cracks, the energy-activated ETICS is watertight even under extreme WDR loads. There is much further work to be done to find out what would be the maximum size of the PV ETICS facade elements before cracks start to appear under different climate conditions.

Earlier studies [17–19] have proven that, sooner or later, cracks do appear and water penetrates the facade surface. Therefore, liquid water was injected directly into the wall to imitate leakages. It was the only way to significantly increase the PCM mortar's moisture content. This study showed that water injected into the wall with diffusion channels managed to escape through the double sheet steel joint below the PV panel. However, this might not be the case for every leakage location, as there were only two places where leaks were simulated. This is another topic where further studies are needed.

The study showed relatively good correlation between the experimental results and the calculated results from the Delphin 6 model. An accurate model makes it possible to conduct calculations for different climate conditions and with various loads and to therefore identify which climates are suitable for En-ActivETICS facades. Thorough hygrothermal calculations should be conducted in the future to assess the performance of the façade in other climates with the calculation model. More test facades should be built to different climate zones and to different buildings (for example, upper corners of the facades of high-rise buildings) to compare the calculated performance with actual performance. Critical climate zones for building the system could be selected via hygrothermal modelling.

Drainage channels with a cross section of 400 mm^2 spaced at every 130 mm are highly recommended for every climate, as the risk of cracks appearing near the joints is high. Regardless of location, craftmanship errors and misapplication of construction technology can happen. If cracks have appeared, water leakage would increase the moisture content of PCM mortar behind the vapour-tight PV as it was simulated in the experiments. Results from drying-out experiments with solar radiation show that even if cracks do appear, the diffusion channels work as a secondary layer of protection.

Moreover, if diffusion channels are used, it makes it possible to apply multiple PV panels side-by-side, as excess moisture exits the wall through channels, not just through the surrounding facade plaster. This will make it possible to apply more PV panels on the facade and would lead to higher electricity production.

One limitation of the study was the relatively short time it took to perform the experiments. The extreme climate experiments were conducted during tens of climate cycles, but the results are not entirely extendable for the lifetime of the real facade system. The aging effect of the materials did not have an influence on the results because of the relatively short amount of time.

Despite the relative success of the experiments described in this study, the technology is still in development. Much more research should be conducted before manufacturing on the industrial scale can be started. Long-term hygrothermal studies should be conducted in real environments in different climate zones in the future. Possibilities to make the construction process of the wall system more effective via pre-fabrication should also be studied. Moreover, the economic aspects and carbon footprint of the system should be analyzed.

5. Conclusions

Creating a facade solution having a vapour-tight barrier as the outermost layer of the facade could result in high levels of moisture content inside the wall, especially if rainwater leaks inside the wall. This study proved that adding an extra layer of safety to the wall with diffusion channels helps the moisture to dry out much faster. Moreover, it was proven with water injection tests that diffusion channels also work as drainage channels. It is also significant that when vertical channels are used, applying PV panels side by side does not inhibit moisture performance as the PCM does not need to dry out only through the facade plaster. It makes the facade solution much more efficient, as there would be fewer PV–facade plaster joints to build on site for each PV panel installed. Overall, the diffusion channels are recommended to ensure the hygrothermal performance of the facade system.

This research resulted in a calibrated Delphin 6 model. Modelled results were calculated for each of the experiments to check the validation accuracy. A validated calculation model makes it possible to conduct long-term hygrothermal performance analyses for different climates in the future.

Overall, the En-ActivETICS facade system seems to be resistant to harsh wind-driven rain and freeze-thaw cycles. No cracks or physical deterioration appeared after 10 accelerated cycles. Facade plaster and PV panels were still fixed well on the thermal insulation and PCM mortar, and there was no immediate sign of frost damage. However, the overall durability still needs to be studied on the test wall under real climate conditions for several years. Before industrial application, several test facades with real building dimensions should also be tested in different climate zones.

It was found that in the case of test walls with diffusion channels, it was not possible experimentally to increase the moisture content of the PCM mortar higher than 0.11 m^3/m^3. Excess water drained out via the channels. Without diffusion channels, the moisture content in PCM mortar increased slightly because of the combined effects of indoor moisture load as well as injected water. Moisture content reached as high as 0.18 m^3/m^3. Both experiments and hygrothermal modelling showed that high moisture content (caused by water leakage) in walls with diffusion channels dried out to 0.08 m^3/m^3 in 10 days of high solar radiation, while it would take at least twice as much without the channels. Therefore, this facade system should be recommended to develop further only if diffusion channels are required.

Author Contributions: Conceptualization, S.I.; Methodology, M.T., S.I. and P.K.; Software, P.K. and M.P.; Validation, M.T.; Investigation, M.T.; Resources, P.K. and M.P.; Data curation, M.T.; Writing—original draft, M.T.; Writing—review & editing, S.I., T.K. and D.H.; Visualization, M.T.; Supervision, T.K.; Project administration, T.K. All authors have read and agreed to the published version of the manuscript.

Funding: This research was funded by the project En-ActivETICS in a framework of M-ERA.NET by ETAG (grant No. 3-4/MOBERA1719029), NCBiR (grant No. M-ERA.NET2/2018/2/2019) & SAS (grant No. M-ERA.NET 2/2018/786/En- ActivETICS), by the Estonian Centre of Excellence ZEBE (grant TK146), by personal research funding (grant PRG483), and by the European Commission through the LIFE IP project BuildEST (Reference: LIFE20 IPC/EE/000010).

Data Availability Statement: Detailed research data is available on demand, by contacting the corresponding author via e-mail.

Conflicts of Interest: The authors declare no conflict of interest. The funders had no role in the design of the study; in the collection, analyses, or interpretation of data; in the writing of the manuscript; or in the decision to publish the results.

References

1. Economidou, M.; Todeschi, V.; Bertoldi, P.; D'Agostino, D.; Zangheri, P.; Castellazzi, L. Review of 50 years of EU energy efficiency policies for buildings. *Energy Build.* **2020**, *225*, 110322. [CrossRef]
2. Basher, M.K.; Nur-E-Alam, M.; Rahman, M.M.; Alameh, K.; Hinckley, S. Aesthetically Appealing Building Integrated Photovoltaic Systems for Net-Zero Energy Buildings. Current Status, Challenges, and Future Developments—A Review. *Buildings* **2023**, *13*, 863. [CrossRef]
3. Maghrabie, H.M.; Abdelkareem, M.A.; Al-Alami, A.H.; Ramadan, M.; Mushtaha, E.; Wilberforce, T.; Olabi, A.G. State-of-the-art technologies for building-integrated photovoltaic systems. *Buildings* **2021**, *11*, 383. [CrossRef]
4. SUPSI. *Standardization, Performance Risks and Identification of Related Gaps for a Performance-Based Qualification in BIPV*; SUPSI: Manno, Switzerland, 2019.
5. City of Tomorrow Probing for PV Façade Systems Made of Lightweight Plastic Modules with Reversible Fittings for New and Old Buildings (PV-FAS_Light + Easy). Available online: https://nachhaltigwirtschaften.at/en/sdz/projects/pv-fas-light-easy.php (accessed on 15 March 2023).
6. Hasan, A.; McCormack, S.J.; Huang, M.J.; Sarwar, J.; Norton, B. Increased photovoltaic performance through temperature regulation by phase change materials: Materials comparison in different climates. *Sol. Energy* **2015**, *115*, 264–276. [CrossRef]
7. González-Peña, D.; Alonso-deMiguel, I.; Díez-Mediavilla, M.; Alonso-Tristán, C. Experimental analysis of a novel PV/T panel with PCM and heat pipes. *Sustainability* **2020**, *12*, 1710. [CrossRef]
8. Vasileva, I.L.; Nemova, D.V.; Vatin, N.I.; Fediuk, R.S.; Karelina, M.I. Climate-Adaptive Façades with an Air Chamber. *Buildings* **2022**, *12*, 366. [CrossRef]
9. Tao, M.; Zhenpeng, L.; Jiaxin, Z. Photovoltaic panel integrated with phase change materials (PV-PCM): Technology overview and materials selection. *Renew. Sustain. Energy Rev.* **2019**, *116*, 109406. [CrossRef]
10. Obergfell, T.; Gölz, S.; Haussmann, T.; Gschwander, S.; Wagner, A. Influence of User Behaviour on the Functioning and Performance of Passive Phase-Change Material Systems after More Than a Decade of Operation. *Buildings* **2022**, *12*, 1797. [CrossRef]
11. Ilomets, S.; Heim, D.; Chodak, I.; Czarny, D.; Kalamees, T. A method to develop energy activated ETICS. *E3S Web Conf.* **2020**, *172*, 21006. [CrossRef]
12. Heim, D.; Wieprzkowicz, A.; Knera, D.; Ilomets, S.; Kalamees, T.; Špitalský, Z. Towards improving the durability and overall performance of pv-etics by application of a pcm layer. *Appl. Sci.* **2021**, *11*, 4667. [CrossRef]
13. Talvik, M.; Ilomets, S.; Kalamees, T.; Klõšeiko, P.; Heim, D.; Wieprzkowicz, A.; Knera, D. Thermal performance of ETICS, energy activated with PCM and PV. *J. Phys. Conf. Ser.* **2021**, *2069*, 012116. [CrossRef]
14. Wieprzkowicz, A.; Heim, D.; Knera, D. Coupled Model of Heat and Power Flow in Unventilated PV/PCM Wall-Validation in a Component Scale. *Appl. Sci.* **2022**, *12*, 7764. [CrossRef]
15. Talvik, M. Hygrothermal Performance of PCM and PV Covered ETICS. Masters' Thesis, Tallinn University of Technology, Tallinn, Estonia, 2021.
16. Talvik, M.; Ilomets, S.; Klõšeiko, P.; Kalamees, T.; Heim, D. Construction process and measuring initial dryout of PCM mortar in Energy Activated ETICS Façade. In Proceedings of the 2022 Buildings XV Conference, Clearwater, FL, USA, 5–8 December 2022.
17. Sulakatko, V. Modelling Construction Process Impact Factors on Degradation of Thin Rendered Facades. Ph.D. Thesis, Tallinn University of Technology, Tallinn, Estonia, 2019.
18. Samuelson, I.; Mjörnell, K.; Jansson, A. Moisture damage in rendered, undrained, well insulated stud walls. In *Third International Symposium on Tunnel Safety, Danish Society of Engineers*; IDA: Copenhagen, Denmark, 2008; pp. 1253–1260.
19. Liisma, E.; Sepri, R.; Raado, L.M.; Lill, I.; Witt, E.D.Q.; Sulakatko, V.; Põldaru, M. Defect analysis of renovated facade walls with ETICS solutions in cold climate conditions. In Proceedings of the CESB 2016-Central Europe towards Sustainable Building, Prague, Czech Republic, 22–24 June 2016; pp. 174–181.
20. Kvande, T.; Bakken, N.; Bergheim, E.; Thue, J.V. Durability of ETICS with rendering in Norway-Experimental and field investigations. *Buildings* **2018**, *8*, 93. [CrossRef]
21. Haghighi, Z.; Dehnavi, M.A.; Konstantinou, T.; van den Dobbelsteen, A.; Klein, T. Architectural photovoltaic applications: Lessons learnt and perceptions from architects. *Buildings* **2021**, *11*, 62. [CrossRef]
22. Kertsmik, K.; Kuusk, K.; Lylykangas, K.; Kalamees, T. Energy & Buildings Evaluation of renovation strategies: Cost-optimal, CO_2 e optimal, or total energy optimal? *Energy Build.* **2023**, *287*, 112995. [CrossRef]
23. Solarfam 100W TPT Flexible Solar Panel-Technical Details. Available online: https://www.solarfam.nl/100w-shingle-tpt-flexible-solar-panel-solarfam.html (accessed on 15 March 2023).
24. Kottek, M.; Grieser, J.; Beck, C.; Rudolf, B.; Rubel, F. World map of the Köppen-Geiger climate classification updated. *Meteorol. Zeitschrift* **2006**, *15*, 259–263. [CrossRef] [PubMed]
25. METER Group. *I.U. EC-5 Manual*; METER Group: Pullman, WA, USA, 2021.

26. Olsson, L. Rain intrusion rates at façade details—A summary of results from four laboratory studies. *Energy Procedia* **2017**, *132*, 387–392. [CrossRef]
27. Esen, V.; Sağlam, Ş.; Oral, B. Light sources of solar simulators for photovoltaic devices: A review. *Renew. Sustain. Energy Rev.* **2017**, *77*, 1240–1250. [CrossRef]
28. Tawfik, M.; Tonnellier, X.; Sansom, C. Light source selection for a solar simulator for thermal applications: A review. *Renew. Sustain. Energy Rev.* **2018**, *90*, 802–813. [CrossRef]
29. Kotkas, J. Comparative Anaalysis of Wind Driven Rain Models for Hygrothermal Modelling in Estonian Climate and Development of Laboratory Test Setup. Department of Civil Engineering. Master's Thesis, Tallinn University of Technology, Tallinn, Estonia, 2015.
30. Grunewald, J. Diffusiver und konvektiver Stoff-und Energie-transport in kapillarporösen Baustoffen. Ph.D. Thesis, TU Dresden, Dresden, Germany, 1996; 220p.
31. Nicolai, A. Modeling and Numerical Simulation of Salt Transport and Phase Transitions in Unsaturated Porous Building Materials. Ph.D. Thesis, Syracuse University, Syracuse, NY, USA, 2008; 251p. [CrossRef]
32. Bauklimatik-Dresden Delphin Documentation. Available online: https://bauklimatik-dresden.de/delphin/documentation.php?aLa=en (accessed on 15 March 2023).
33. Volkova, K.; Põldaru, M.; Ilomets, S.; Kalamees, T.; Talvik, M.; Heim, D. The effect of temperature, humidity and mechanical properties on crack formation on external thin plasters of ETICS. In Proceedings of the International Building Physics Conference, Copenhagen, Denmark, 25–27 August 2021.
34. Rubitherm Rubitherm PCM RT Series. Diverse Organic PCM. Available online: https://www.rubitherm.eu/index.php/produktkategorie/organische-pcm-rt (accessed on 2 March 2023).

buildings

MDPI

Article

Influence of Piloti Forms on Wind Comfort of Different Building Group Layouts by Large Eddy Simulation

Yueyun Hu [1], Congchuan Hu [2], Guangdong Liu [2], Xiaofang Shan [1,3], Qinli Deng [1,3,*], Zhigang Ren [1,3] and Qianyu Tang [4]

1 School of Civil Engineering and Architecture, Wuhan University of Technology, Wuhan 430070, China
2 Luneng Group Co., Ltd., No. 5 Chaowai Street, Chaoyang District, Beijing 100023, China
3 Hainan Institute, Wuhan University of Technology, Sanya 572000, China
4 Building Energy Research Center, Tsinghua University, Beijing 100084, China
* Correspondence: deng4213@whut.edu.cn; Tel.: +86-27-87651786

Abstract: This paper studies the influence of different piloti rates (0%, 20%, 40%, 60%, 80%, 100%) on outdoor wind comfort for three building groups, i.e., determinant type, point type, and enclosure type. LES (Large Eddy Simulation) is used to simulate the wind environment of three clusters at six different piloti rates. This paper mainly studies the effect of piloti rate on wind speed at pedestrian level (1.5 m). The outdoor wind environment was analyzed using the average wind speed ratio, and outdoor wind comfort was evaluated using the comfortable wind ratio. The following results were obtained: (1) The piloti setting has little influence on the overall wind speed in the target area, and even an inappropriate piloti rate setting may reduce the overall average wind speed in the target area. (2) A comprehensive comparison of the three building layouts shows that the comfortable wind ratio of the determinant layout is the highest when the piloti ratio is 80%. The results of this study can provide architects and urban planners with reference for piloti and urban layout settings.

Keywords: outdoor wind environment; wind comfort; urban form; large eddy simulation; piloti rate

Citation: Hu, Y.; Hu, C.; Liu, G.; Shan, X.; Deng, Q.; Ren, Z.; Tang, Q. Influence of Piloti Forms on Wind Comfort of Different Building Group Layouts by Large Eddy Simulation. *Buildings* **2023**, *13*, 234. https://doi.org/10.3390/buildings13010234

Academic Editor: Francesco Ricciardelli

Received: 9 December 2022
Revised: 6 January 2023
Accepted: 11 January 2023
Published: 13 January 2023

1. Introduction

The safety and comfort of pedestrians around urban buildings largely depend on the wind environment at the height of the pedestrian floor [1–4]. In order to solve this problem, people have done a lot of research on pedestrian wind environments [5–9]. To clarify what urban layout, building shape, vegetation, building density, and other factors can make the wind environment of the pedestrian floor safe and comfortable [10–12]. Therefore, it is of great significance for architectural designers and decision makers to improve pedestrian level wind environment and create good pedestrian wind comfort.

In recent years, in order to improve the wind environment at urban pedestrian levels and enhance the comfort of the wind environment, people have adopted some strategies [13–16].

Cui [17] quantitatively evaluated the pedestrian wind environment of a U-shaped canyon by conducting particle image test experiments in wind tunnels. The results show that the wind speed of a U-shaped canyon at pedestrian height is generally lower than that of a parallel canyon, especially in a parallel wind direction. Weerasuriya [18] tested the influence of twisted wind characteristics on pedestrian altitude wind speed in a boundary layer wind tunnel. Taking Tsuen Wan, Hong Kong, as the research object, we found that the wind speed of the twisted wind largely depends on the size and direction of the yaw angle; especially in nearby areas with low densities of buildings, the local wind circulation is significantly affected by the twisted wind. Through a wind tunnel experiement, He [19] studied the influence of irregularity (fragmentation, angularity, and curvature) and permeability on the wind environment of the pedestrian floor. The results show that the wind environment at pedestrian level is negatively correlated with urban irregularity. There is a positive correlation between the wind performance of the pedestrian layer and urban permeability. Ikegaya [20] used the computational fluid dynamics (CFD) method to study the wind field of the pedestrian floor around the building with an aspect ratio of 1:1:2. It is proposed that the evaluation method of peak factor can better evaluate the wind environment on a pedestrian floor.

Piloti design plays a huge role in improving the microclimate around buildings [21]. In hot summers, a piloti design can not only provide shade for people but also improve the ventilation effect around the building [22]. Du [23] studied the shapes of "I", "L", "U", and "□" of four common building structures through CFD, used the average wind speed ratio and the average wind speed change rate to identify wind comfort, and quantitatively evaluated the improvement brought by the "lift-up" design. The results show that the piloti design can improve the wind comfort of the built environment, and its influence degree is closely related to the incident wind direction. Among them, the wind comfort under the oblique wind direction is better than the other two wind directions. Huang [24] compared the differences in meteorological parameters and response heat perception between the piloti part of the piloti building and the area directly exposed to outdoor sunlight based on field measurement and subject questionnaire. The results show that most people report feeling more comfortable in the shade of a piloti building. Meanwhile, through data analysis, it is concluded that solar radiation and wind speed are the two main factors affecting outdoor thermal comfort. Sha [25] used CFD to study the influence of piloti design with different wind directions on pedestrian wind comfort. The results show that when the wind direction is between 30° and 45°, the open-air setting is more effective reducing the concentration of pollutants than at 0° and 15°. The piloti design can enhance the dilution of pollutant concentrations in all winds. Tse [26] studied the influence of piloti design and size on the surrounding wind environment through a wind tunnel experiment. The results show that the piloti height is the parameter that has the greatest influence on the surrounding wind environment. Chen [27] used the CFD method to explore the impact of building height and upstream buildings on piloti buildings. The results show that the piloti building is taller or shorter than the upstream building, which can make the surrounding wind environment better. However, increasing building height and removing upstream buildings are not necessarily beneficial to the surrounding wind environment of the piloti buildings.

Computational wind engineering simulation research originated around 1963, when Smagorinsky [28] developed one of the first successful LES methods. With the improvement of computing power, large eddy simulation is becoming more and more popular in wind engineering. Muralam [29] and Yu Da-Hai [30] applied the LES model to study the flow around the blunt body. The results show that although the LES model can accurately reflect the complex characteristics of flow around a blunt body, it is difficult to apply in engineering due to the large amount of calculation. Shah KB [31] found that LES could well predict the phenomenon of flow around blunt bodies and provide important information for wind engineering researchers. Selvan [32] used the LES model to study the full-scale wind tunnel experiment, and the results showed that the prediction of the mean value was in good agreement with the measured results. Kravchenko AG [33] studied the large-eddy numerical simulation of subcritical flow around a cylinder and proposed some key problems to appropriately simulate the numerical simulation of flow around a blunt body. Tutar [34] used large eddy simulation to conduct a numerical study on two parallel buildings and presented the distribution characteristics of the flow field of the parallel buildings. Lim HC [35] used LES to study a rectangular building located in a turbulent boundary layer, and the results showed that a reasonable incoming flow and computational domain boundary can greatly improve the coincidence between LES and test results. In general, LES performs better than other turbulence modeling methods but requires much more computing resources.

Existing wind environment studies are mainly focused on the layout of buildings without piloti or with the piloti wind environment of a single building or a row of buildings.

It is generally ignored that the diverse and complex forms of urban blocks will increase the uncertainty of the effect of piloti design on improving the wind environment. In addition, there is a lack of research on the wind comfort of pedestrian floors based on the building overhead rate in urban blocks. This study considers the complexity of the form of urban blocks, and the target area of the study is set in the middle of the architectural complex. At the same time, this study is not conducted on a single building or a single row of buildings but on a small group of buildings so as to ensure the credibility of the research results. In this study, six different overhead rates were set up to study the influence of different piloti rates on the surrounding wind environment, and it was found that the piloti rate was the best for pedestrian wind comfort. At the same time, three commonly used building layouts are selected, and the comparison between these three building layouts is also made, and it is obtained that the building layout has the best piloti rate and pedestrian wind comfort. This study provides some suggestions for urban planners to mitigate the increasingly serious urban heat island effect.

2. Methodology

2.1. CFD Turbulence Models

The 3D Navier-Stokes (NS) equation is used for flow computations. The technique of LES with the Smagorinsky model is used for turbulence modeling. The continuity and momentum equations in tensorial notation are as follows:

Continuity Equation:

$$\frac{\partial \overline{U_i}}{\partial X_i} = 0 \tag{1}$$

Momentum Equation:

$$\frac{\partial \overline{U_i}}{\partial t} + \frac{\partial \overline{U_i}\overline{U_j}}{\partial X_j} = -\frac{\partial \overline{P}}{\partial X_j} + 2\frac{\partial}{\partial X_i}(\nu + \nu_{sgs})\overline{S_{ij}} \tag{2}$$

The variable 'ν' in Equation (2) is the kinematic viscosity of fluid, whereas 'ν_{sgs}' is the turbulent kinematic viscosity given as:

$$\nu_{sgs} = (C_s\Delta)^2\sqrt{2\overline{S_{ij}}\overline{S_{ij}}} \tag{3}$$

'C_s' is the Smagorinsky constant taken as Csgs = 0.167 for the current work, and 'Δ' is the cube root of the volume of a cell used in the Smagorinsky model, which is given as:

$$\Delta = \sqrt[3]{\Delta x\ \Delta y\ \Delta z} \tag{4}$$

Similarly, $\overline{S_{ij}}$ in Equation (2), the shear rate tensor computed as:

$$\overline{S_{ij}} = \frac{1}{2}\left(\frac{\partial \overline{U_i}}{\partial X_j} + \frac{\partial \overline{U_i}}{\partial X_i}\right) \tag{5}$$

2.2. Assessment Approach

2.2.1. Mean Wind Speed Ratio

In this study, the average wind speed was used to evaluate the outdoor wind environment and wind comfort. In order to make the study more general, the normalized average wind speed is used in this study. The mean velocity ratio (MVR) is calculated as follows.

$$MVR = \frac{U_N}{U_{ref}} \tag{6}$$

where, U_N is the average wind speed at any point on the pedestrian floor, U_{ref} is the average wind speed at the entrance height of 1.5 m, U_{ref} = 1.528 m/s.

2.2.2. Comfortable Wind Ratio

In this study, the comfortable wind ratio was used as the evaluation standard for outdoor comfortable wind. The so-called comfortable wind ratio refers to the area occupied by the comfortable wind in the target area over the area of the target area. However, because the area occupied by comfortable wind is not easy to calculate, a point selection process is carried out for the target area. Therefore, the comfortable wind ratio in this study is the ratio of the number of points in the comfortable wind speed interval of the target area to the total number of points in the target area.

2.2.3. Outdoor Comfortable Wind Evaluation Criteria

The influence of outdoor wind environments on human wind perception is mainly manifested in two aspects: the influence of wind on human behavior and the influence of wind on human thermal comfort. Wind speed and the fluctuation of wind speed are the key factors affecting people's feeling of wind, but the ambient air temperature, humidity, people's clothes, and solar radiation factors also affect people's feeling of wind. Taking all the factors into consideration will be a very complicated problem. Many international studies have proposed the upper limit of wind speed from the perspective of human behavior safety, but this method can only evaluate the uncomfortable problems caused by strong winds, and few studies have defined the wind speed range in terms of thermal comfort. In 1985, according to the influence of temperature on wind perception, Japanese scholars Yasunari Murakawa and Yasunari Morikawa proposed the evaluation criteria for wind speed from the perspectives of safety and comfort [36], as shown in Table 1.

Table 1. Evaluation principle of velocity considering temperature influence.

Evaluation Range	Temperature Range (°C)		
	<10	10~25	>25
Range of breezes that cause thermal discomfort in the human body (m/s)	——	——	<0.7
Human comfortable wind speed range (m/s)	<1.3	<1.5	0.7~1.7
Excessive range between comfortable and uncomfortable winds (m/s)	1.3~2.0	1.5~2.3	1.7~2.9
Range of strong winds that cause human discomfort (m/s)	>2.0	>2.3	>2.9

2.3. *Turbulence Model Validation*

2.3.1. Case Description

The experimental data for wind field turbulence model validation in this study refers to the wind tunnel test of the Japan Building Association AIJ [37]. As shown in Figure 1a, the cube represents a single building, and its size is 0.2 m (D1) × 0.2 m (W1) × 0.2 m (H). Figure 1b shows the 120 measurement points in the X-Y plane. Based on these 120 points, the simulation results are compared with experimental data, and the correlation coefficient is obtained.

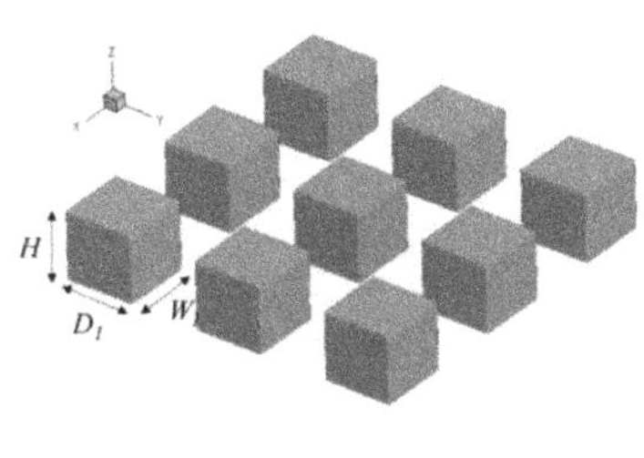

(a)

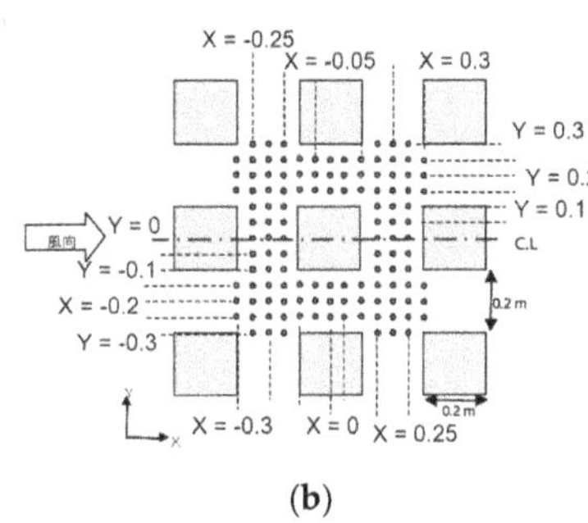

(b)

Figure 1. Wind tunnel model and schematic diagram of measuring points.(**a**) A wind tunnel experimental physical model. (**b**) Measuring point azimuth diagram.

As recommended by the Japanese AIJ guide, the computational domain entrance is 5H away from the windward section of the building (H is the height of the building, which is 0.2 m in this model), and the boundary of the computational domain on both sides is 5H away from the building. The top computing domain boundary is 10H away from the building, and the outlet boundary computing domain is 15H away from the rear end of the leeward building. The blocking rate for this arrangement is 1.8%.

2.3.2. Boundary Conditions

The inlet boundary condition is the exponential rate function obtained by fitting the experimental data measured in the wind tunnel experiment, and the equation of the inlet wind speed is $\frac{u_Z}{u_H} = \left(\frac{Z}{H}\right)^{\alpha}$, where u_H is the reference wind speed at the building height (Z = 0.2 m), u_H = 3.654 m/s; u_z is the wind speed at the entrance Z height; α is ground roughness, α = 0.28.

2.3.3. Wind Environment Verification Result

Two lines, x = −0.25 m and y = 0.25 m, were selected for comparison and verification between simulated data and wind tunnel tests. The verification results are shown in Figure 2. In Figure 2, the y-axis is the wind speed ratio, and the x-axis is the coordinate of the layout of measuring points in Figure 1b. For the line x = −0.25 m, the coordinate on the x-axis in Figure 2a is the coordinate point of y in Figure 1b. For the line y = 0.25 m, the coordinate on the x-axis in Figure 2b is the coordinate point of x in Figure 1b. When x = −0.25 m, the Spearman correlation is 0.929 and the confidence level is 99%. While y = 0.25, the M-S Spielman correlation was 0.912, with a significance level of 99%. Therefore, the simulation results can be considered credible.

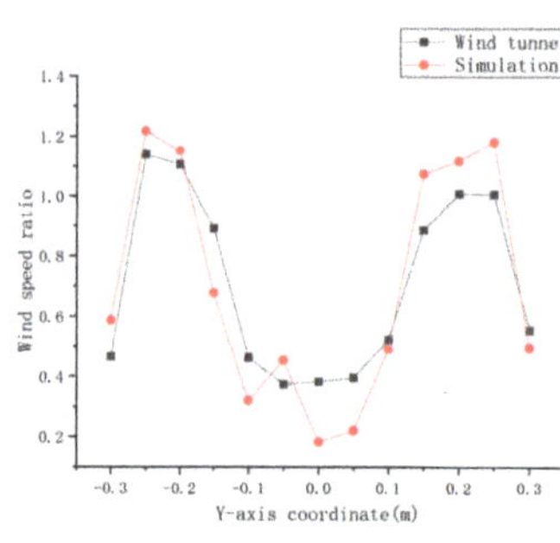

(a)

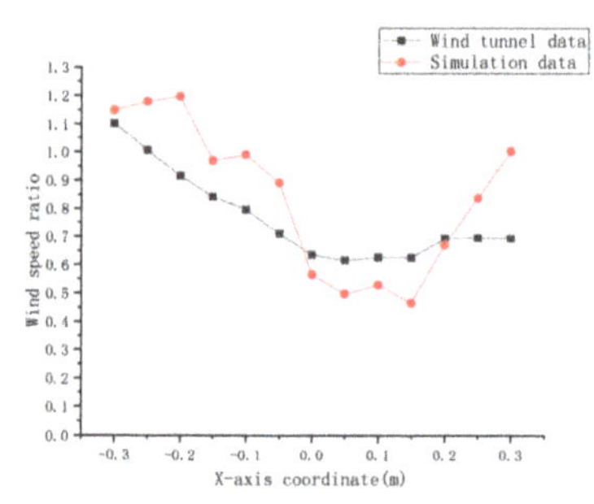

(b)

Figure 2. Comparison between wind tunnel and simulated data. (**a**) Location of the county: x = −0.25 m. (**b**) Location of the county: y = 0.25 m.

3. Building Configuration Description

3.1. Building Configuration

The analytical model in this study is determined by referring to three common residential layouts, i.e., determinant cluster, point cluster, and enclosing cluster, in Wuhan, China.

As shown in Figure 3, the architectural complex designed by this research institute consists of 88 single buildings in 8 rows and 11 columns. The size of the computational domain is 1882 × 1204 × 462 m, and the design size of the entire architectural complex is 1000 × 700 m. The distance between the front, left, and right of the entire building complex and the edge of the computing domain is 252 m (6H, H = 42 m), and the distance between the rear and the edge of the computing domain is 630 m (15H), which is in line with the recommended setting in the AIJ guide. The maximum blocking rate of the model in this study is 3.6%, which is slightly higher than the 3% blocking rate recommended by the guideline, but as long as the blocking rate is less than 10% [38], the result can also be considered reliable.

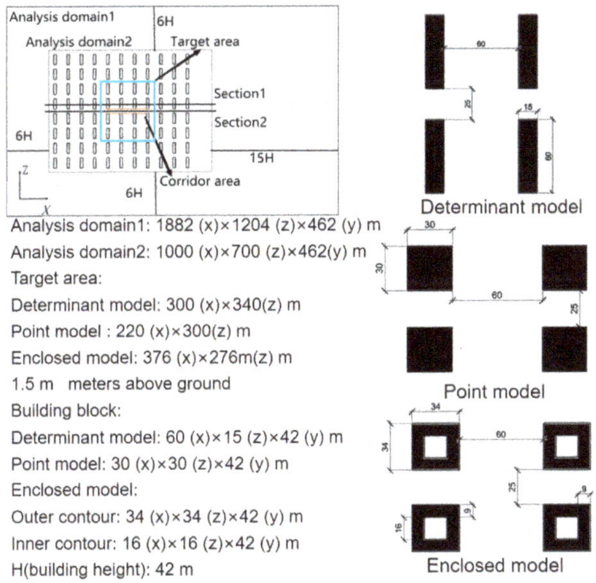

Figure 3. Architectural layout diagram.

The dimensions of a single building are shown in Figure 3. The distance between the front and back of the building is 60 m, and the distance between the left and right sides of the building is 25 m. The building size of the determinant group was 60 m (x) × 15 m (z) × 42 m (y); the point group building size was 30 m (x) × 30 m (z) × 42 m (y). The dimensions of the surrounding building are 34 m (x) × 34 m (z) × 42 m (y), and the dimensions of the inner building are 16 m (x) × 16 m (z) × 42 m (y). Although each model is different in size, its footprint is the same.

In order to eliminate the uncertainty of wind environment effect of piloti design in complex urban blocks, 16 single buildings in the center of an architectural complex were selected as the target area. The target areas of determinant, point, and enclosed types are 300 × 340 m, 220 × 300 m, and 376 × 236 m, respectively. Although the target area is different in size, its position in the whole building group is the same. There are four rows of buildings in the front row, three rows of buildings in the back row, and two rows of buildings on the left and right. At the same time, the pedestrian height of the middle passage and piloti area in the target area are studied separately. In order to study the direction of the wind in the target area, the vector diagrams of the middle section of the target area and the middle section of the building in the target area are studied. As shown in Figure 3, the middle section of the target area is Section 1. The middle section of the building in the target area is Section 2. For the determinant model, point model, and enclosed model, Section 1 is the Z = 602 m section, and Section 2 are Z = 644.5 m, Z = 629.5 m, and Z = 631.5 m, respectively.

Table 2 shows the introduction to each case. The piloti rate of each model is 0%, 20%, 40%, 60%, 80%, and 100%. The area of different groups in this study is the same. The piloti form studied is that the piloti part begins in the middle and gradually spreads to both sides. As shown in Table 1, the so-called piloti ratio is the ratio of the horizontal area of 3 m of the building to the floor area of the building. As shown in Figure 4, the so-called piloti area refers to the part of the building raised from the ground to allow the air to flow through.

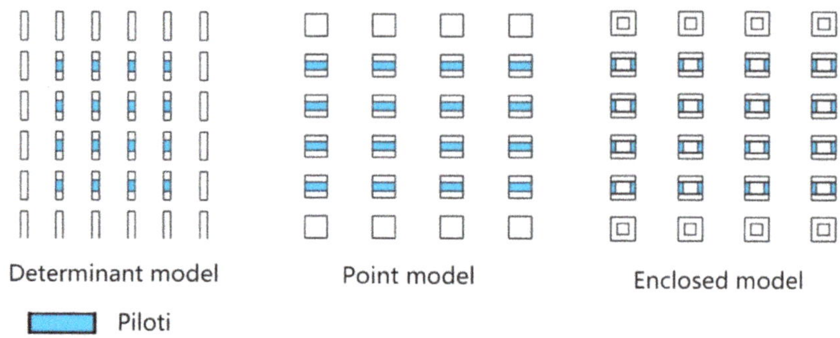

Figure 4. Building model piloti diagram.

Table 2. Analysis cases.

Case	Floor Space (m^2)	Model	Piloti Ratio	Piloti Area (m^2)
case1	900	Determinant	0%	0
case2	900	Determinant	20%	180
case3	900	Determinant	40%	360
case4	900	Determinant	60%	540
case5	900	Determinant	80%	720
case6	900	Determinant	100%	900
case7	900	Point	0%	0
case8	900	Point	20%	180
case9	900	Point	40%	360
case10	900	Point	60%	540
case11	900	Point	80%	720
case12	900	Point	100%	900
case13	900	Enclosed	0%	0
case14	900	Enclosed	20%	180
case15	900	Enclosed	40%	360
case16	900	Enclosed	60%	540
case17	900	Enclosed	80%	720
case18	900	Enclosed	100%	900

3.2. Boundary Condition

The software used in this study is Openfoam-v8.

The boundary condition type of inlet velocity is codeFixedValue, and the power law air inlet velocity is adopted. The formula is $U = U_S\left(\frac{Z}{Z_S}\right)^{\alpha}$, where US is the summer wind speed in Wuhan, China [39], U_S = 2.7 m/s, Zs = 10 m, and the ground roughness α is set to 0.3. The exit boundary condition is pressureInletOutletValue. The wall boundary condition is set to noSlip. The inlet flow set is a laminar flow inlet.

The pressure boundary condition type for the entrance and wall is set to zeroGradient. The outlet boundary condition type is set to totalPressure, which corresponds to the velocity in the inlet boundary condition.

The viscosity boundary conditions of the inlet, outlet, and wall are type calculated.

The total calculation time step is 1200 s. 1200 s is calculated from the inlet wind speed blowing across the entire computing domain to allow adequate fluid development. At the same time, the number of co was set, the maximum set is not more than 1.

The time term of the NS equation is approximated using the Euler difference scheme, while the convection, gradient, and LaPlace terms are approximated using Gauss linear approximations. The precision of the solution of the NS equation is 10^{-6}.

The turbulence condition used in this paper is $I(Z) = 0.1\left(\frac{Z}{Z_S}\right)^{-\alpha-0.05}$, where $\alpha = 0.3$, $Z_S = 462$ m, and Z is the height.

3.3. Grid Introduction

Figures 5 and 6 show the grid picture of the x-y and x-z sections when the piloti ratio of the determinant model is 0%. The author densified the target area studied and adopted a relatively sparse grid layout far away from the building model. The boundary layer grid of the first floor of the buildings in the target area is 1.6×10^{-3} (y+ < 8). The grid number of all models is shown in Table 3.

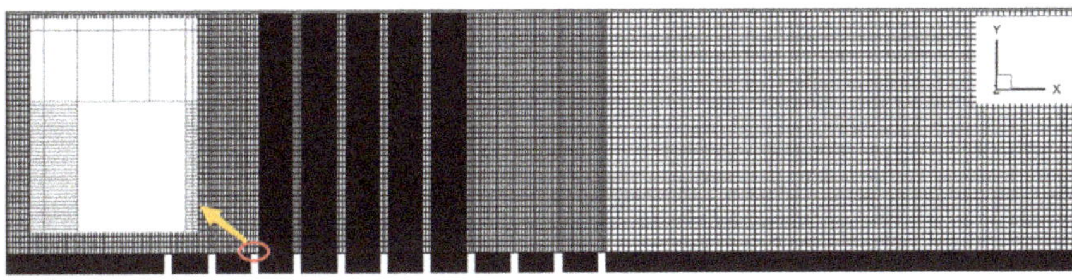

Figure 5. Schematic diagram of the x-y section grid.

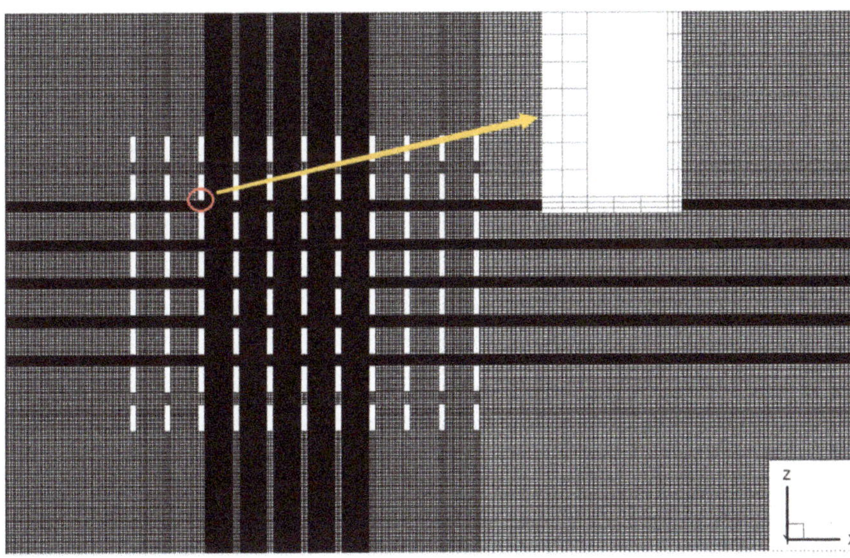

Figure 6. Schematic diagram of the x-z section grid.

Table 3. Grid information.

Piloti Rate	0%	20%	40%	60%	80%	100%
Grid number of the determinant model (10^4)	2304	1607	1600	1625	1625	1597
Grid number of the point model (10^4)	2305	1671	1653	1668	1671	1644
Grid number of the enclosed model (10^4)	2306	1715	1746	1752	1749	1715

4. Results and Discussion

4.1. The Effects of Piloti Design in the Determinant Group

4.1.1. The Effects of Piloti Design on Pedestrian Layer Wind Speed in the Target Area

As shown in Figure 7, the overall average wind speed ratio of target areas with different piloti rates presents a downward trend as a whole, with slight, but small, fluctuations in the middle. When the piloti ratio is 100%, the reason why the average wind speed ratio suddenly rises is that there are no obstacles at the bottom, which leads to the increase of the average wind speed ratio. Therefore, for the determinant model, the increase in piloti rate has a negative effect on the average wind speed of pedestrian height in the target area.

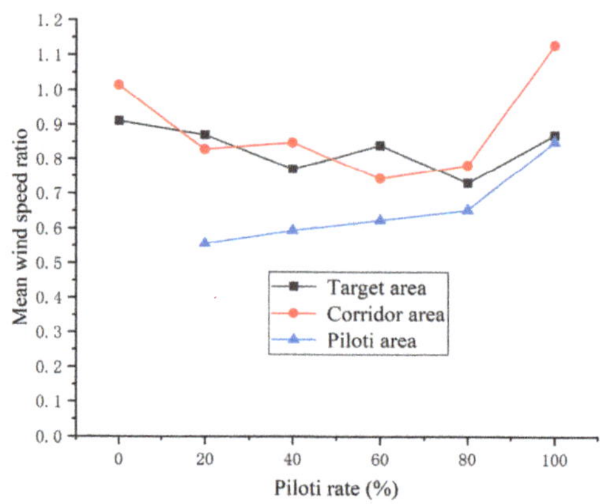

Figure 7. Variation of the mean wind speed ratio in the determinant model.

Combined with Figures 8 and 9, it can be known that the mainstream wind speed in the target area is countercurrent wind. The regional average wind speed ratio does not increase with the increase of the building piloti rate because the target area is located in the middle part of the building complex and there are obstacles in front of it. Therefore, when the wind blows into the target area, the wind speed in the target area has been attenuated, so the wind speed in the target area is small. However, there are no obstacles around the whole building complex, so the wind speed is relatively high. In the target area of the building complex, the wind speed outside the building complex is large while the wind speed in the target area is small, so the wind outside the building complex will flow to the inside of the building complex, and the inside of the target area is affected by the side wind. Furthermore, as shown in Figure 8, when the side wind enters, the wind direction from the side changes from the downwind direction to the opposite wind direction due to the obstruction of buildings. In the target area, the wind speed of the countercurrent wind is higher than that of the downcurrent wind, so most of the wind direction in the target area changes.

4.1.2. The Effects of Piloti Design on the Wind Speed at the Pedestrian Level of the Aisle in the Target Area

As shown in Figure 7, the average pedestrian wind speed ratio in the middle aisle of the target area generally presents a downward trend. Combined with Figure 10, it can be seen that the dominant wind in the middle corridor of the target area is countercurrent wind, while some wind direction blows upward, resulting in the decline of wind speed on the pedestrian floor. When the piloti ratio is 100%, the increase in the average wind speed ratio is due to the fact that there is no obstacle to block it, which leads to the increase in the average wind speed.

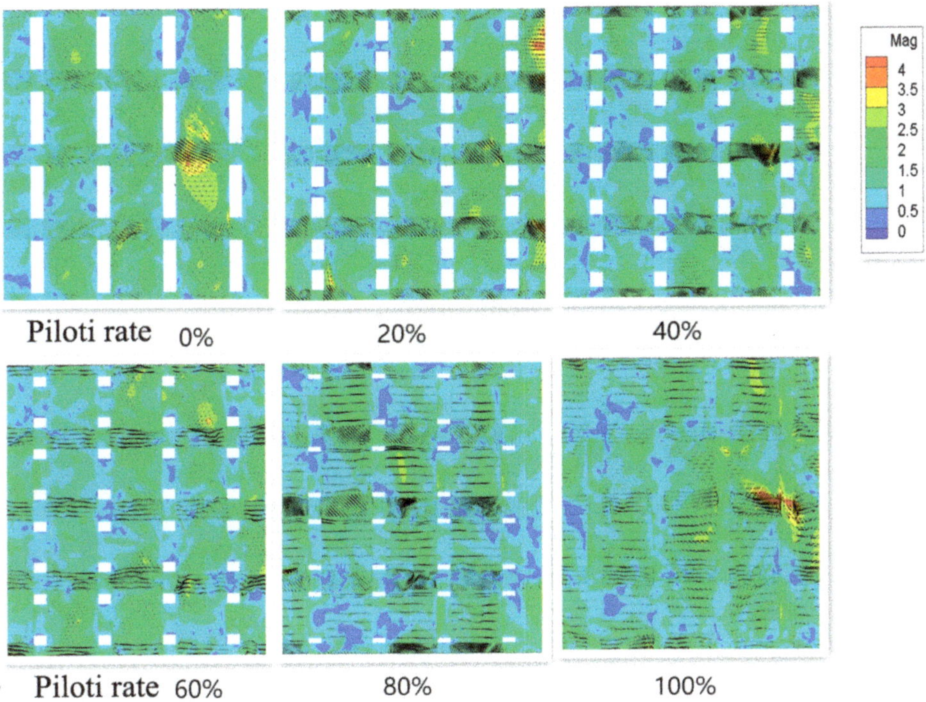

Figure 8. Vector diagram of the target area of a pedestrian layer.

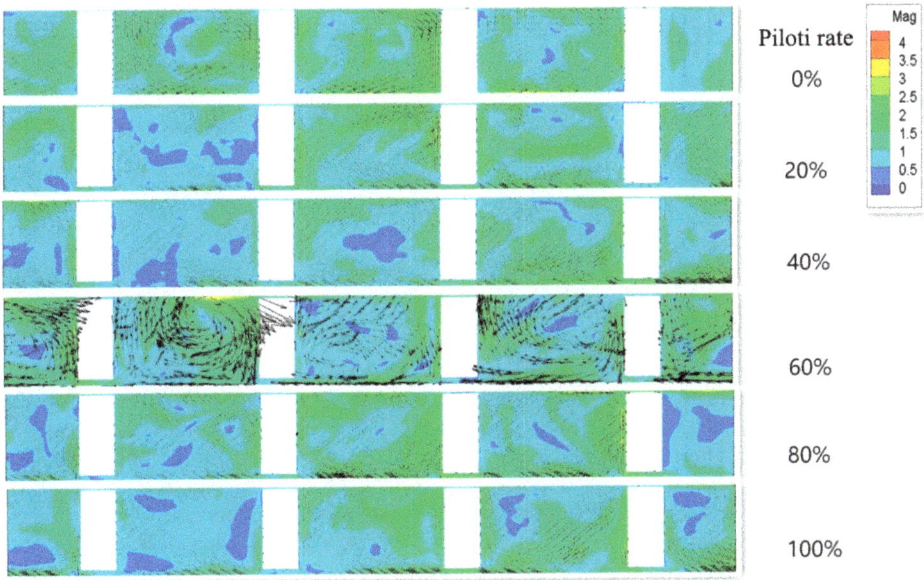

Figure 9. Vector diagram of the target area of Section 1 (Determinant model).

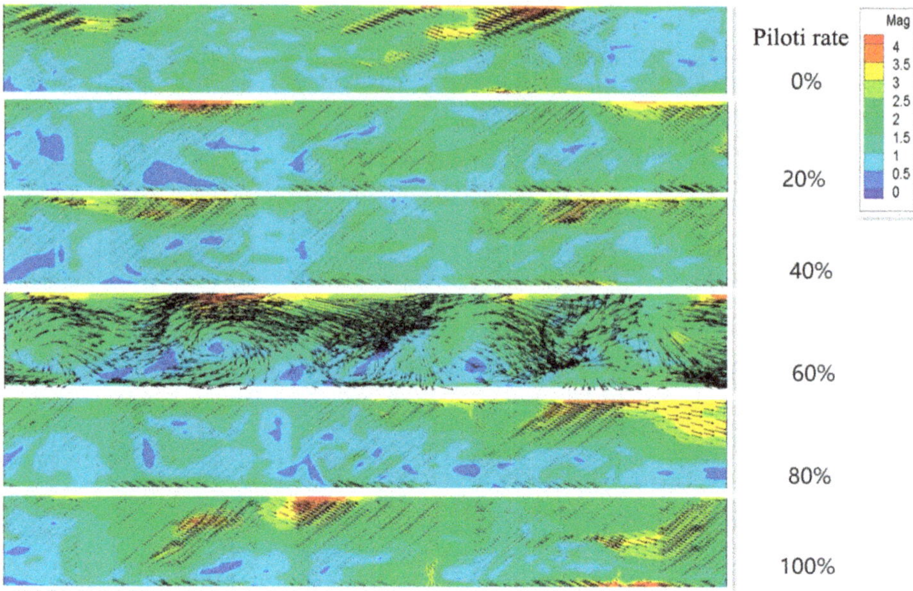

Figure 10. Vector diagram of the target area of Section 2 (Determinant model).

4.1.3. The Effects of Piloti Design on Wind Speed of the Pedestrian Layer in the Piloti Area

As shown in Figure 7, with the increase in piloti rate, the average wind speed ratio of the pedestrian floor height in piloti areas also gradually increases. It is highest when the piloti rate is 100% because it is fully piloti and there is no obstruction because the average wind speed is at a maximum. Therefore, for the determinant model, the increase in the piloti rate can increase the average wind speed of pedestrian height in the piloti space.

4.1.4. The Effects of Piloti Design on Pedestrian Layer Wind Comfort in the Target Area

The hottest month in Wuhan is July, with an average temperature of 28.4 °C [39]. Therefore, its comfortable wind range is 0.7–1.7 m/s.

The target area of the study was sampled every 0.5 m, and 355,705, 368,105, 380,009, 391,913, 403,817, and 415,721 points were selected for the piloti rate ranging from 0% to 100%.

The bar chart of the ratio and number of comfortable wind points on pedestrian floors with different piloti rates in the target area is shown below. It has been discovered that, in the target area, the proportion of comfortable wind increases as the piloti rate increases. However, when the piloti rate is 100%, the proportion of large wind speeds increases, leading to a decline in the comfortable wind ratio.

At the same time, it can be seen from Figure 11 that when the piloti rate increases from 0% to 20%, the comfortable wind ratio increases by 1.66%, with a small increase in the ratio range. However, from 20 to 40 percent, the comfortable wind ratio increased significantly, by 4.96 percent. From 40% to 60% and from 60% to 80%, the comfortable wind ratio increased by 1.76% and 1.04%, respectively.

Therefore, it can be concluded that in the summer, for the determinant group, the comfortable wind ratio increases with an increase in the piloti rate, but if the piloti rate exceeds 80%, because of the increase in the large wind speed, the comfortable wind ratio decreases. At the same time, it can be concluded that when the piloti rate is between 20% to 40%, the comfortable wind ratio increases the most.

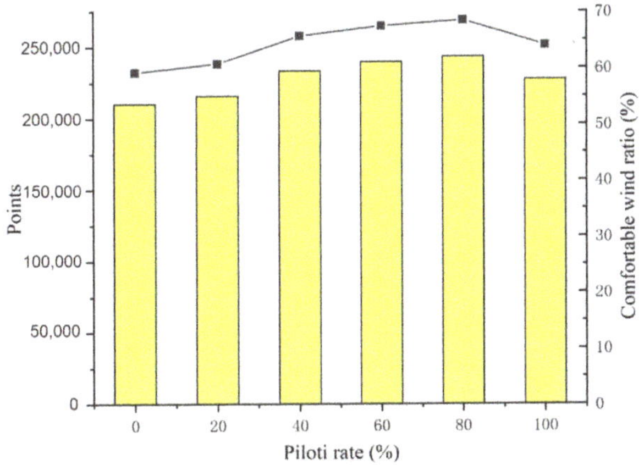

Figure 11. The changing pattern of a comfortable wind ratio under different piloti rates (Determinant model).

4.2. The Effects of Piloti Design in the Point Group

4.2.1. The Effects of Piloti Design on Pedestrian Layer Wind Speed in the Target Area

It can be seen from Figure 12 that the overall wind speed ratio of target areas under different piloti rates fluctuates, but the fluctuation is small. The maximum difference between two piloti rates does not exceed 0.07 except when the piloti rate is 100%. The maximum wind speed ratio occurs on layouts with a 100% piloti ratio, but this is because there are no obstacles under the target area and wind flow is not blocked, resulting in a higher average wind speed.

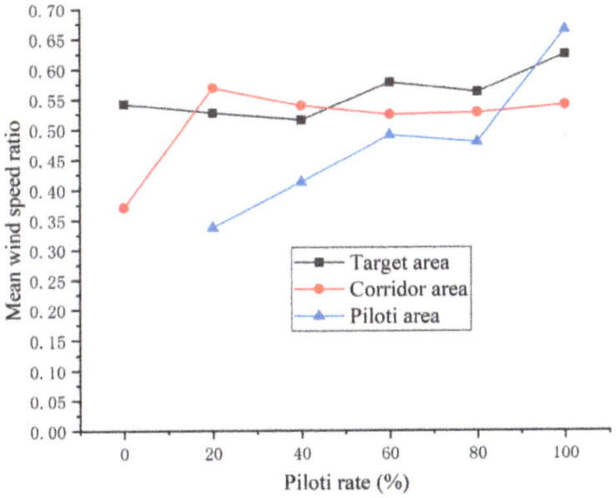

Figure 12. Variation of mean wind speed ratio in the point model.

Combined with Figures 13 and 14, it can be seen that the mainstream wind direction in the target area is countercurrent wind. The average wind speed ratio of the target area does not increase with the increase in the piloti rate of the building, and even slightly decreases when the piloti rate is 20% and 40%. This is because there is an opening at the bottom, resulting in the increased wind speed of the downstream wind is greater than that of the upstream wind. When the piloti rate is 60%, the increase is due to the further increase of the bottom opening, which leads to a larger increase in the upstream wind speed than the downstream wind speed.

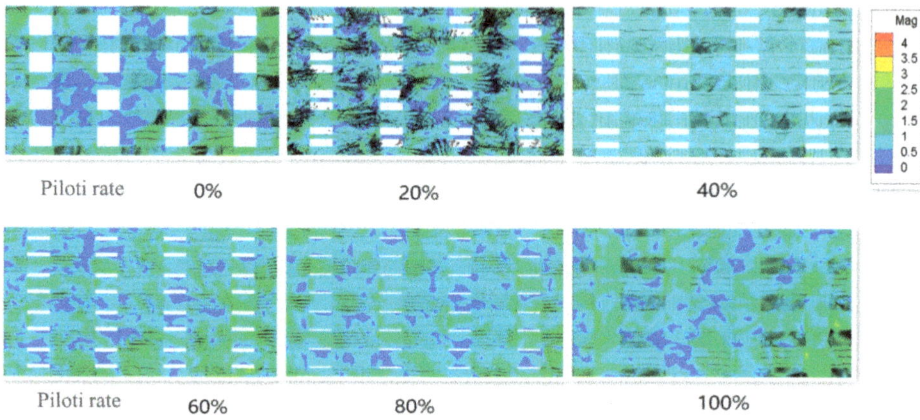

Figure 13. Vector diagram of the target area.

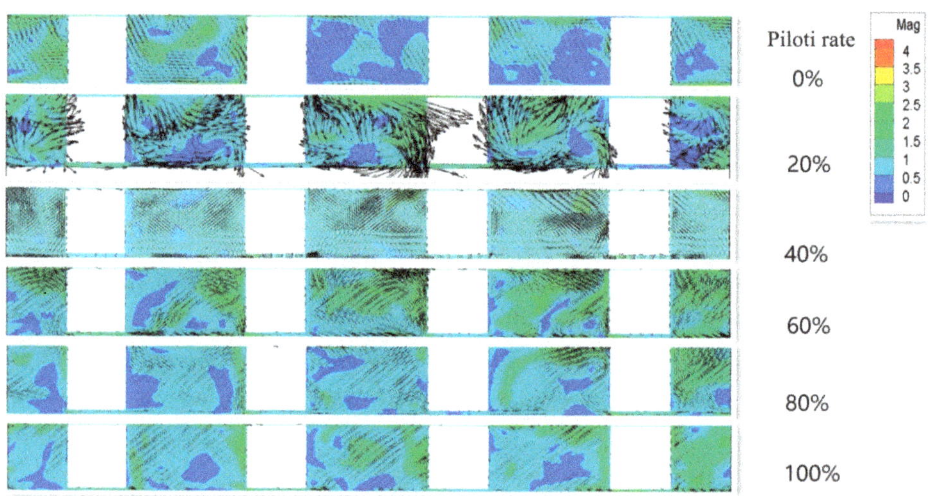

Figure 14. Vector diagram of the target area of Section 1 (Point model).

4.2.2. The Effects of Piloti Design on the Wind Speed at the Pedestrian Level of the Aisle in the Target Area

It can be seen from Figure 12 that the average wind speed ratio of the corridor in the target area is the minimum under the condition of no piloti, reaches the maximum when the piloti rate is 20%, and then begins to fluctuate gently, but the fluctuation is small. Combined with Figure 15, the main wind direction of the target area is still countercurrent. This shows that, for point layout, building piloti on the first floor can greatly improve the wind speed of the middle corridor.

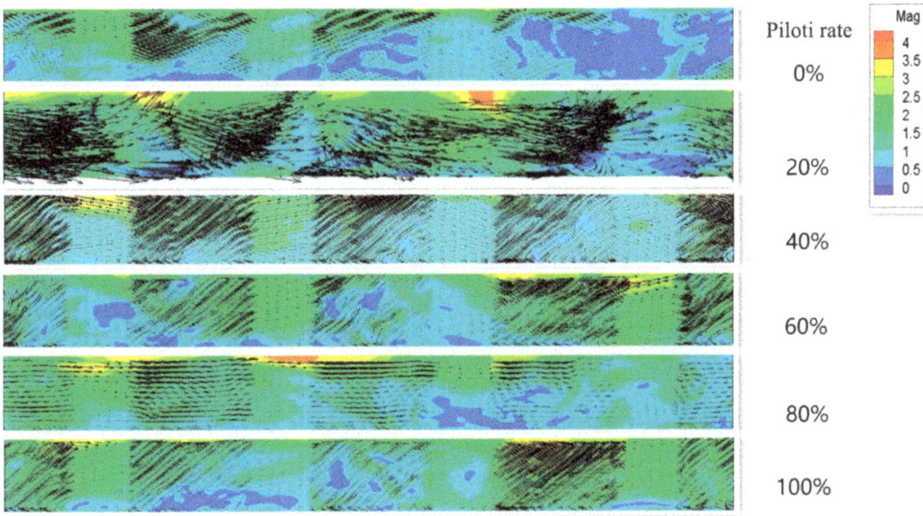

Figure 15. Vector diagram of the target area of Section 2 (Point model).

4.2.3. The Effects of Piloti Design on Wind Speed of the Pedestrian Layer in the Piloti Area

As shown in Figure 12, the average wind speed ratio at a height of 1.5 m inside the piloti space presents a trend of gradual increase. It can be concluded that for the point model, with the increase in piloti rate, the average wind speed of pedestrian floors in the piloti area also increases gradually. When the piloti ratio was 60% to 80%, the average wind speed ratio decreased slightly, which may be because at a 60% piloti ratio, the average wind speed of downstream wind and countercurrent wind increased by the same amplitude, so the average wind speed did not change much. When the piloti ratio is 100%, the sudden rise in the average wind speed ratio is because there is no blockage at the bottom, so the average wind speed ratio increases.

4.2.4. The Effects of Piloti Design on Pedestrian Layer Wind Comfort in Target Area

The target area of the study was sampled every 0.5 m, and 263,906, 276,593, 288,305, 300,017, 311,729, and 323,441 points were selected for the piloti rate ranging from 0% to 100%.

It can be seen from Figure 16 that in the target area, the comfortable wind ratio presents an overall upward trend with the increase in piloti rate. When the piloti rate is 100%, the comfortable wind ratio reaches its maximum.

However, when the piloti ratio was 40%, the comfortable wind ratio showed a decreasing trend compared with 20%. Combined with Figure 12, it can be seen that for the layout with a 40% piloti ratio, its average wind speed ratio in the target area is the smallest among all point models. This is due to the increase in the speed of the countercurrent wind, which leads to a decrease in the overall wind speed. Therefore, when the piloti ratio is 40%, the comfortable wind ratio decreases.

For the point model, the proportion of comfortable wind is highest when the piloti ratio is 100%. However, when the piloti rate is 100%, it is impossible to achieve, and when the piloti rate is 80%, the comfortable wind ratio is only 1.23% smaller than when the piloti rate is 100%. Therefore, it is also appropriate to choose a slightly smaller piloti rate.

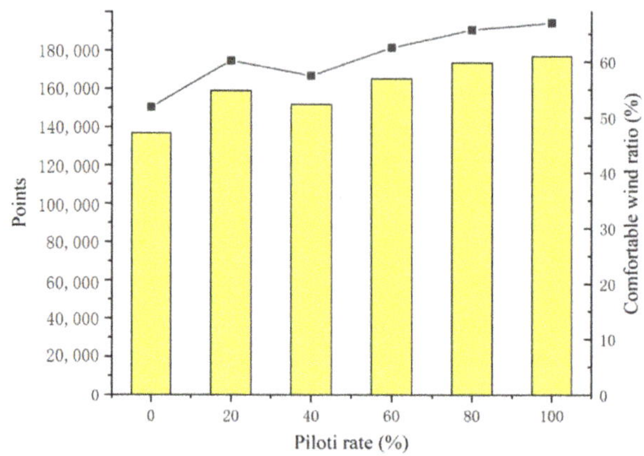

Figure 16. The changing pattern of a comfortable wind ratio under different piloti rates (Point model).

4.3. The Effects of Piloti Design in the Enclosed Group

4.3.1. The Effects of Piloti Design on Pedestrian Layer Wind Speed in the Target Area

Figure 17 shows that the average wind speed of pedestrian floors in the target area decreases first and then gradually flattens. This is due to the change in the building piloti rate. However, it fluctuates slightly when it is flat, and the difference in average wind speed ratio between different piloti rates is less than 0.035. This indicates that for the enclosed model, the building piloti rate has little influence on the average wind speed of pedestrian floors in the target area. Even when the piloti rate is 20%, this arrangement is not conducive to the average wind speed of the pedestrian layer in the target area.

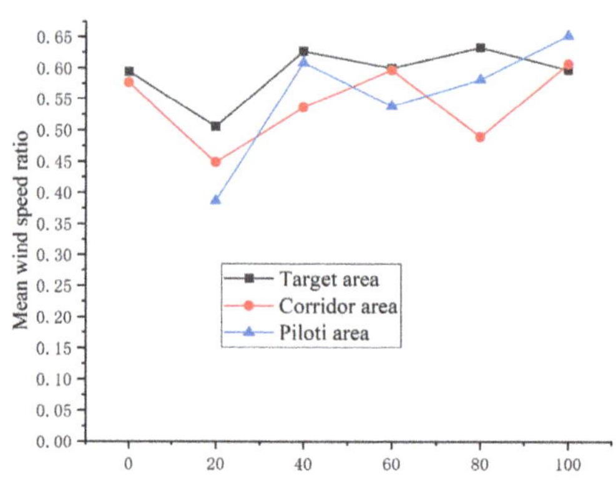

Figure 17. Variation of mean wind speed ratio in the enclosed model.

According to Figures 18 and 19, the mainstream wind speed in the target area is countercurrent wind. When the piloti rate is 20%, the decrease of the average wind speed is because the wind direction is partly upward, which leads to the decrease of the average wind speed in the target area. When the piloti ratio is 100%, it is lower than when it is 40%, 60%, and 80%, which is also because part of the wind direction is upward.

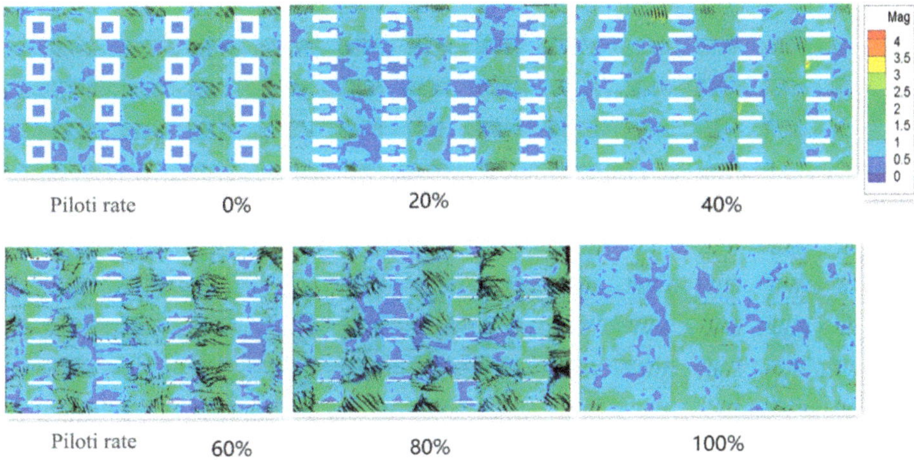

Figure 18. Vector diagram of the target area of the pedestrian layer.

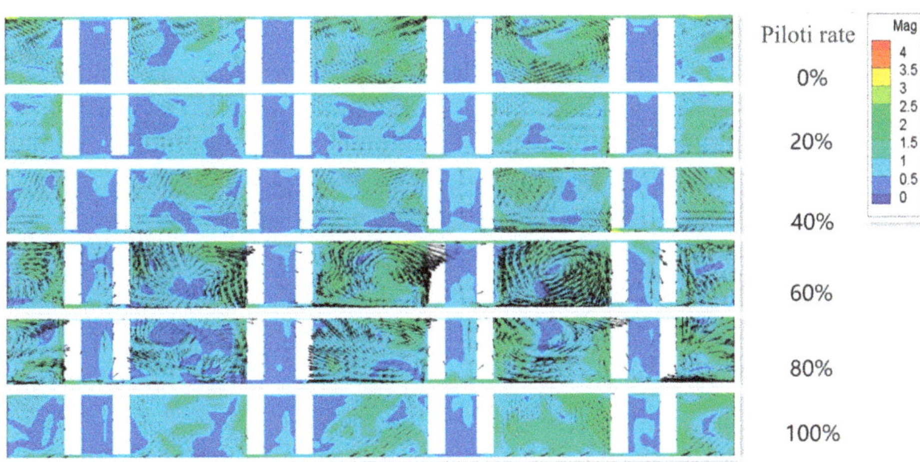

Figure 19. Vector diagram of the target area of Section 1 (Enclosed model).

4.3.2. The Effects of Piloti Design on the Wind Speed at the Pedestrian Level of the Aisle in the Target Area

As shown in Figure 17, the average wind speed ratio of the pedestrian floor in the middle aisle of the target area shows a trend of fluctuation, and the fluctuation range is large. The maximum average wind speed ratio of the middle passage at a height of 1.5 m in the target area is 100%, which is 0.607.

When the piloti ratio is 20%, the average wind speed ratio is the lowest. At the same time, according to Figure 20, the wind flows upward and the wind speed ratio decreases because of the low piloti rate. The average wind speed ratio of the target area is the lowest when the piloti ratio is 20%.

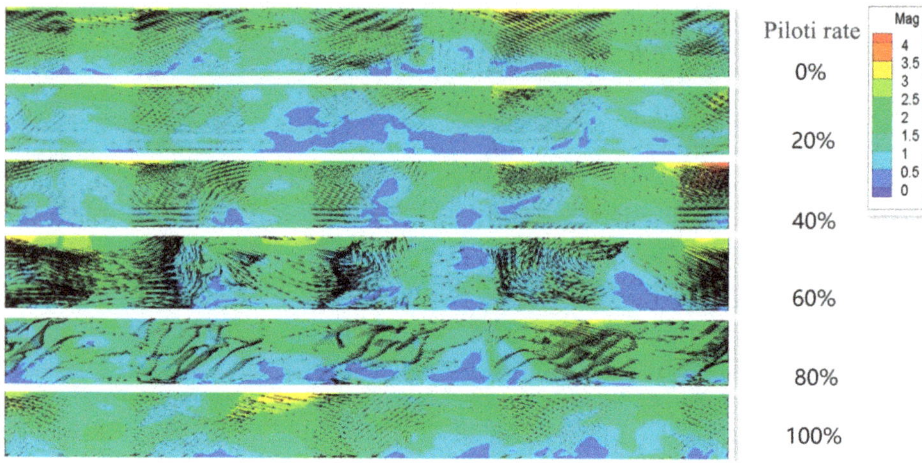

Figure 20. Vector diagram of the target area of Section 2 (Enclosed model).

4.3.3. The Effects of Piloti Design on Wind Speed of the Pedestrian Layer in the Piloti Area

It can be seen from Figure 17 that, with the continuous increase of piloti rate, the average wind speed ratio at the height of 1.5 m in the piloti space presents an overall trend of increase. When the piloti ratio is 100%, the average wind speed ratio reaches its maximum, which is 0.652. Compared with the piloti rate of 20%, there is a great improvement.

4.3.4. The Effects of Piloti Design on Pedestrian Layer Wind Comfort in the Target Area

The target area of the study was sampled every 0.5 m, and 285,985, 298,753, 312,259, 325,921, 336,961, and 348,001 points were selected for the piloti rate ranging from 0% to 100%.

It can be seen from Figure 21 that for the enclosed model, the increase in piloti rate has a certain effect on improving wind comfort in the target area of the enclosed layout, but the effect is small. Even if the piloti rate is not set correctly, it may reduce outdoor comfort.

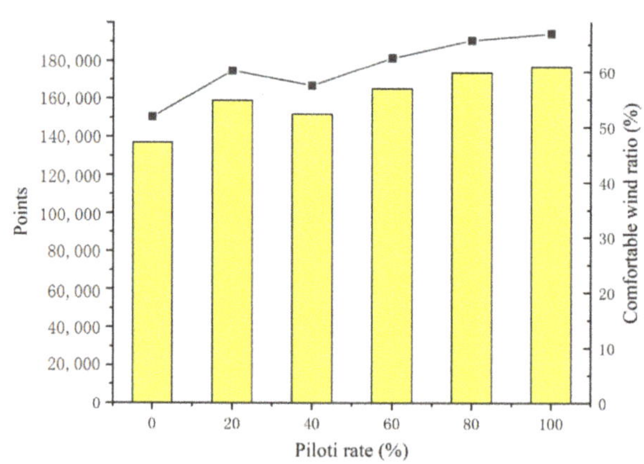

Figure 21. The changing pattern of a comfortable wind ratio under different piloti rates (Enclosed model).

When the piloti ratio is 20%, the comfortable wind ratio is the lowest. In combination with the vector diagram, this is due to the double effect of the side-current wind and a large part of the wind flowing upwards. Because of these two factors, when the piloti rate is 20%, the overall average wind speed ratio is smaller, which leads to a smaller comfortable wind ratio in the target area.

When the piloti ratio is 60%, the comfortable wind ratio is 65.3%. Compared with no piloti, the improvement range is 2.41%. This is a significant improvement over the 20% reduction in piloti rates and the less than 1% improvement in piloti rates of 40%, 80%, and 100%.

Therefore, it can be concluded that for the enclosed model, outdoor pedestrian wind comfort reaches its maximum when the piloti rate is 60%.

4.4. Comparisons between Different Building Configurations

The piloti setting has little influence on the overall wind speed in the target area, and even an inappropriate piloti rate setting may reduce the overall average wind speed in the target area. For the determinant layout, with the increase in piloti rate, the average wind speed of pedestrian height in the target area is adversely affected. For point and enclosed layouts, setting the correct piloti rate can improve the overall mean wind speed of the area, requiring the piloti rate to be greater than 40%.

For point layouts, the piloti setting can greatly improve the average wind speed of the center aisle. This is not the case with determinant and enclosing layouts. For the determinant layout, the piloti setting reduces the average wind speed of the center aisle. The enclosed layout is the same, but when the piloti ratio is 60%, the average wind speed in the corridor is slightly increased.

At the same time, with the increase in piloti rate, the average piloti space wind speed of the three building layouts increases accordingly. This indicates that the greater the piloti ratio, the greater the wind speed inside the piloti space.

When there is no piloti condition, the determinant comfortable wind ratio is 59.14%. The point comfortable wind ratio is 51.84%; the enclosed comfortable wind ratio is 62.89%. The enclosed building layout is better than the other two kinds of building layout when there is no piloti. The point layout had the worst wind comfort ratio and was much lower than the other two layouts. For the determinant model, the point model, and the enclosed model, the maximum comfortable wind ratios in the target area are 68.44%, 66.99%, and 65.3% when the piloti ratio is 80%, 100%, and 60%, respectively. The wind comfort ratio is highest when the row layout and piloti ratio are both at 80%. Compared with the no piloti condition, the piloti rate of determinant, point, and enclosure models are improved by 9.3%, 15.15%, and 2.41%, respectively. The comfortable wind ratio is most obvious in the ground-floor piloti point-to-point layout. For the enclosed model, the improvement effect of the bottom frame is poor.

5. Conclusions

This paper mainly studies the influence of different piloti ratios on outdoor wind comfort for three typical building layouts. A total of six different piloti rates were set: 0%, 20%, 40%, 60%, 80%, and 100%. Three common residential building layouts in Wuhan, China were selected as the research objects, which are determinant type, point type, and enclosed type. There were 18 sets of models. The average wind speed was compared among the target area, the middle corridor of the building, and the pedestrian layer of the building's piloti space. The outdoor wind comfort was analyzed by the comfortable wind ratio.

The results show that: (1) The piloti setting has little influence on the overall wind speed in the target area, and even an inappropriate piloti rate setting may reduce the overall mean wind speed in the target area. (2) For point layout, the piloti setting can improve the average wind speed in the middle passage. However, for the determinant and enclosure layout, the average wind speed in the corridor area will be reduced. (3) With the increase in piloti rate, the average wind speed of the piloti area can be improved. (4) A comprehensive comparison of the three building layouts shows that the comfortable wind ratio of the determinant layout is the highest when the piloti ratio is 80%.

The results of this study can provide architects and urban planners with reference for piloti and urban layout design. The reasonable piloti rate setting can significantly improve outdoor wind comfort as well as outdoor pedestrian wind comfort in densely built cities.

Author Contributions: Methodology, Q.D.; software, X.S.; validation, Y.H., C.H. and G.L.; formal analysis, Z.R.; resources, Q.D.; data curation, Y.H.; writing—original draft preparation, Y.H.; writing—review and editing, Y.H.; supervision, Q.D. and Q.T.; funding acquisition, Q.D., X.S. and Q.T. All authors have read and agreed to the published version of the manuscript.

Funding: This research was funded by the Hainan Province Science and Technology Special Fund, ZDKJ2021024 and the Sanya Science and Education Innovation Park of Wuhan University of Technology, 2021KF0002 and 2021KF0004.

Data Availability Statement: Not applicable.

Acknowledgments: This research is supported by Luneng Group Co., Ltd.'s "Research Project on Relevant Upgrade Technology Solutions to Respond to the New Green Building Evaluation Standards".

Conflicts of Interest: The authors declare no conflict of interest.

References

1. Wang, W.; Wang, X.; Ng, E. The coupled effect of mechanical and thermal conditions on pedestrian-level ventilation in high-rise urban scenarios. *Build. Environ.* **2021**, *191*, 107586. [CrossRef]
2. Zou, J.; Yu, Y.; Liu, J.; Niu, J.; Chauhan, K.; Lei, C. Field measurement of the urban pedestrian level wind turbulence. *Build. Environ.* **2021**, *194*, 107713. [CrossRef]
3. Zou, J.; Liu, J.; Niu, J.; Yu, Y.; Lei, C. Convective heat loss from computational thermal manikin subject to outdoor wind environments. *Build. Environ.* **2021**, *188*, 107469. [CrossRef]
4. Cao, S.-H.; Ming, P.-P.; Zhao, X. Fuzzy comprehensive evaluation of human thermal comfort in simulating natural wind environment. *Build. Environ.* **2021**, *188*, 107447. [CrossRef]
5. Tominaga, Y.; Shirzadi, M. Wind tunnel measurement of three-dimensional turbulent flow structures around a building group: Impact of high-rise buildings on pedestrian wind environment. *Build. Environ.* **2021**, *206*, 108389. [CrossRef]
6. Kikumoto, H.; Choi, W.; Ooka, R. Development of probabilistic assessment framework for pedestrian wind environment using Bayesian technique. *Build. Environ.* **2021**, *187*, 107419. [CrossRef]
7. Du, Y.; Mak, C.M.; Ai, Z. Modelling of pedestrian level wind environment on a high-quality mesh: A case study for the HKPolyU campus. *Environ. Model. Softw.* **2018**, *103*, 105–119. [CrossRef]
8. Wu, Y.; Zhan, Q.; Quan, S.J. Improving local pedestrian-level wind environment based on probabilistic assessment using Gaussian process regression. *Build. Environ.* **2021**, *205*, 108172. [CrossRef]
9. Weerasuriya, A.U.; Zhang, X.; Lu, B.; Tse, K.T.; Liu, C.H. A Gaussian Process-Based emulator for modeling pedestrian-level wind field. *Build. Environ.* **2021**, *188*, 107500. [CrossRef]
10. Ricci, A.; Guasco, M.; Caboni, F.; Orlanno, M.; Giachetta, A.; Repetto, M.P. Impact of surrounding environments and vegetation on wind comfort assessment of a new tower with vertical green park. *Build. Environ.* **2022**, *207*, 108409. [CrossRef]
11. Kabošová, L.; Chronis, A.; Galanos, T.; Kmet', S.; Katunský, D. Shape optimization during design for improving outdoor wind comfort and solar radiation in cities. *Build. Environ.* **2022**, *226*, 109668. [CrossRef]
12. Gao, H.; Liu, J.; Lin, P.; Li, C.; Xiao, Y.; Hu, G. Pedestrian level wind flow field of elevated tall buildings with dense tandem arrangement. *Build. Environ.* **2022**, *226*, 109745. [CrossRef]
13. Hu, H.; Chen, Q.; Qian, Q.; Zhou, X.; Chen, Y.; Cai, Y. Field investigation for ambient wind speed and direction effects exposure of cyclists to PM2.5 and PM10 in urban street environments. *Build. Environ.* **2022**, *223*, 109483. [CrossRef]
14. Takebayashi, H. Effects of air temperature, humidity, and wind velocity distribution on indoor cooling load and outdoor human thermal environment at urban scale. *Energy Build.* **2022**, *257*, 111792. [CrossRef]
15. Yang, Q.; Xu, X.; Lin, Q.; Tamura, Y. Generic models for predicting pedestrian-level wind around isolated square-section high-rise buildings. *J. Wind Eng. Ind. Aerodyn.* **2022**, *220*, 104842. [CrossRef]
16. Shirzadi, M.; Tominaga, Y. Multi-fidelity shape optimization methodology for pedestrian-level wind environment. *Build. Environ.* **2021**, *204*, 108076. [CrossRef]
17. Cui, D.; Hu, G.; Ai, Z.; Du, Y.; Mak, C.M.; Kwok, K. Particle image velocimetry measurement and CFD simulation of pedestrian level wind environment around U-type street canyon. *Build. Environ.* **2019**, *154*, 239–251. [CrossRef]
18. Weerasuriya, A.U.; Tse, K.T.; Zhang, X.; Li, S.W. A wind tunnel study of effects of twisted wind flows on the pedestrian-level wind field in an urban environment. *Build. Environ.* **2018**, *128*, 225–235. [CrossRef]
19. He, Y.; Liu, Z.; Ng, E. Parametrization of irregularity of urban morphologies for designing better pedestrian wind environment in high-density cities—A wind tunnel study. *Build. Environ.* **2022**, *226*, 109692. [CrossRef]
20. Ikegaya, N.; Kawaminami, T.; Okaze, T.; Hagishima, A. Evaluation of exceeding wind speed at a pedestrian level around a 1:1:2 isolated block model. *J. Wind Eng. Ind. Aerodyn.* **2020**, *201*, 104193. [CrossRef]
21. Du, Y.; Mak, C.M.; Li, Y. Application of a multi-variable optimization method to determine lift-up design for optimum wind comfort. *Build. Environ.* **2018**, *131*, 242–254. [CrossRef]
22. Li, J.; Wang, W.; Jin, H.; Li, Y.; Bu, N. Thermal responses of people exhibiting high metabolic rates when exercising in piloti spaces in hot and humid areas. *J. Build. Eng.* **2022**, *48*, 103930. [CrossRef]

23. Du, Y.; Mak, C.M.; Liu, J.; Xia, Q.; Niu, J.; Kwok, K.C.S. Effects of lift-up design on pedestrian level wind comfort in different building configurations under three wind directions. *Build. Environ.* **2017**, *117*, 84–99. [CrossRef]
24. Huang, T.; Niu, J.; Xie, Y.; Li, J.; Mak, C.M. Assessment of "lift-up" design's impact on thermal perceptions in the transition process from indoor to outdoor. *Sustain. Cities Soc.* **2020**, *56*, 102081. [CrossRef]
25. Sha, C.; Wang, X.; Lin, Y.; Fan, Y.; Chen, X.; Hang, J. The impact of urban open space and 'lift-up' building design on building intake fraction and daily pollutant exposure in idealized urban models. *Sci. Total Environ.* **2018**, *633*, 1314–1328. [CrossRef] [PubMed]
26. Tse, K.T.; Zhang, X.; Weerasuriya, A.U.; Li, S.W.; Kwok, K.C.S.; Mak, C.M.; Niu, J. Adopting 'lift-up' building design to improve the surrounding pedestrian-level wind environment. *Build. Environ.* **2017**, *117*, 154–165. [CrossRef]
27. Chen, L.; Mak, C.M. Integrated impacts of building height and upstream building on pedestrian comfort around ideal lift-up buildings in a weak wind environment. *Build. Environ.* **2021**, *200*, 107963. [CrossRef]
28. Smagorinsky, J. General Circulation Experiments with the Primitive Equations. *Mon. Weather Rev.* **1963**, *91*, 99–164. [CrossRef]
29. Murakami, S.; Mochida, A. On turbulent vortex shedding flow past 2D square cylinder predicted by CFD. *J. Wind Eng. Ind. Aerodyn.* **1995**, *54–55*, 191–211. [CrossRef]
30. Yu, D.; Kareem, A. Numerical simulation of flow around rectangular prism. *J. Wind Eng. Ind. Aerodyn.* **1997**, *67–68*, 195–208. [CrossRef]
31. Shah, K.B.; Ferziger, J.H. A fluid mechanicians view of wind engineering: Large eddy simulation of flow past a cubic obstacle. *J. Wind Eng. Ind. Aerodyn.* **1997**, *67–68*, 211–224. [CrossRef]
32. Panneer Selvam, R. Computation of pressures on Texas Tech University building using large eddy simulation. *J. Wind Eng. Ind. Aerodyn.* **1997**, *67–68*, 647–657. [CrossRef]
33. Kravchenko, A.G.; Moin, P. Numerical studies of flow over a circular cylinder at ReD = 3900. *Phys. Fluids* **2000**, *12*, 403. [CrossRef]
34. Tutar, M.; Oguz, G. Large eddy simulation of wind flow around parallel buildings with varying configurations. *Fluid Dyn. Res.* **2002**, *31*, 289–315. [CrossRef]
35. Lim, H.C.; Thomas, T.G.; Castro, I.P. Flow around a cube in a turbulent boundary layer: LES and experiment. *J. Wind Eng. Ind. Aerodyn.* **2009**, *97*, 96–109. [CrossRef]
36. Murakami, S.; Morikawa, Y. Criteria for assessing wind-induced discomfort considering temperature effect. *J. Archit. Plan. Environ. Eng.* **1985**, *358*, 9–17.
37. Tominaga, Y.; Mochida, A.; Yoshie, R.; Kataoka, H.; Nozu, T.; Yoshikawa, M.; Shirasawa, T. AIJ guidelines for practical applications of CFD to pedestrian wind environment around buildings. *J. Wind Eng. Ind. Aerodyn.* **2008**, *96*, 1749–1761. [CrossRef]
38. Blocken, B.; Gualtieri, C. Ten iterative steps for model development and evaluation applied to Computational Fluid Dynamics for Environmental Fluid Mechanics. *Environ. Model. Softw.* **2012**, *33*, 1–22. [CrossRef]
39. Nation Meteorological Information Center. *China's Special Meteorological Data Set for Thermal Environment Analysis*; China Architecture and Building Press: Beijing, China, 2005.

buildings MDPI

Article

Production of Low-Cost, High-Strength Concrete with Waste Glass as Fine Aggregates Replacement

Ali Jahami [1,*], Jamal Khatib [1] and Rabab Raydan [2]

[1] Department of Civil and Environmental Engineering, Beirut Arab University, Beirut 1105, Lebanon
[2] Faculty of Engineering, University of Porto, 4100 Porto, Portugal
* Correspondence: a.jahmi@bau.edu.lb

Abstract: The world of building materials is constantly and rapidly developing. New technologies are needed to reduce the cost of producing these materials and to ensure better efficiency when the materials are used in various engineering projects. One of these materials is high-strength concrete. This paper investigates the production of low-cost, high-strength concrete by partially replacing fine aggregates (FA) with waste glass sand (WGS). Four concrete mixes were considered in this study with varying percentages of WGS (0%, 25%, 50%, and 75%). For each mix, cubic, cylindrical, and beam specimens were cast to study the workability and different mechanical properties of concrete-like density, elasticity modulus, compressive strength, ultrasonic pulse velocity (UPV), split tensile strength, and flexural strength. In addition, the cost of each mix was calculated to evaluate the cost reduction efficiency of concrete with WGS compared to normal concrete. Results showed that the workability of concrete enhanced as the percentage of WGS increased. In terms of concrete mechanical properties, it was shown that the elasticity modulus, compressive strength, split tensile strength, and flexure strength for a concrete mix with 50% WGS as FA replacement was increased by 7%, 27%, 9%, and 50%, respectively. Also, it was concluded that the presence of WGS in concrete mixes reduced the production cost by up to 30% for a 75% replacement level. The authors recommended the usage of 50% WGS as the optimum replacement percentage for low-cost, high-strength concrete.

Keywords: sustainability; waste management; concrete; mechanical properties; workability

Citation: Jahami, A.; Khatib, J.; Raydan, R. Production of Low-Cost, High-Strength Concrete with Waste Glass as Fine Aggregates Replacement. *Buildings* **2022**, *12*, 2168. https://doi.org/10.3390/buildings12122168

Academic Editors: Xin Ren, Yaolin Lin and Teng Shao

Received: 8 November 2022
Accepted: 5 December 2022
Published: 8 December 2022

1. Introduction

Using natural sand as fine aggregates in concrete production has been the conventional element used in the industry for several decades. However, this excessive use can result in the depletion of locally available sand [1,2]. Current practices that consider sustainability as a priority are agreed on finding ways to reuse the waste in order to prevent environmental consequences and save natural resources without compromising the quality of the concrete [3–9]. One of the alternative materials that can substitute the fine natural aggregates in concrete is recycled glass. Using glass as sand replacement helps in preserving natural sand while reducing the amount of glass waste generated [1,5,10].

The key parameters that affect the efficiency of using glass as sand replacement are related to the glass's physical and chemical properties, which include size, gradation, shape, specific gravity, and resistance to alkali-aggregate reactions [4]. The finer the glass particles, the better the mechanical and durability performance is [11,12]. Using glass sand at replacement levels between 5% to 15% resulted in a positive slump value and compaction factor [13–15], in addition to having a considerable improvement in the concrete density attributed to the micro-filling effect. However, for replacement levels beyond 20%, a loss of workability is realized [12,13]. The slump of concrete decreases as the percentage of sand glass replacement content increases [5,15]. Some studies attribute this decrease to the sharp and larger surface area of the sand glass [1,10,12,15].

From the main concerns that lie behind the use of waste glass sand in concrete is the Alkali-Silica Reaction (ASR), which occurs between amorphous silica present in the glass composition and alkali in cement, resulting in excessive expansion when moisture is present and thus affects the durability of concrete [11]. Studies show that using waste glass sand up to 20% replacement levels did not show substantial ASR expansion [6,16]. Also, since ASR reactions are attributed to various parameters, including particle size, well-graded glass send can serve to increase the density due to the micro-filling effect and thus lower the risk of ASR expansion [12]. Although the inclusion of waste glass sand might lead to an increase in cost, it serves as an eco-friendly option that obtains acceptable concrete characteristics and a reduced cost of waste management [15]. However, it is recommended by some studies that up to 10% replacement levels of low-grade and uneconomic glass fines can be used to substitute natural sand in concrete [17]. This serves to reduce the consumption of large amounts of natural resources and contribute to the benefit of the economy and environment [18,19].

Several studies conducted on the use of waste glass as sand replacement in concrete have agreed that the effective substitution of sand up to between 10 to 30% exposes no detrimental long-term effects on the quality of concrete [4–15]. The best percentage of replacement in common with most studies is at 20% sand replacement level [1,4,6,7,12,13].

Several studies have been conducted on the use of waste glass (WG) in ultra-high-performance concrete (UHPC) as a binder and fine aggregates (FA) replacement [20–27]. Soliman and Tagnit-Hamou [20] investigated the replacement of silica fume (SF) in UHPC with waste glass powder (WGP). The replacement percentages adopted in this study were 30%, 50%, 70%, and 100%. It was demonstrated that replacing SF in UHPC by 30–50% led to enhancing concrete compressive strength by 15% and 8%, respectively. Moreover, better fresh concrete properties were achieved as the percentage of WGP increased.

Another study was conducted by Jiao et al. [22] on concrete incorporating WG as FA replacement. The study investigated different replacement percentages of WG (25%, 50%, 75%, and 100%) to recommend the best replacement percentage in terms of concrete mechanical properties. It was noticed that concrete with 75% WG as FA replacement had the highest compressive strength among all specimens. However, incorporating waste glass in concrete showed a negligible effect on both flexure and tensile strength.

Chu et al. [23] studied the partial replacement of paste (cement + silica fume + fly ash) in concrete by WGP. Three replacement percentages of WGP were considered as follows: 5%, 10%, and 15%. In terms of workability, results showed that concrete with 10% WGP had the best workability in terms of slump value and flow diameter. Regarding concrete mechanical properties, results exhibited that at a 10% replacement level, both compressive and flexure strength increased significantly (50% increase for compressive strength and 30% increase for flexure strength) compared to the control mix.

Based on the literature review conducted, it can be concluded that most of the previous studies conducted on sand replacement by the waste glass in concrete focused on the mechanical properties and didn't consider the cost issue. Some studies were conducted on the cost analysis of UHPC with waste glass as cement replacement, but also, they are few. Therefore, this study aimed to assess the use of waste glass as fine aggregate replacement in low-cost, high-strength sustainable concrete mixtures under standard curing conditions. The workability and mechanical properties of concrete were assessed in this study by performing a slump flow test, density test, elasticity modulus test, ultrasonic pulse velocity (UPV) test, compressive strength test, split tensile test, and flexural strength test. A theoretical analysis was conducted on the findings, and some fitting relations were derived and compared with ACI code equations.

2. Methods

2.1. Mix Design

The properties of various constituents used in the mix design, including Waste Glass Sand (WGS), Quartz Sand (QS), Quartz Powder (QP), and Silica Fume (SF), are shown in Table 1. The cement used has tri-calcium aluminate content of less than 8%, di-calcium and tri-calcium aluminate combined content of more than 65%, low alkali content, and moderate fineness ranging from 3000 to 4500 cm^2/kg. The low content of tricalcium aluminate reduces the water demand and improves overall workability. The waste glass sand is ground in the lab with the aid of a compressive strength machine to reduce glass particle sizes (to 300 μm and 150 μm), as shown in Figure 1.

Table 1. Properties and Composition of the Materials Used.

Properties	Identification	QS	WGS	SF	QP	Cement
Chemical Composition (%)	Silicon Dioxide (SiO_2)	99.43	71.10	98.86	99.20	22.8
	Iron Oxide (Fe_2O_3)	0.02	0.51	0.14	0.01	4.43
	Aluminum Oxide (Al_2O_3)	0.18	2.15	0.12	0.10	4.01
	Calcium Oxide (CaO)	0.11	11.10	0.57	0.21	65.32
	Magnesium Oxide (MgO)	0.01	1.30	0.20	0.17	2.07
	Sodium Oxide (Na_2O)	-	14.07	0.35	0.11	0.07
	Potassium Oxide (K_2O)	0.03	0.28	0.21	0,81	0.56
Physical Properties	Packing Density (kg/m^3)	1536.31	1579.62	-	-	-
	Mean Particle Diameter (μm)	300	300	-	-	-
	Max. Particle Diameter (μm)	600	600	-	-	-
	Crushing Values (%)	94.72	96.71	-	-	-

Figure 1. (**a**) Compressive Strength Machine; (**b**) Crushed Waste glass sand.

The mixtures were made with different percentages of replacement of sand by WGS with the following proportions (0%, 25%, 50%, and 75%), as shown in Table 2. The nomenclature used for each mix consisted of the term WGS followed by the replacement level of waste glass sand. Therefore, the four mixes were named WGS0 (control mix), WGS25 (25% replacement level), WGS50 (50% replacement level), and WGS75 (75% replacement level). All mixes were cast in molds and were normally cured by water at 20 °C till the testing day.

Table 2. Mix Design.

Materials (kg/m^3)	WGS0	WGS25	WGS50	WGS75
Cement	802	802	802	802
SF	223	223	223	223
Water	246	246	246	246
QS	962	722	481	241
WGS	0	241	481	722
QP	241	241	241	241
HRWRA [1]	30.75	30.75	30.75	30.75

[1] High-Range Water-Reducing Admixture.

Oven-dried aggregates were sieved with the help of the mechanical shaker through a stack of sieves to consider a specific particle size for each aggregate type. A particle size of 300 μm was chosen for both QS and WGS (particles retained in sieve #50), while a particle size of 150 μm was chosen for QP (particles retained in sieve #100), as shown in Figure 2.

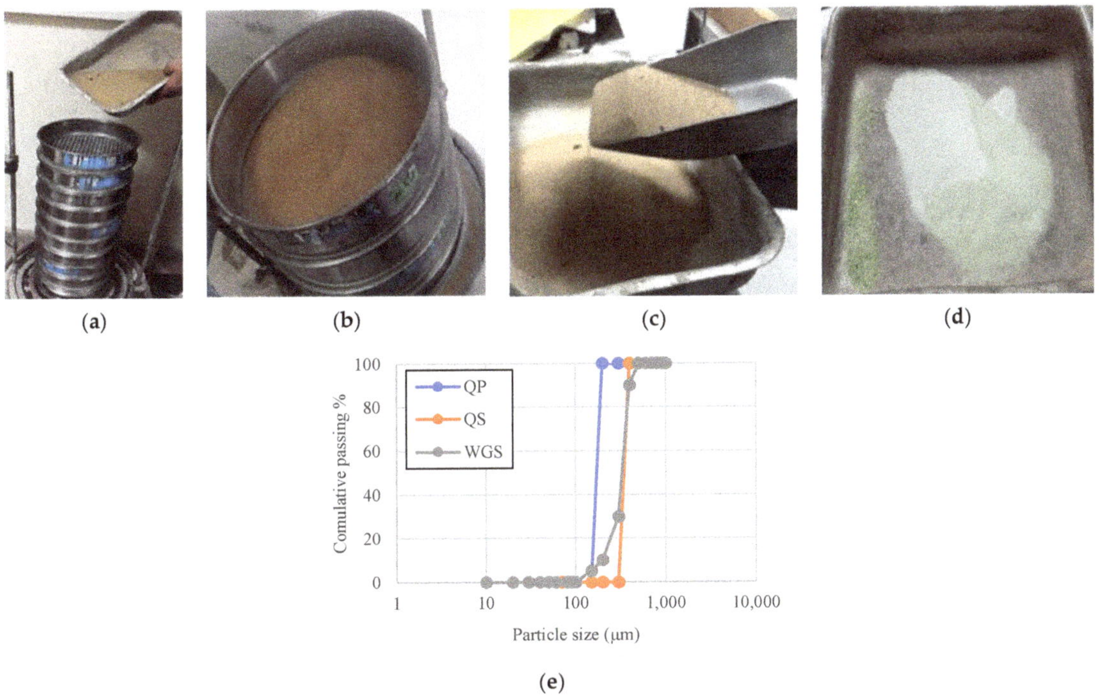

Figure 2. (**a**) Mechanical Shaker; (**b**) Sieving Apparatus; (**c**) Sieved Aggregates; (**d**) Waste glass sand; (**e**) Gradation curve for QP, QS, and WGS.

2.2. Mixing and Casting Procedure

The dry components like Cement, Silica fume (SF), Quartz powder (QP), Quartz sand (QS), and Waste glass sand (WGS) were added to the mixer for three minutes at a low speed. Then, the High-Range Water-Reducing Admixture (HRWRA) and water were added. The diluted HRWRA in half the amount of mixing water was added gradually to the mixture after two minutes of the mixing period, and the remaining quantity of water and HRWRA was added in another two minutes of the mixing. After the preparation of the mixture, the slump flow was measured to check the workability of the mixes. All mixes were de-molded and cured in a water bath in a room at a normal temperature of 20 °C until the day of testing. For each mix, 9 cubes of (100 × 100 × 100 mm) dimensions were cast for compressive strength and UPV tests. Also, 9 cylinders of 150 mm diameter and 300 mm height were cast for elasticity modulus, compression, and split tensile tests. As for the flexure test, 3 prisms of 40 × 40 × 160 mm were cast. Figure 3 depicts the concrete cubes and prisms after casting.

Figure 3. Molded Concrete Specimens.

2.3. Testing

Figure 4 shows the various strength tests performed on the different specimens. The slump flow test was performed according to ASTM 1437 [28]. The elasticity modulus test was done according to ASTM C469 [29] requirements. The compressive strength test was done in accordance with test standards ASTM C109 [30] after a curing duration of 7, 14, and 28 days. The load increased gradually until reaching failure. The density test was done according to ASTM C188 [31]. The flexural test was done according to the ASTM C348 [32] standards, and the splitting tensile test was done according to ASTM C496 [33], where the concrete cylinders were placed horizontally along its length, and a vertical load (P) was applied gradually until the specimen reaches failure. The Ultrasonic pulse velocity (UPV) was performed according to ASTM C597 [34].

(a) (b) (c) (d)

Figure 4. (**a**) Compressive Strength Test; (**b**) Flexural Strength Test; (**c**) Split Tensile Test; (**d**) UPV Test.

3. Results and Discussion

3.1. Slump Flow

Figure 5a below shows the slump flow assessment for the mixtures. Based on the experimental visualization, the control mix gave the lowest diameter for the flow table test measured as 165 mm and increased with the increase in replacement levels, gradually reaching 175 mm for 75 % percentage of replacement which indicated the best workability. This is attributed to the low waste glass sand absorption ability. Figure 5b shows the flow diameter for the various mixtures prepared.

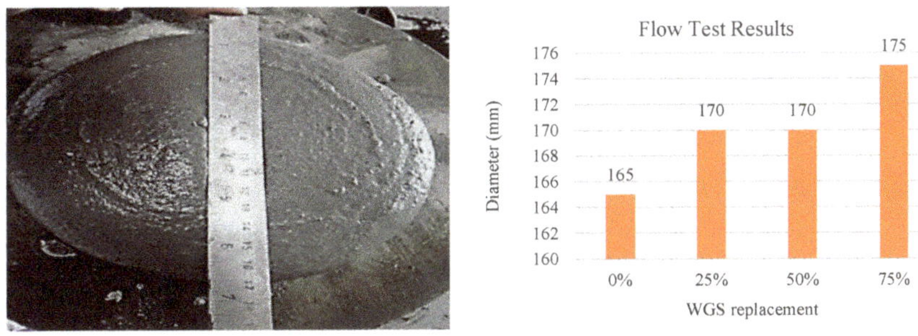

(**a**) (**b**)

Figure 5. (**a**) Flow Test Measurement; (**b**) Flow Test Results.

Other studies showed a similar trend for workability to this study [22,35]. For example, the study done by Jiao et al. [22] showed that concrete specimens with 100% WGS as a sand replacement had the best workability compared to other replacement percentages (0%, 25%, 50%, 75%, and 100%). Moreover, flowability test results done by Liu et al. [35] demonstrated that concrete with 75% WGS as a sand replacement had the best workability among all replacement percentages. However, the study conducted by Tamanna et al. [1] showed different behavior, where increasing the replacement percentage of WGS led to decreasing slump values. This was justified by the sharp edges and rough texture of the WGS, which led to the reduction of slump values.

3.2. Density

The dry density of concrete was determined for all mixes by measuring the weight/area ratio as shown in Figure 6a for the cubic specimen of (100 mm × 100 mm × 100 mm). It was noticed that as the percentage of waste glass sand replacement increased, the density experienced a decrease relating to the fact that glass powder is lighter than sand affecting the overall aggregate volume and thus decreasing the density. When the percentage of waste glass sand replacement reached 50% and above, the concrete turned to lightweight concrete with a (density < 1850 kg/m^3) as observed in Figure 6. Comparable results were determined for the specimens at 14 and 28 days.

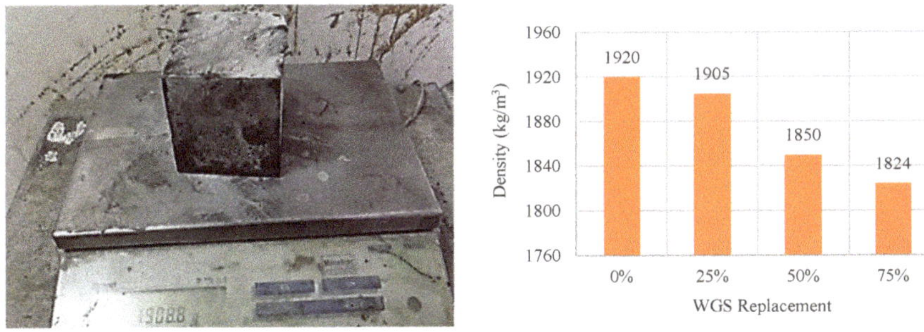

(**a**) (**b**)

Figure 6. (**a**) Weight Measurement; (**b**) Density of Cubic Specimens at 7 days.

Compared to previous work done on similar topics, the density showed to decrease as the percentage of WGS increased [1,22]. However, in some studies, when the WGS used had a higher density than the quartz sand, the density of concrete showed an increase as the replacement percentage increased [35].

3.3. Compressive Strength

The use of waste glass sand as aggregate replacement contributes to the improvement of the concrete microstructure. It contributes to a beneficial reaction with cement hydrates relating to the fineness and waste glass sand particle properties. As shown in Figure 7 below, the 50% glass replacement was determined as the most efficient percentage in having the highest compressive strength. At 7 days, the 25% glass replacement gave 3.44% higher strength than the control mix, while the 50% glass replacement gave an 18.22% increase in compressive strength. It was noticed that as the waste glass sand replacement levels reached 75%, the bonding between the aggregates was weakened, and the WGS aggregate could not fully play the advantage of high crushing strength, leading to a slight decrease in the compressive strength.

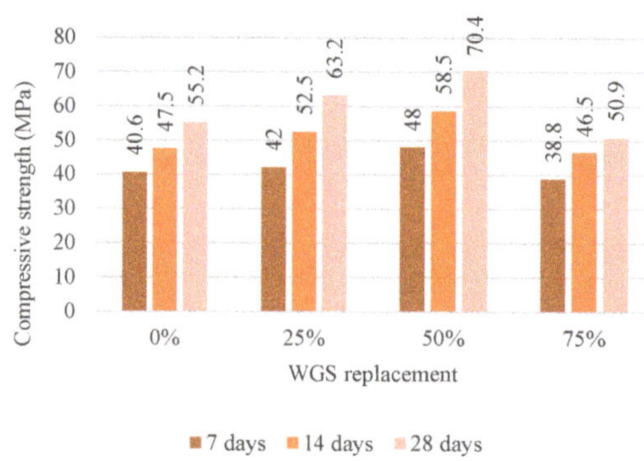

Figure 7. Compressive Strength Results at 7, 14, and 28 days for cubic specimens.

A similar trend is realized for the compressive strength of cylindrical specimens, where 50% glass replacement led to the highest compressive strength at 28 days. Figure 8 depicts the compressive strength of cubic specimens versus cylindrical specimens. The compressive strength of cylindrical specimens was less than the compressive strength of cubic specimens. The ratio between cylindrical and cubic strength ranged from 0.85 to 0.89, according to Figure 8.

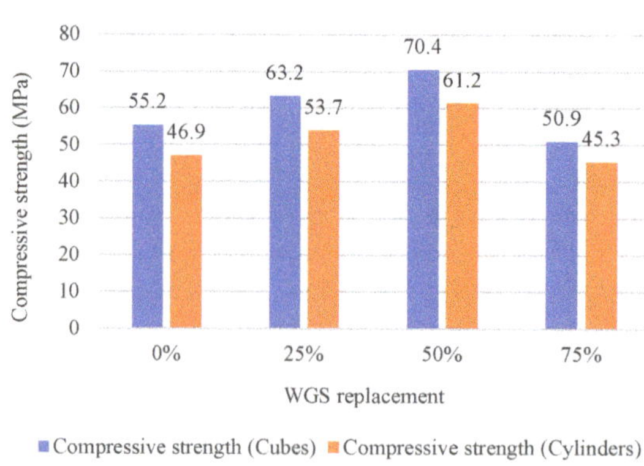

Figure 8. Compressive Strength Results at 28 days for cubic vs. cylindrical specimens.

The best replacement percentage for the highest compressive, according to previous studies, ranges from 20% to 75%, depending on the type and shape of WGS used in the replacement procedure [1,22]. However, in some studies, the presence of WGS led to decreasing the compressive strength of concrete [35]. This was justified by the high porosity of WGS concrete, which led to a negative effect on the compressive strength.

3.4. Ultrasonic Pulse Velocity (UPV)

The UPV test was conducted along at (7 days, 14 days, and 28 days) for the different concrete mixtures. It is noticed that as the waste glass sand replacement percentage increased, the UPV increased, indicating better concrete quality with low void content. From the results presented in Figure 9, it is indicated that the 50% replacement level with waste glass sand showed the best performance and exhibited continuous improvement with the concrete age.

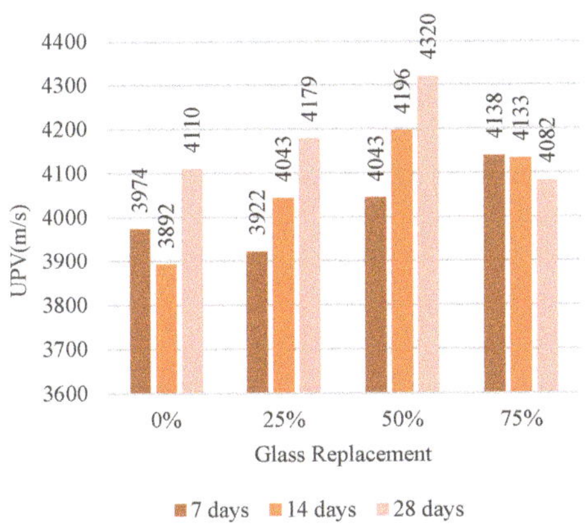

Figure 9. UPV Results at 7, 14, and 28 Days.

Based on the UPV findings of this study, the fitting curve that relates UPV to cylindrical compressive strength (f′c) for concrete containing waste glass as a fine aggregate replacement was presented. It has the formula:

$$Y = 0.247e^{0.0013x} \quad (1)$$

where "Y" is the compressive strength of concrete in MPa, and "x" is the ultrasonic pulse velocity (UPV) in m/s. The (R^2) factor for the fitting curve formula is 0.97, which indicates high accuracy. Figure 10 depicts the experimental data along with the fitting curve.

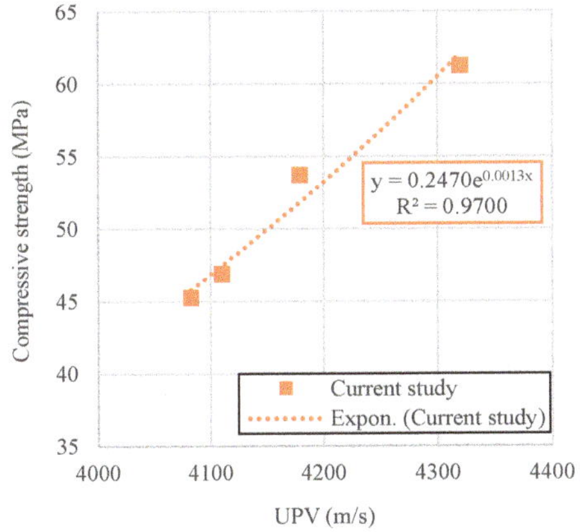

Figure 10. Fitting curve for UPV and the corresponding compressive strength.

3.5. Modulus of Elasticity

Substituting aggregates with Waste glass sand, according to some research studies, tends to have a positive influence on the elasticity modulus when used at small percentages. This can lead to a similar or slightly higher modulus of elasticity compared to control specimens. The results in Figure 11 show that as the percentage of WGS in concrete increased, the elasticity modulus slightly increased up to a 50% replacement level, whilst a 75% replacement level resulted in a reduction in the modulus of elasticity by around 12% compared to the control specimen. In summary, the elasticity modulus tends to show a dramatic decrease with an increasing percentage beyond 50%.

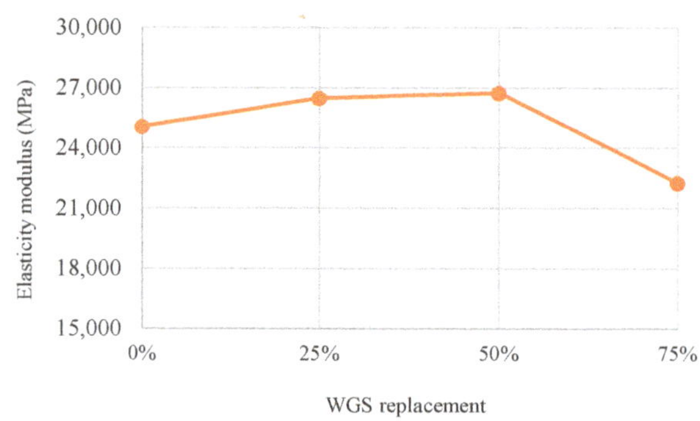

Figure 11. Elasticity Modulus vs. Glass Sand Replacement Percentage.

3.6. Split Tensile Strength

The split tensile results of the mixtures shown in Figure 12 showed that cylinders having 50% glass sand replacement levels obtained the highest tensile strength with a 9% improvement in comparison with the control mix. Beyond 50% replacement levels, the tensile strength experienced a decreasing manner that may be attributed to the fragility of concrete in tension because of its very brittle nature and the ineffectiveness of glass powder replacement as the improvement was slightly changing. As noticed, when the glass sand replacement levels increased, the concrete cracked and reached its ultimate strength earlier.

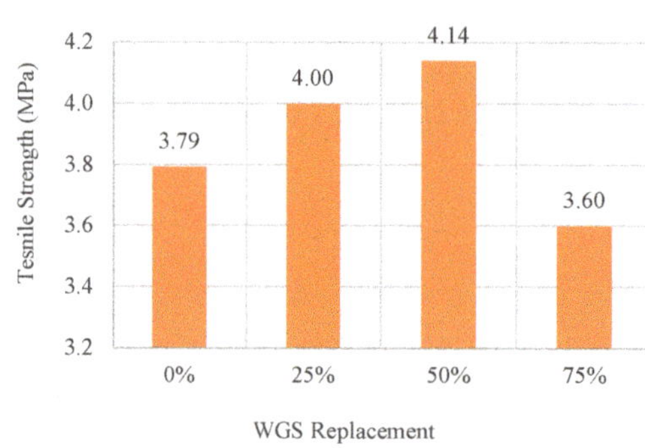

Figure 12. Split tensile strength at 28 days.

Table 3 shows the ratio between the split tensile strength and compressive strength (f_t/f'_c) at 28 days. It can be noted that as the percentage of WGS increased, the (f_t/f'_c) ratio decreased. This is valid until reaching 50% replacement, where values beyond 50% replacement lead to increasing this factor. This is due to the significant drop in compressive strength for specimen WGS75 which led to a higher (f_t/f'_c) ratio.

Table 3. The ratio between Split Tensile Strength and Compressive Strength.

Specimen	f_t [1] (MPa)	f'_c [2] (MPa)	f_t/f'_c
WGS0	3.79	46.9	0.081
WGS25	4.00	53.7	0.074
WGS50	4.14	61.3	0.068
WGS75	3.60	45.3	0.079

[1] Split tensile strength, [2] Cylindrical compressive strength.

Previous research showed a similar trend regarding the split tensile strength [1,22], but they didn't agree on the best replacement percentage of WGS that resulted in the highest tensile strength, and some of them concluded that replacing WGS with different types of sands, e.g., quartz sand and river sand, led to different increasing percentages in the tensile strength [22].

The ACI code [36] derived a relation between the split tensile strength and cylindrical compressive strength for normal concrete. The equation was plotted in Figure 13 and used as a reference to be compared with the fitting curve from this study that relates the same parameters. The fitting curve has the formula:

$$Y = 0.725x^{0.4257} \quad (2)$$

where "Y" is the split tensile strength of concrete in MPa, and "x" is the compressive strength of concrete in MPa. The (R^2) factor for the fitting curve is 0.92, which indicates high accuracy. It can be shown, according to Figure 13, that split tensile strength values for concrete with waste glass were higher than normal concrete for the same compressive strength values. This could be due to the high enhancement of split tensile strength for WGS concrete compared to normal concrete.

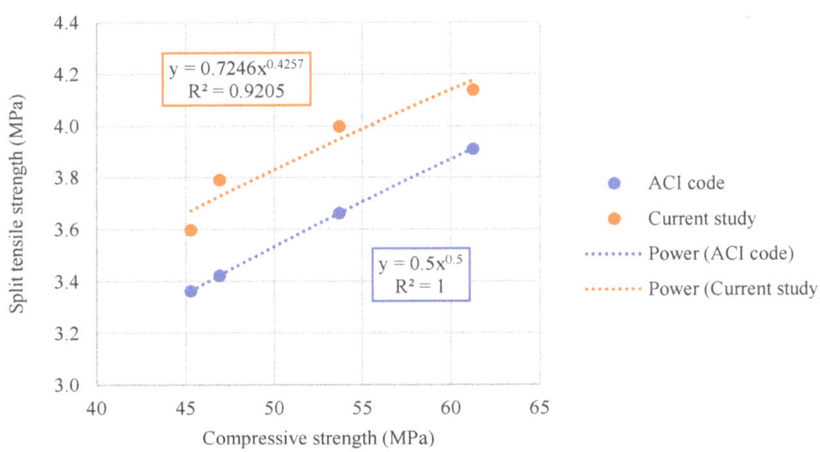

Figure 13. Fitting curve for the split tensile strength compared to ACI code equation.

3.7. Flexure Strength

The flexural strength results presented in Figure 14 showed that specimens with 50% glass sand replacement showed the highest flexural strength of 10.56 MPa. This is 50% higher than the flexure strength of the control specimen. As the replacement percentage went beyond 50%, the flexure strength showed a significant reduction. This lies within the range of several previous studies conducted on concrete with WGS [1,22]. However, some other studies indicated a reduction in flexure strength as the percentage of WGS increased due to the high porosity of concrete incorporating WGS [35].

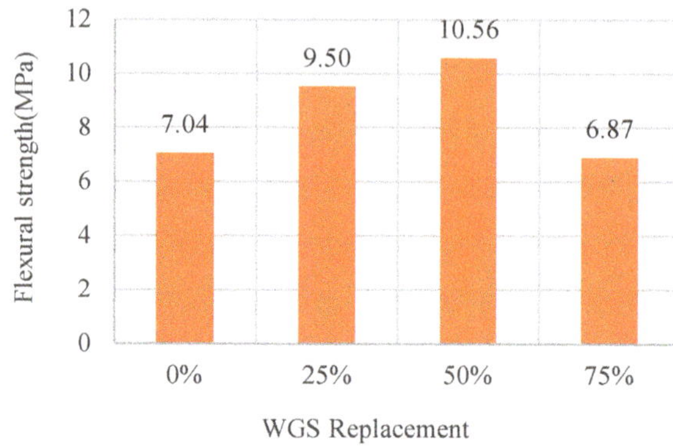

Figure 14. Split tensile strength at 28 days.

Similar to split tensile strength, the fitting curve was presented for the flexure strength of concrete with WGS as follows:

$$Y = 0.02x^{1.53} \tag{3}$$

where "Y" is the flexure strength of concrete in MPa, and "x" is the compressive strength of concrete in MPa. The (R^2) factor for the fitting curve is 0.95, which indicates high accuracy. It can be shown, according to Figure 15, that flexure strength values for concrete with waste glass were higher than normal concrete for the same compressive strength values. This is similar to the trend realized for the split tensile strength for WGS concrete.

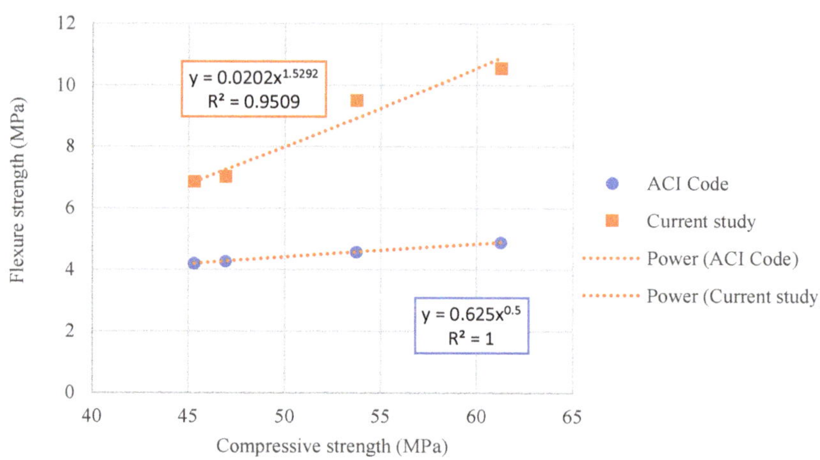

Figure 15. Fitting curve for the flexure strength compared to ACI code equation.

3.8. Economic Study

In addition to using glass powder as an eco-friendly replacement for natural sand, it also plays a role in reducing the overall cost of the mixture. After summing up the total cost per mixture, as shown in Table 4, it was noticed that as the percentage of glass replacement of sand increased, the cost decreased, reaching a maximum reduction of 8% when 75% of glass powder was used.

Table 4. Cost analysis details for each concrete mix.

Material	**Cost ($/m^3)**			
	WGS0	**WGS25**	**WGS50**	**WGS75**
Cement	32.08	32.08	32.08	32.08
SF	33.45	33.45	33.45	33.45
Water	1.25	1.25	1.25	1.25
QS	20.00	15.00	10.00	5.00
WGS	0.00	2.00	4.00	6.00
QP	5.00	5.00	5.00	5.00
HRWRA	30.75	30.75	30.75	30.75
$\sum$	122.53	119.53	116.53	113.53

Another way to present the cost analysis results is by determining the Performance/Cost (P/C) ratio, which determines the strength gained per each 1 USD paid for preparing 1 m^3 of each mix. Three mechanical properties of concrete were considered in this analysis: compressive strength, split tensile strength, and flexure strength. According to Figure 16, the P/C ratio for compressive strength for the control specimen was 0.383 MPa/$/m^3. This ratio increased as the replacement percentage of WGS increased to reach 0.449 MPa/$/m^3, 0.526 MPa/$/m^3, and 0.339 MPa/$/m^3 for replacement levels 25%, 50%, and 75%, respectively.

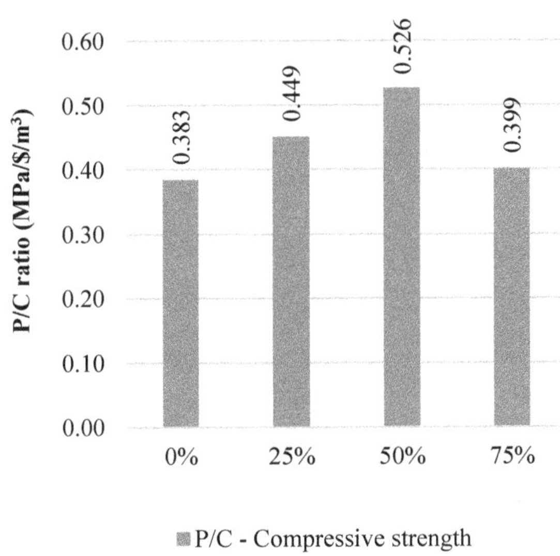

Figure 16. P/C ratios for compressive strength.

Moreover, a similar trend was realized for the split tensile strength and flexure strength, where the P/C ratio increased from 0.031 MPa/$/m^3 for specimen WGS0 to reach a maximum value of 0.036 MPa/$/m^3 for specimen WGS50. As for the P/C ratios for flexure strength, it increased from 0.057 MPa/$/m^3 for WGS0 to reach a maximum value of 0.091 MPa/$/m^3 for WGS50. Although the P/C ratio for split tensile and flexure strength decreased at replacement level 75%, it was still higher than the P/C ratio of control specimen WGS0, as depicted in Figure 17.

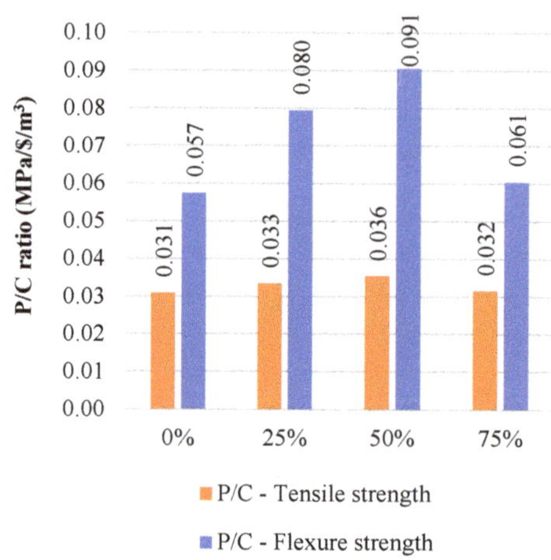

Figure 17. P/C ratios for split tensile and flexure strengths.

This indicates that the utilization of WGS as sand replacement is possible economically, speaking up to 75% replacement level. However, the authors recommend the 50% replacement level since it corresponds to maximum enhancement in the P/C ratio (37% for compressive strength, 16% for split tensile strength, and 60% for flexure strength).

4. Conclusions

This study focuses on evaluating the mechanical properties of concrete containing waste glass as sand replacement. Based on the study findings, the following conclusions can be drawn:

1. In terms of density, results showed a slight decrease as the percentage of WGS increased in concrete. The density was decreased by 5% for the specimen with a 75% replacement level compared to the control specimen. Other studies reached the same conclusion regarding the density due to the lighter weight of WGS compared to QS. However, for other types of WGS where the weight of WGS is higher than the replaced sand, the density showed to slightly increase with the increase of replacement level.
2. Regarding concrete workability, slump flow results showed a slight increase in the distribution diameter as the replacement percentage of WGS increased, which indicates better workability. Specimen WGS75 showed the best workability among the concrete mixes, which is similar to the findings of other studies that adopted the same replacement level. However, in other studies, it was shown that the shape of glass particles (the presence of sharp edges) could have an adverse effect on workability.
3. In terms of material stiffness, results showed that the specimen with 50% waste glass sand showed the highest elasticity modulus among all specimens. The elasticity modulus was shown to increase by 7% for specimen WGS50. However, as the replacement percentage reached 75%, the elasticity modulus showed to decrease by 15% compared to the control specimen. The authors recommend the 50% replacement level for the best material stiffness.
4. In terms of compressive strength, results showed that concrete specimens with a WGS replacement level of 50% had the highest compressive strength compared to the control specimen. The compressive strength was increased by 27% for specimen WGS50 compared to specimen WGS0. This was confirmed by other similar studies, which recommended a replacement level between 20–75% for the highest compressive strength. Moreover, UPV results showed the same trend where specimen WGS50 had the highest values compared to other specimens, which indicates higher compressive strength at a 50% replacement level.

5. In terms of tensile strength, both split tensile and flexure tests showed that concrete with 50% WGS had the best performance in tension. Split tensile and flexure strengths for specimen WGS50 were 9% and 50% higher than control specimens, respectively. This was confirmed by other studies, which recommended replacement levels between 25% to 75% for the highest tensile and flexure strength.
6. In terms of production cost, replacing sand with WGS led to decreasing the production cost by up to 8% for a replacement percentage of 75%. As for the performance/cost (P/C) ratio, it was concluded that as the replacement level of WGS increased, the P/C ratio for compressive, tensile, and flexure strength increased, which means that using WGS as sand replacement is economically feasible. However, the authors recommend a replacement level of 50% since it led to the highest P/C ratios (37% for compressive strength, 16% for split tensile strength, and 60% for flexure strength). This will help in reducing the overall construction cost since high-strength concrete is known for its high production cost

Finally, for future research, the authors recommend studying the effect of using WGS with different shapes (cylindrical, discoidal, spherical, abutting—burr) as sand replacement on the mechanical behavior of concrete. Also, it is worth studying the effect of different grain sizes of WGS on the mechanical performance of concrete.

Author Contributions: Conceptualization, A.J. and J.K.; methodology, A.J. and J.K.; formal analysis, A.J. and R.R.; investigation, A.J. and R.R.; resources, J.K.; data curation, A.J. and R.R; writing—original draft preparation, A.J. and R.R; writing—review and editing, A.J. and J.K.; visualization, J.K.; supervision, J.K.; project administration, J.K. and A.J. All authors have read and agreed to the published version of the manuscript.

Funding: This research received no external funding.

Institutional Review Board Statement: Not applicable.

Informed Consent Statement: Not applicable.

Conflicts of Interest: The authors declare no conflict of interest.

References

1. Tamanna, N.; Tuladhar, R.; Sivakugan, N. Performance of recycled waste glass sand as partial replacement of sand in concrete. *Constr. Build. Mater.* **2020**, *239*, 117804. [CrossRef]
2. Ashish, D.K. Feasibility of waste marble powder in concrete as partial substitution of cement and sand amalgam for sustainable growth. *J. Build. Eng.* **2018**, *15*, 236–242. [CrossRef]
3. Małek, M.; Łasica, W.; Jackowski, M.; Kadela, M. Effect of waste glass addition as a replacement for fine aggregate on properties of mortar. *Materials* **2020**, *13*, 3189. [CrossRef] [PubMed]
4. Meddah, M.S. Use of waste window glass as substitute of natural sand in concrete production. *IOP Conf. Ser. Mater. Sci. Eng.* **2019**, *603*, 032011. [CrossRef]
5. Olofinnade, O.M.; Ede, A.N.; Ndambuki, J.M.; Ngene, B.U.; Akinwumi, I.I.; Ofuyatan, O. Strength and microstructure of eco-concrete produced using waste glass as partial and complete replacement for sand. *Cogent Eng.* **2018**, *5*, 1483860. [CrossRef]
6. Kazmi, D.; Williams, D.J.; Serati, M. Waste glass in civil engineering applications—A review. *Int. J. Appl. Ceram. Technol.* **2020**, *17*, 529–554. [CrossRef]
7. Harrison, E.; Berenjian, A.; Seifan, M. Recycling of waste glass as aggregate in cement-based materials. *Environ. Sci. Ecotechnol.* **2020**, *4*, 100064. [CrossRef]
8. Khatib, J.; Jahami, A.; Elkordi, A.; Baalbaki, O. Structural performance of reinforced concrete beams containing plastic waste caps. *Mag. Civ. Eng.* **2019**, *7*, 73–79.
9. Bawab, J.; Khatib, J.; Jahami, A.; Kenai, S.; Elkordi, A. Effect of using cathode-ray tubes (CRT) waste glass on concrete properties. *BAU J.-Sci. Technol.* **2020**, *2*, 3.
10. Ferdous, W.; Manalo, A.; Siddique, R.; Mendis, P.; Zhuge, Y.; Wong, H.S.; Lokuge, W.; Aravinthan, T.; Schubel, P. Recycling of landfill wastes (tyres, plastics and glass) in construction—A review on global waste generation, performance, application and future opportunities. *Resour. Conserv. Recycl.* **2021**, *173*, 105745. [CrossRef]
11. Sawant, V.G. Experimental Investigation of Waste Glass Powder as the Partial Replacement of Sand in Making Concrete. *Iconic Res. Eng. J.* **2018**, *1*, 7–9.
12. Ahmad, J.; Zhou, Z.; Usanova, K.I.; Vatin, N.I.; El-Shorbagy, M.A. A Step towards Concrete with Partial Substitution of Waste Glass (WG) in Concrete: A Review. *Materials* **2022**, *15*, 2525. [CrossRef] [PubMed]

13. Warnphen, H.; Supakata, N.; Kanokkantapong, V. The Reuse of Waste Glass as Aggregate Replacement for Producing Concrete Bricks as an Alternative for Waste Glass Management on Sichang Island. *Eng. J.* **2019**, *23*, 43–58. [CrossRef]
14. Atoyebi, O.D.; Gana, A.J.; Longe, J.E. Strength assessment of concrete with waste glass and bankoro (Morinda Citrifolia) as partial replacement for fine and coarse aggregate. *Results Eng.* **2020**, *6*, 100124. [CrossRef]
15. Tiwari, A.; Singh, S.; Nagar, R. Feasibility assessment for partial replacement of fine aggregate to attain cleaner production perspective in concrete: A review. *J. Clean. Prod.* **2016**, *135*, 490–507. [CrossRef]
16. Collivignarelli, M.C.; Cillari, G.; Ricciardi, P.; Miino, M.C.; Torretta, V.; Rada, E.C.; Abbà, A. The production of sustainable concrete with the use of alternative aggregates: A review. *Sustainability* **2020**, *12*, 7903. [CrossRef]
17. Wang, T.; San Nicolas, R.; Kashani, A.; Ngo, T. Sustainable utilisation of low-grade and contaminated waste glass fines as a partial sand replacement in structural concrete. *Case Stud. Constr. Mater.* **2022**, *16*, e00794. [CrossRef]
18. Khatib, J.M.; Jahami, A.; Elkordi, A.; Abdelgader, H.; Sonebi, M. Structural assessment of reinforced concrete beams incorporating waste plastic straws. *Environments* **2020**, *7*, 96. [CrossRef]
19. Khatib, J.; Jahami, A.; El Kordi, A.; Sonebi, M.; Malek, Z.; Elchamaa, R.; Dakkour, S. Effect of municipal solid waste incineration bottom ash (MSWI-BA) on the structural performance of reinforced concrete (RC) beams. *J. Eng. Des. Technol.* **2021**. [CrossRef]
20. Soliman, N.A.; Tagnit-Hamou, A. Partial substitution of silica fume with fine glass powder in UHPC: Filling the Micro Gap. *Constr. Build. Mater.* **2017**, *139*, 374–383. [CrossRef]
21. Soliman, N.A.; Tagnit-Hamou, A. Development of ultra-high-performance concrete using glass powder—Towards ecofriendly concrete. *Constr. Build. Mater.* **2016**, *125*, 600–612. [CrossRef]
22. Jiao, Y.; Zhang, Y.; Guo, M.; Zhang, L.; Ning, H.; Liu, S. Mechanical and fracture properties of ultra-high performance concrete (UHPC) containing waste glass sand as partial replacement material. *J. Clean. Prod.* **2020**, *277*, 123501. [CrossRef]
23. Chu, S.H.; Li, L.; Shen, P.L.; Lu, J.X.; Poon, C.S. Recycling of waste glass powder as paste replacement in Green UHPFRC. *Constr. Build. Mater.* **2022**, *316*, 125719. [CrossRef]
24. Salahaddin, S.D.; Haido, J.H.; Wardeh, G. The behavior of UHPC containing recycled glass waste in place of cementitious materials: A comprehensive review. *Case Stud. Constr. Mater.* **2022**, *17*, e01494. [CrossRef]
25. Yoo, D.-Y.; You, I.; Zi, G. Effects of waste liquid–crystal display glass powder and fiber geometry on the mechanical properties of ultra-high-performance concrete. *Constr. Build. Mater.* **2021**, *266*, 120938. [CrossRef]
26. Abellán-García, J. Tensile behavior of recycled-glass-UHPC under direct tensile loading. *Case Stud. Constr. Mater.* **2022**, *17*, e01308. [CrossRef]
27. Kou, S.C.; Xing, F. The effect of recycled glass powder and reject fly ash on the mechanical properties of fibre-reinforced ultrahigh performance concrete. *Adv. Mater. Sci. Eng.* **2012**, *2012*, 263243. [CrossRef]
28. Standard Test Method for Flow of Hydraulic Cement Mortar. Available online: https://www.astm.org/c1437-20.html (accessed on 16 October 2022).
29. Standard Test Method for Static Modulus of Elasticity and Poisson's Ratio of Concrete in Compression. Available online: https://www.astm.org/c0469_c0469m-22.html (accessed on 16 October 2022).
30. Standard Test Method for Compressive Strength of Hydraulic Cement Mortars (Using 2-in. or [50-mm] Cube Specimens). Available online: https://www.astm.org/c0109_c0109m-20.html (accessed on 16 October 2022).
31. Standard Test Method for Density of Hydraulic Cement. Available online: https://www.astm.org/c0188-17.html (accessed on 16 October 2022).
32. Standard Test Method for Flexural Strength of Hydraulic-Cement Mortars. 2021. Available online: https://www.astm.org/c0348-21.html (accessed on 16 October 2022).
33. Standard Test Method for Splitting Tensile Strength of Cylindrical Concrete Specimens. Available online: https://www.astm.org/standards/c496 (accessed on 16 October 2022).
34. Standard Test Method for Pulse Velocity Through Concrete. 2016. Available online: https://www.astm.org/c0597-16.html (accessed on 16 October 2022).
35. Liu, T.; Wei, H.; Zou, D.; Zhou, A.; Jian, H. Utilization of waste cathode ray tube funnel glass for ultra-high performance concrete. *J. Clean. Prod.* **2020**, *249*, 119333. [CrossRef]
36. American Concrete Institute. *Building Code Requirements for Structural Concrete (ACI 318-14): An ACI Standard: Commentary on Building Code Requirements for Structural Concrete (ACI 318R-14), an ACI Report*; American Concrete Institute: Farmington Hills, MI, USA, 2014.

buildings

MDPI

Article

An Experimental and Numerical Study on Fire Spread in a Furnished Room

Małgorzata Król [1,*] and Aleksander Król [2]

[1] Faculty of Energy and Environmental Engineering, Silesian University of Technology, Konarskiego 18, 44-100 Gliwice, Poland
[2] Faculty of Transport and Aviation Engineering, Silesian University of Technology, Krasińskiego 8, 40-019 Katowice, Poland
* Correspondence: malgorzata.krol@polsl.pl; Tel.: +48-322-372-867

Abstract: The main objective of this research was to examine the development of fire in a furnished room in a typical high-rise building. This work was part of research on the fire safety of building occupants. It included two controlled fires in which a standard sofa in a room was set on fire. Several thermocouple trees were arranged in the test room and the temperature was continuously recorded. Additionally, each fire test was videotaped for further analysis. Since an unexpected forepeak of the temperature course was observed, special attention was paid to explain this phenomenon. For this purpose, numerical models of fire development in a furnished room were built using the well-recognized software package, ANSYS Fluent and Fire Dynamic Simulator (FDS). The numerical research was focused on fire spread over a single piece of furniture, the sofa. The data recorded in real experiments were used to tune and validate the numerical models. The results of the Fluent numerical simulation were consistent with the recorded experimental data and proved that after the initial growth, there was a critical phase of fire development in which the fire might almost snuff or flare again. Meanwhile, the FDS results, despite being generally in accordance with the experiment, did not reproduce this critical phase of fire spreading.

Keywords: compartment fire; real fire; thermocouples; CFD

Citation: Król, M.; Król, A. An Experimental and Numerical Study on Fire Spread in a Furnished Room. *Buildings* **2022**, *12*, 2189. https://doi.org/10.3390/buildings12122189

Academic Editors: Yaolin Lin, Xin Ren and Teng Shao

Received: 4 November 2022
Accepted: 8 December 2022
Published: 10 December 2022

1. Introduction

The development of a fire in an enclosed space poses a huge risk to anyone within. Engineers are constantly working to improve the safety of people in buildings [1–4]. This relates to the development of nonflammable materials and the continuous improvement of fire ventilation and extinguishing systems. It also applies to the location of combustible materials in the vicinity of particularly tall buildings [5]. There is also ongoing research on the proper modeling of the fire development phenomenon to reliably reproduce air and smoke flows and structural fire resistance [6]. The fire risk index for buildings was also created to define the level of fire threat in buildings [7].

When considering a fire developing in a compartment, it should be remembered that its course is determined by the availability of air and fuel. One commonly uses the terms that the fire is controlled by fuel supply (fuel limited) or air supply (oxygen limited) [8]. In the literature, the terms Regime I and Regime II, respectively [9], can also be found. Regime I is characterized by the vents being small enough that they allow for the compartment to fill with smoke. The air supply is limited. In Regime II, the vents are sufficiently large, so that the smoke evacuates the enclosure and the air flows inside. Complex heat and mass transfer processes dominate over Regime II [10]. This makes the description of Regime II complicated, and it is difficult to define the direct link between ventilation, temperature and burning rate, as it can be done for Regime I.

Numerical analyses are increasingly used in all these activities, but some difficulties arise when connecting these analyses with tests and trials under real fire conditions. However, numerical analyses seem to be a good tool for testing various types of fire engineering solutions. However, it should be noted that the numerical models created during such analyses must be validated using real fires.

Research involving the mapping of fire development using fire zone model [11] computational fluid dynamics (CFD) programs has been carried out for many years [10,12,13]. The Fire Dynamics Simulator (FDS) is often used in these analyses. Analysis of the research conducted so far shows that modeling fire development in rooms faces many problems. The large number of parameters determining the course of such a fire makes it difficult to represent a fire numerically.

Modeling the temperature distribution in a room where a fire develops is difficult. Stern-Gottfried points out that the temperature distribution in a room is not uniform. Locally, the maximum temperature can be 75% higher than the average temperature for a given room [14]. The Dalmarnock Fire Test One was used for their analyses. The same studies were also used to validate the numerical model developed in the Fire Dynamics Simulator (FDS). Extensive measurements allowed for an analysis of the measured results against the results of numerical analyses. As part of the study, the authors noted the difficulty of modeling near the fire and the relatively good agreement of the results away from the fire [15].

Research on fires developing in confined spaces also concerns the impact of fire on the building structure. Gupta studied the thermal characteristics of fire spread in a fully developed fire, a growing fire, and a traveling fire. This was possible because, unlike the previous research, Gupta conducted his research in a very large open plan compartment. According to their research, each fire spread mode induced significant and characteristic spatial heat distributions [16].

Research on the development of fires in confined spaces often includes studies on the development of flames outside openings on the façade of a building. These studies are important to understand the causes and course of a fire that develops along the façade of the building [17–19]. The windows on the façade of a building can cause the flames to spread outside the room onto the façade [20,21]. However, windows can also affect the course of a fire inside the room. This is related to the intense air inflow into the room forced by a fire plume. The effect of incoming air into a building due to wind is particularly evident in the case of a tall building fire [22], but it can also be used to create good conditions on the refuge floor [23]. Airflow can disturb the natural stratification in the room and cause hot smoke to mix with cold air. This can result in the disappearance of the hot and cooler layers [24]. This research not only concerns windows on the building façade, it also includes investigations of doors fitted with an air curtain. Research has been conducted on how the effect of an air curtain jet influences the temperature in the upper layer of a compartment fire [25].

Experimental investigations and numerical simulations of small room fires have already been performed in the literature [26]. Chen performed the measurements in a room built in a laboratory, which allowed them to monitor multiple parameters during the development of a fire. However, the conditions of the experiment were significantly different from those of a real fire. The measurements were compared with the results of numerical analyses in which the Fire Dynamic Simulator was used. The results were highly comparable. Material properties and oxygen limit settings in the FDS software were tested to explore their influence on the tendency of heat release rate. A.H. Majdalani carried out similar research on the development of fires. He built a room model in the laboratory and examined the principal characteristics of two unique behaviors of a fully developed compartment fire. Experimentation and computational modeling were used to explore, compare and contrast the characteristics of these two behaviors [27].

Research has also been conducted on the development of fires in real buildings. The experiment was carried out in a warehouse building, where a fire was planned on a shelf on a rack. The ignition, fire spread and combustion characteristics were studied. At the same time, CFD analyses were performed with the use of FDS. The results demonstrated a high degree of agreement between the experimental results and the CFD data. However, attention was drawn to the limited possibilities of modeling using FDS [28].

Similarly, Bystrom described interesting studies. They performed a fire experiment in a concrete building at a low ambient temperature. Research has shown that a compartment can be divided into two layers during a fire. However, the temperature in the upper layer is not uniform. The differences in temperature values were related to the distance from the fire source. A series of numerical analyses using FDS were also performed, and the outside temperature was assumed. The study demonstrated that the power of a fire would be greater at a higher outside temperature [29].

Mackay described compartment fire behavior training combined with numerical analyses using FDS. The main objective of the research was the training of firefighters. The results of these studies were also used to validate the numerical model [30]. In addition to the FDS program, the CFX program was used in the numerical analyses of the development of compartment fires. Hasib used the results of an experimental fire generated in an enclosure during the growth period to validate the predictions of CFX [31]. Fire studies conducted in the compartment are sometimes limited to the analysis of flame spread over solid fuels. He led the research, the main goal of which was to compare the surface and internal heat transfers to understand the mechanism underlying the degree of fuel packing [32,33]. Another study looked at the heat and mass transport process that defines the dynamics of combustion and flame spread in the case of wood crib fires in large rooms [34].

In the present study, fire tests were carried out in a room, which was enclosed in panels with high fire resistance. Six thermocouple trees were placed inside the room to measure the temperature during the tests. Additionally, the fire tests were recorded on video. The work is based mainly on the temperature distributions recorded in two fire tests which differed in the moment when the firefighting operation began. When analyzing the time dependence of the temperature, an unexpected course of the fire development was revealed: a clear forepeak was observed for almost all temperature curves. Therefore, numerical models were built to investigate this phenomenon. The well-recognized general purpose software packages Ansys Fluent and FDS were used. This former choice was justified by the possibility of a detailed investigation of model interdependencies at a very basic level [35]; the latter one is commonly used in fire engineering. In addition, the use of two software packages made it possible to compare the results achieved by them and their capabilities. The problems encountered when performing numerical analyses are presented in the manuscript, along with solutions that can help other researchers with numerical modeling of fire development in a room.

2. Materials and Methods

The main aim of the research was to test different configurations of a staircase smoke removal system, which is a crucial aspect of safe evacuation in high-rise buildings. The system consisted of an air supply fan with adjustable capacity and roof discharging vents, among others. In addition to these analyses, the test room was equipped with measurement devices, which allowed for the detailed analyses of compartment fire development. The experiments required the permanent presence of a fire brigade to ensure the appropriate safety level for the staff and material assets.

The test room was prepared on the 5th floor of a 9-story abandoned office building (Figure 1). The room was connected to the staircase through a short corridor with a remotely controlled door.

Figure 1. View of the building with the test room marked (on the west façade).

2.1. Experimental Setup

The test room was 5.7 × 5.8 × 2.5 m and had four windows. The rightmost window was built up with a cover from the inside. The three other glassed windows were tight. There was a door on the opposite side of the room from the windows. During both tests, the door connecting the test room to the corridor was open. Room partitions were covered with fireproof boards. The fireproof boards were made of silicate cement with a thickness of 50 mm and the following parameters: density 450 kg/m^3, specific heat 950 J/kg·K and thermal conductivity 0.083 W/m·K. There was a sofa and a table in the room. The furniture was the same in each test.

Six thermocouple trees were distributed throughout the test room, as shown in the figure (Figure 2). There were three to eight thermocouples in the individual trees. K-type thermocouples with beads of 3 mm diameter were used. The sheath was made of heat resistant steel and was insulated inside with magnesium oxide. The measurement accuracy for the thermocouples was ±2.2 °C or 0.75% in the range of 0–1200 °C. The NI-9213 Temperature Input Modules recorded temperature measurements [36]. Signal Express National Instruments software was used for data acquisition. The positions of the thermocouples in individual trees are shown in Table 1.

Table 1. Summary of thermocouple trees.

Number of the Tree	The Heights of the Thermocouples on the Trees [mm] (Measured from the Floor)
1	1000, 1600, 1800, 2000, 2200, 2400
2	1600, 1800, 2000, 2200, 2400
3	400, 1000, 1400, 1800, 2000, 2200, 2400, 2600
4	1000, 1600, 1800, 2000, 2200, 2400
5	400, 1400, 1600, 1800, 2000, 2200, 2400
6	400, 2400, 2600

The ignition source no. 4 from BS 5852:2006 was used to start the fires [37]. This standard ignition source is described in detail by Gupta [34]. The parameters of the ignition source are as follows: bar length 40 mm, bar width 6.5 mm, number of bars 10, bar weight 8.5 g, number of bar layers in two directions 5, and sterile gauze dimensions 40 × 40 mm. 1.4 mL of isopropyl alcohol was applied to the sterile gauze before starting the test. The igniter was placed in the middle of the sofa, as shown in Figure 3.

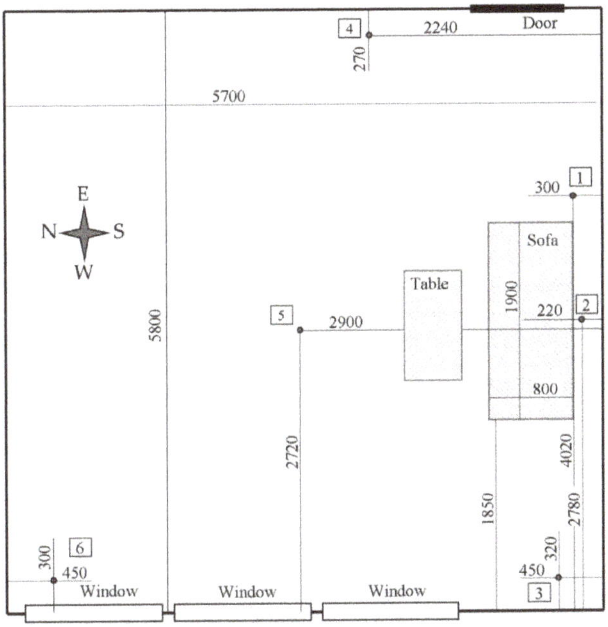

Figure 2. The test room (distances given in mm).

Figure 3. The location of the ignition source on the sofa at the beginning of a fire.

2.2. Experiment Results

Two full-scale fires were performed in the test room. The meteorological conditions during both fire tests are presented in Table 2.

Table 2. The meteorological conditions during both fire tests.

	Fire Test 1	Fire Test 2
Air temperature 2 m above the ground, °C	18	10
Air temperature 100 m above the ground (approx. 5 m above the roof of the building), °C	16	8
Wind speed 10 m above ground (10 min average), m/s	2.3	3.8
Wind speed 100 m above ground (10 min average), m/s	3.7	7.0
Wind direction	west	west

The beginning of the fire was similar in both tests. The temperature inside the room was 17 °C, and the ignition source successfully initiated the fire development. First, the sofa backrest lit up. Then the flame grew larger and larger forming a conical burning region at the backrest. Next, the room was gradually filled with smoke and hot combustion gases and the fire still developed until it covered the entire sofa. This can be seen in a series of images presented in Figure 4.

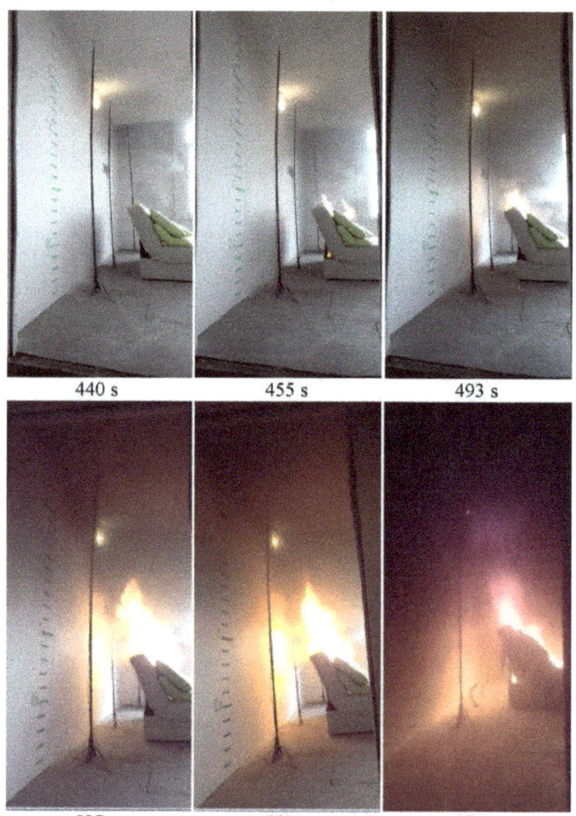

Figure 4. Subsequent stages of fire development (the first test). Time in seconds from the moment of ignition.

During the first test, as shown in Table 2, the outside temperature was significantly higher, but should not have impacted on the conditions inside the room significantly. However, it was observed (Figures 5 and 6) that the temperature above the sofa started to rise almost three minutes earlier than in the second test fire. This was probably caused by different storage conditions of the sofa prior to the experiment, such as packaging methods and room humidity. The first phase of the fire, which did not change the temperature above the sofa, lasted about seven minutes. Then, the temperature started to increase and, in the tenth minute, the temperature above the back of the sofa (at 1.6 m) reached 600 °C. Then, it dropped and started to rise again, reaching 800 °C. At that moment, the decision was made to start the firefighting operation. The entire set of temperature distributions is presented in Figure 5.

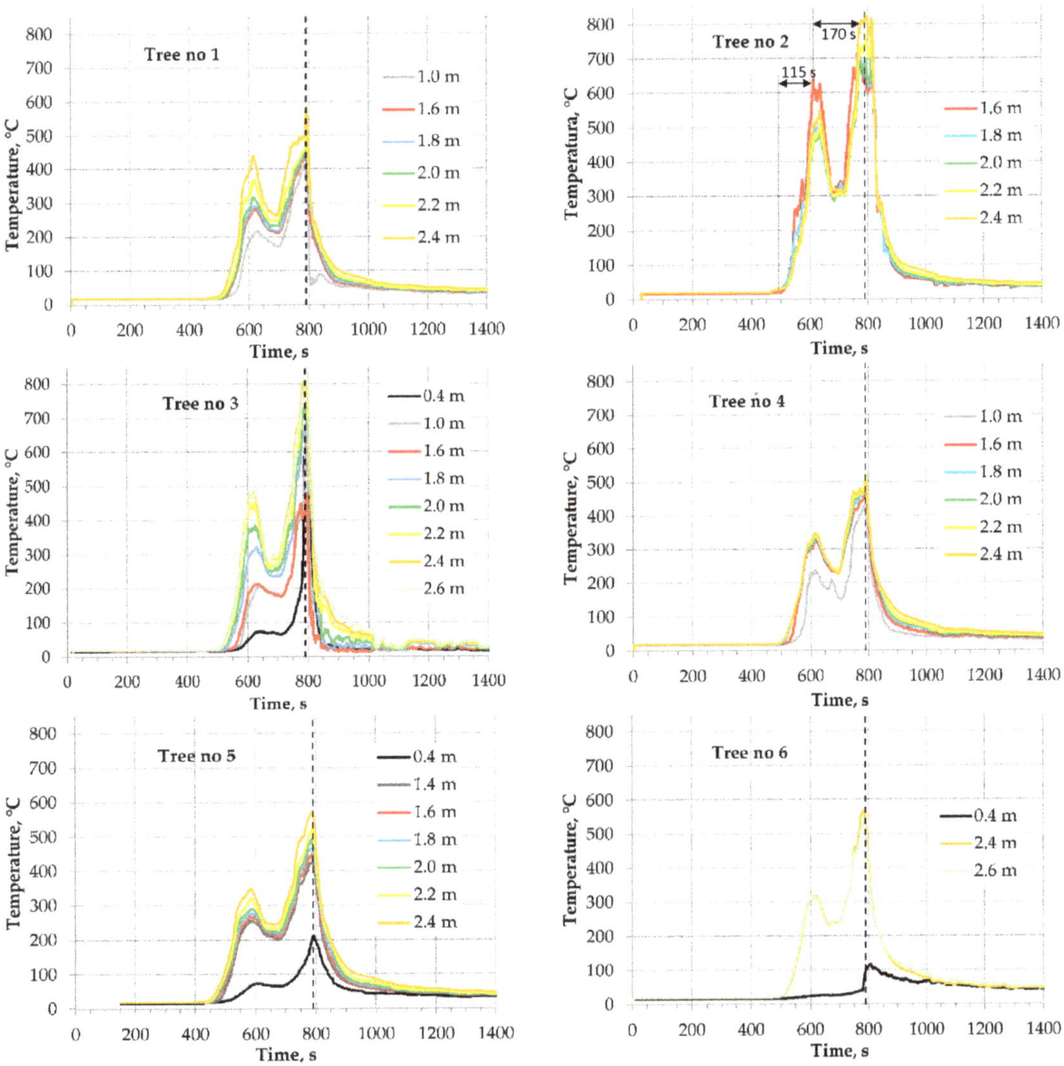

Figure 5. Temperatures during the first test. Dashed lines mark the beginning of the firefighting operations (790 s).

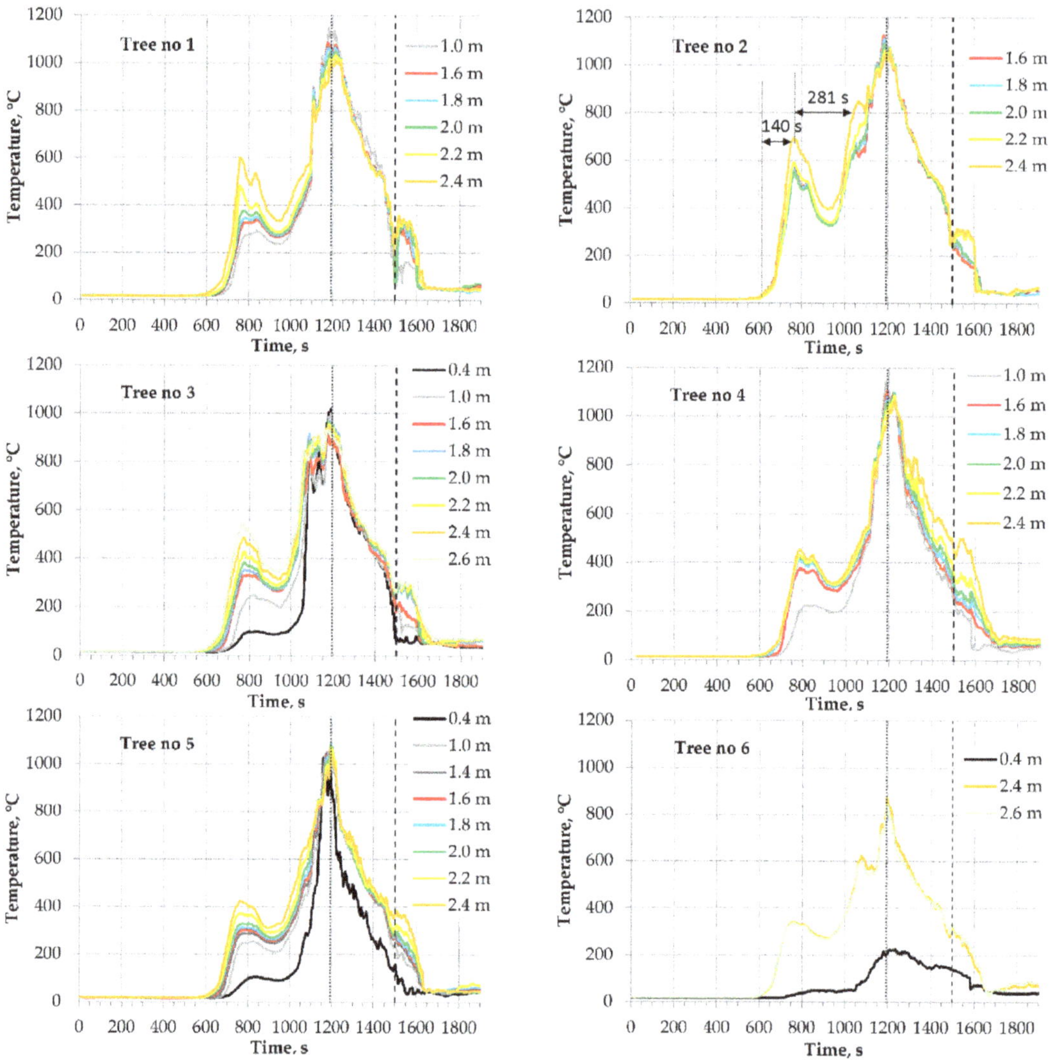

Figure 6. Temperatures during the second test. Dashed lines mark the beginning of the firefighting operations (1500 s), dotted lines mark the moment windows fell out (1193 s).

In the temperature distribution charts for individual thermocouple trees, it is clear that the highest temperature was achieved on the second and third trees. This suggests that smoke and hot air accumulated above the sofa and in the corner of the room by the window. Cold air flowing through the open door probably caused the temperatures on the first thermocouple tree to be lower. The temperatures further down the room, at some distance from the source of the fire, were also lower. In the case of the fifth and sixth trees, the temperature did not reach 600 °C. The course of temperature variation was the same in all trees. The graphs appear to show the stratification of air in the room, which manifested in lower temperatures at low heights, and high temperatures were almost the same at the higher points. This is especially visible for thermocouple trees away from the sofa (trees 5 and 6).

After the first test, the room was cleaned and returned to its original state. This included replacing windows because they may have been weakened due to the heat.

During the second fire test, the first phase of the fire lasted almost ten minutes. Then, the flames spread rapidly in two minutes. Above the back of the sofa (at 1.6 m), the temperature increased to 700 °C. Next, the temperature dropped and then started to rise again. So, the odd course of fire spread was observed again; there were moments when the fire dimmed slightly. Eventually, the temperature reached over 1100 °C in the twentieth minute. Such a high temperature caused the windows to fall out, resulting in a rapid temperature drop. The diagrams of the temperature distribution during the second fire test for all thermocouple trees are shown in Figure 6. Five minutes after the windows had fallen out, the firefighting operation began, as shown in Figure 7. As can be seen, the room reached flashover and the flames were ejected from the window openings. This was confirmed during the room inspection, after the fire was completely extinguished—all equipment items were completely burnt.

When examining the temperature data, it is clear that until the maximum temperature is reached, the shapes of the curves are similar to those of the first fire. After the initial period of growth, there is a decrease and then an increase to the maximum values. Once the maximum temperature is reached for all measuring points, the temperature drops. However, in the second fire test, it is not caused by the intentional extinguishing action, but by the windows' disintegration and the cold air inflow. As a result, the temperature did not drop in the same way as in the first fire. Instead, the temperature dropped slowly after the windows fell out, and the test room cooled down freely until the extinguishing action began.

As was mentioned earlier, the fire developed slower in the second fire test; this is also visible when comparing the time between the first increase of temperature to the forepeak. This was analyzed for the thermocouple tree no. 2: the temperature at the forepeak is almost the same in both tests (650–700 °C), but the time was 115 s in the first test and 140 s in the second one. Similarly, the time between the forepeak and the moment when 800 °C was reached is shorter in the first test (170 s vs. 281 s). These differences occurred despite the same experimental setup in both tests.

In both fire tests, a significant nonuniformity of the temperature distribution was observed, which could be expected because only the growth phase was examined. Two groups of thermocouple trees are notable: the first group contains trees located just close to the fire source (1 and 2) and the tree in the near corner (3), where the hot gases accumulated. The second group consists of distant trees (4 and 6) and the tree in the middle of the room (5), where the free spread of hot gases occurred. It was particularly visible in the fire growth phase: the measured temperature was almost independent of the height for the first group. Meanwhile, the temperature recorded by the second group indicated a clear stratification. This stratification almost ceased for the developed fire, excluding tree 6. Although, even for the developed fire, the maximum recorded temperature was distributed very unevenly.

Figure 7. Windows breaking during the second test fire.

2.3. Numerical Model

The numerical model was built using ANSYS Workbench and solved using ANSYS Fluent. Embedding the combustible material burning model in the ANSYS Fluent environment opens the possibility of a detailed numerical study of the heat flow and combustion products with high reliability, which is characteristic of this tool [38]. It allows for the selection of different turbulence models, which results in accurate flow reproduction. Due to the available combustion models, it is possible to model fire development under different air supply conditions [39].

Room dimensions and material properties were kept the same as the actual ones. The walls, floor and ceiling were modeled as concrete with real thickness. The ANSYS Fluent shell conduction feature was applied, and an ambient temperature of 10 °C was assumed at the outer surfaces (as it actually was for the second real fire). The windows were modeled similarly, where the actual thermal properties of the glass panes were applied. Items in the room were reproduced as well. Since during real fire tests the door was open, it was modeled as a 'pressure outlet' boundary condition.

Conditions inside the room changed significantly as the fire was developing. In the beginning, the amounts of carbon dioxide and water vapor were negligible, so the optical density of the gases was low. Then, as the fire was growing upwards, significant amounts of those combustion products appeared, which resulted in a high optical density, especially in the upper part of the room. Hence, the radiative heat exchange should be modeled considering all details. That is why the discrete ordinate (DO) model of radiation was used. It solves the full general equation ruling the radiative transport for an absorbing, emitting and scattering medium for each finite volume cell [40]. The DO model is able to cope with a wide spectrum of optical densities, includes soot effects on radiation and is the most accurate model available. These advantages are paid for by significantly higher demands of computational resources. Additionally, the thermal radiation in the fire simulations was regarded as grey [41]. This is because soot acts as the main source and sink of thermal radiation, and its properties are not particularly sensitive to wavelength.

The k-ω SST turbulence model, commonly regarded as reliable, was used for flows modeling. The time step duration was set as adaptive, and it varied between 0.01 s to 0.1 s.

Since the simulations covered just a few minutes of fire development, it was assumed that only the sofa and the table were composed of burning materials. The rest of the room equipment did not start to burn in such a short period. The model of both pieces arranged in the room is shown in Figure 8.

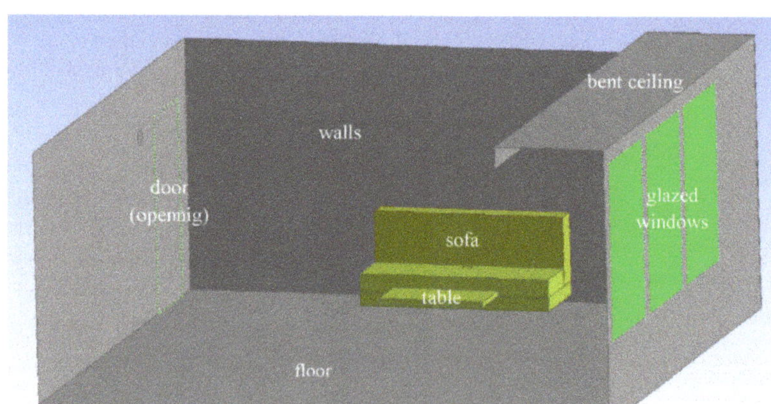

Figure 8. A Fluent model of the partially furnished compartment.

2.3.1. Modeling the Fire Spreading over a Furniture Item

Since a compartment fire is considered, one must not assume in advance that the fire is just fuel controlled. Thus, the fire was modeled using a non-premixed combustion model, which assumes that the final and intermediate combustion products and released energy depend only on the local composition of the gas mixture and temperature [42,43]. The gas mixture is in an equilibrium state or very close to it, and the model can be effectively described by the phrase "what is mixed is burned". So, there is a basic assumption for this approach that chemical combustion reactions are immediate compared to the flows. This simplifies the model and allows a quantity called mean mixture fraction to be kept for all elements. This quantity expresses the mass fraction of considered elements that come from fuel. If a flow is turbulent, diffusion coefficients for all compounds are equal, then the values of mixture fractions are the same for each element and can be replaced by a single value. Therefore, it is enough to solve only one additional equation describing the transport of mean mixture fraction.

To speed up the calculation, preliminary calculations are performed, and a probability density function (PDF) table is created. It is a look-up table containing the details of combustion chemistry for assumed fuel and oxidant composition [44].

As the oxygen abundance directly impacts on the combustion process and this approach takes into account the combustion chemistry, it is suitable for fires controlled by ventilation and by fuel as well. So, this combustion model can model fires under different ventilation regimes and is particularly useful to simulate compartment fires.

2.3.2. Single-Cell Level Fire Model

The combustion process of each piece of equipment was modeled using a set of user defined functions (UDF). This is an ANSYS Fluent feature, which allows for freely expanding the software's capabilities. A UDF is a procedure written in C, which is linked to a numerical model.

Burning pieces of the room equipment were modeled as fluid, porous zones. Their properties were set to imitate the actual materials accurately. Porosity (X_p) expresses the fraction of a material volume filled by a fluid (air). It is used to compute the effective thermal conductivity as the volume average of the thermal conductivities of fluids and solids, which constitute the porous material (λ_{eff}, λ_f and λ_s denote effective thermal conductivity, the fluid thermal conductivity and the solid thermal conductivity, respectively):

$$\lambda_{eff} = X_p\lambda_f + (1 - X_p)\lambda_s \quad (1)$$

For flows in porous media, the pressure drop (∇p) is commonly proportional to the flow velocity ($\vec{u}$). This dependence is described by Darcy's law (where μ denotes dynamic viscosity):

$$\nabla p = \mu \overline{\overline{D_v}} \cdot \vec{u} \quad (2)$$

$\overline{\overline{D_v}}$ is the matrix of viscous resistances, which may be different in different directions. Hence, the flow nature depends on the fluid properties and structure of the porous material. In the considered case, all materials are assumed to be isotropic, so the equation governing the flow simplifies and contains only a single scalar value of viscous resistance (R_v, m^{-2}):

$$\nabla p = \mu\, R_v \cdot \vec{u} \quad (3)$$

Since the mass burns out, the porous properties of combustible materials were also controlled by the dedicated UDFs. The values of the viscous resistance ($R_v(t)$) and the porosity ($X_p(t)$) changes as the material burns. It was assumed that these values changed linearly with the mass loss ($m(t)/m_0$). Viscous resistance dropped from an initial value specific to the material (R_{v0}) to zero, which corresponds to the complete burnout:

$$R_v(t) = \frac{m(t)}{m_0} R_{v0} \tag{4}$$

Porosity increased from an initial value specific to the material (X_{p0}) to unity:

$$X_p(t) = 1 + (X_{p0} - 1)\frac{m(t)}{m_0} \tag{5}$$

In a similar way, the radiation absorption coefficient for volumes of combustible items was treated. This value for polyurethane foam (α_{foam}) was adopted according to data published by Glicksman et al. [45]. The radiation absorption coefficient for a gases mixture (α_{mixt}) was automatically calculated by Fluent software based on the mixture compound and, finally, the effective value the radiation absorption coefficient was expressed as a weighted mean:

$$\alpha(t) = \frac{m(t)}{m_0}\alpha_{foam} + \left(1 - \frac{m(t)}{m_0}\right)\alpha_{mixt} \tag{6}$$

A set of variables called user defined memory (UDM) was assigned for each computational cell. These variables stored the initial amount of fuel, the current amount of fuel and a clock storing the moment of the previous iteration.

In the presented approach, the pyrolysis process was modeled as an emission of gaseous fuel at a rate dependent on the temperature. Since during the pyrolysis of polyurethane foam, methane and light olefins prevail among the products [46], two fuels were examined: methane and acetylene. The preliminary tests showed negligible differences, so eventually, acetylene was selected because it mimics the pyrolysis products composition in a better way. The initial amount of fuel was adjusted to obtain the actual heat of combustion of the given material.

Pyrolysis was modeled as a simple, single-step reaction. Hence, theoretically, the rate of the combustible material mass (m) loss, which is equal to the amount of emitted gaseous fuel, can be expressed as a function of temperature (T), as follows (c_0 denotes a pre-exponential factor, E_a denotes the activation energy per molecule and k_B is the thermodynamic Boltzmann constant):

$$\frac{\mathrm{d}m}{\mathrm{d}t} = -mc_0 \exp\left(-\frac{E_a}{k_B T}\right) \tag{7}$$

The values of parameters c_0 and E_a are not commonly known for most materials. Moreover, real materials are usually not homogeneous, so a number of pyrolysis processes occur when a material is heated. Hence, even apparently similar materials may differ significantly in pyrolysis details. A simplified approach was introduced here by assuming a homogeneous pyrolysis and the parameters were fit using the literature data on the pyrolysis of polyurethane foam and wood. This issue is discussed later in the paper. Despite the simplification, this approach was difficult to implement because it required applying a very short time step, otherwise the numerical instability appeared.

Therefore, a further simplified relationship was adopted to speed up the simulation. Considering that pyrolysis occurs in a relatively narrow temperature range, Formula (7) was expanded in a power series, and only the first order terms were considered. It was assumed that pyrolysis began at a specified temperature (T_{p0}), then its rate linearly increased with the temperature:

$$\frac{\mathrm{d}m}{\mathrm{d}t} = \begin{cases} 0 & T \leq T_{p0} \\ -mc_1(T - T_{p0}) & T > T_{p0} \end{cases} \tag{8}$$

Finally, Formula (8) was used to determine the amount of released gaseous fuel. The values of Tp_0 and c_1 were fit as previously. The preliminary simulations showed there was almost no difference between both approach's results, but the calculation involving the latter formula was completed much faster.

Hereby, the simulation of fire spreading over a combustible item appeared largely independent on the detailed assumptions of the pyrolysis process. This can be briefly explained by analyzing a single cell of the computational domain constituting the combustible item. It is heated in different ways (radiation, convection and conduction) and then, over a given temperature threshold, the pyrolysis starts. In a short time, the cell temperature reaches a high value and all its mass is converted into the gaseous fuel. Hence, in a larger time scale, it is not the accurate rate of pyrolysis that is important but its integral over time.

Since pyrolysis itself is an endothermic reaction, the energy needed for material decomposition must be supplied to sustain the reaction. This energy is called heat of reaction and is a program parameter adopted in accordance with the literature data [47]. This amount of energy is taken into account in the total energy balance.

For volumes that corresponded to combustible pieces of the room equipment, the 'source terms' option was enabled. This allowed for the implementation of Formulas (7) and (8) via UDFs as fuel and mass sources. Since the process of pyrolysis is an endoenergetic one, it needs a specified amount of energy per mass unit of the processed material to be absorbed. When the non-premixed combustion model is used, a fuel stream enters the domain at a given temperature. Hence, it was necessary to add the specified amount of energy to mimic the fuel release at the actual temperature. Both issues were incorporated into the model via another UDF, which established an energy source of a relevant intensity.

The soot yield was adopted as 0.15, which is in accordance with experimental data [48]; however, some sources reported a slightly higher value [49,50].

The ignition was modeled by setting a high temperature (900 K) for a small volume of the sofa seat just next to the backrest.

2.3.3. Tuning the Model Parameters

The model assumptions were validated by relating them to the results of TGA experiments (thermogravimetric analysis) using a polyurethane foam sample. The sample in the form of a thin slice (0.03 m diameter, 0.001 m height) was put at the bottom of a vessel filled with a neutral gas (nitrogen). The bottom of the vessel was heated with a constant temperature rise rate of 10 °C/60 s. Since the Biot number of the sample was equal to $Bi = 0.75$, which is less than unity, the sample could be regarded as lumped. Hence, the temperature was uniform within the whole sample volume, and the pyrolysis process occurred homogeneously.

Eventually, for Formula (7), the values of activation energy $E_a = 2.15 \times 10^{-19}$ J and pre-exponential factor $c_0 = 5 \times 10^9$ 1/s were found, which are typical values for polyurethane pyrolysis [51]. Meanwhile, for Formula (8), the parameters T_{p0} and c_1 were fit to 240 °C and 1.029×10^{-5} 1/s·K, respectively.

The foam density was 40 kg/m^3. The initial value of the porosity was set to 0.94 and the viscous resistance to 2,200,000 m^{-2}, which are typical for standard polyurethane foam [52]. The results were compared with different experimental literature data on polyurethane foam pyrolysis. Since there are many kinds of such foam, the available data differ to some degree, but generally, they can be regarded as consistent, as seen in Figure 9.

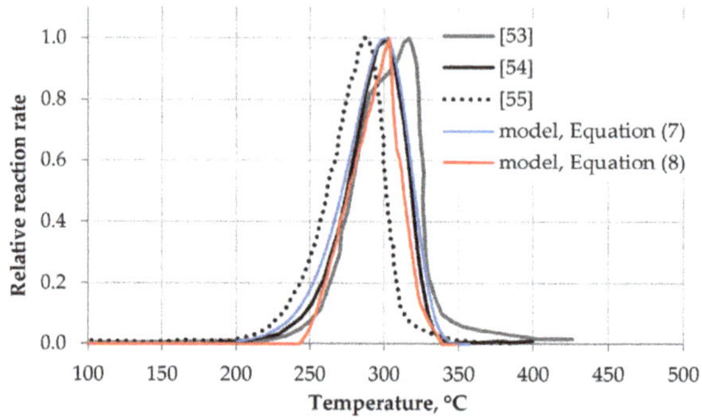

Figure 9. Comparison of modeled pyrolysis rate with experimental data [53–55].

2.3.4. Modeling of Temperature Measurement

Since the data obtained by carried out simulations had to be compared with the experimental results, the process of temperature measurement by thermocouple beads was also modeled. This issue is very important because the difference between the real temperature of gases and the temperature indicated by a thermocouple in a non-equilibrium state may be significant. The problem was solved by placing virtual thermocouples of different bead diameters in every cell of the computational domain. This required additional UDM slots and UDF plug-ins but allowed for accurate temperature recording [56]. Several temperature measurement points were included to model the thermocouple trees. Although virtual thermocouples were placed in every cell of the computational domain, the data were recorded every 5 s only for the actual thermocouple positions.

2.3.5. Mesh Sensitivity Analysis

Since the obtained results may depend on the applied numerical mesh, two different meshes were assessed. Both applied meshes were created according to the work presented by Węgrzyński et al. [48]. Since the fire plume impinges and starts to spread horizontally when in the above area, just beneath the ceiling, both meshes included inflation layers there. This helped to keep the accuracy of the flow image in this region. The burning pieces (the sofa and the table) were modeled using regular mesh, which contained only hexahedral cells. This allowed for a dense fragmentation with a relatively low number of mesh elements. The mixed mesh, containing mostly hexahedrons, was used for the whole room space, since such an approach allows for smooth transitions among elements of different sizes. The parameters of the applied meshes are shown in Table 3.

Table 3. Parameters of the applied meshes.

Mesh	No. of Elements	No. of Nodes	No. of Inflation Layers	Edge Length	
				Room	Burning Items
Normal	470,362	118,188	10	0.150	0.050
Coarse	319,898	81,418	8	0.200	0.075

For comparison purposes, the temperature distributions close by and just above the sofa (trees 1 and 2) for both meshes are shown in Figure 10. The curves are quite similar despite some fluctuations, indicating that the model can be regarded as mesh independent.

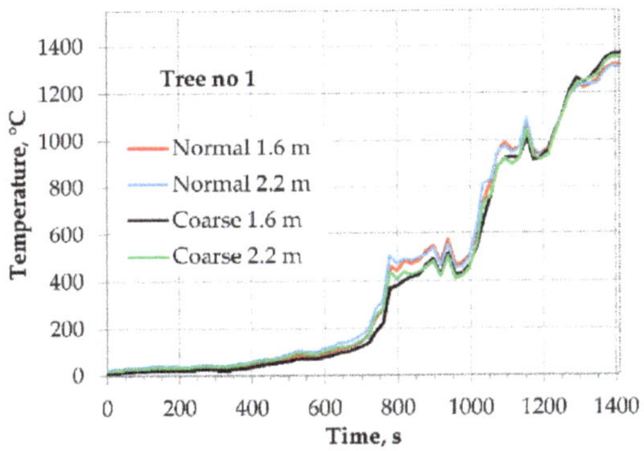

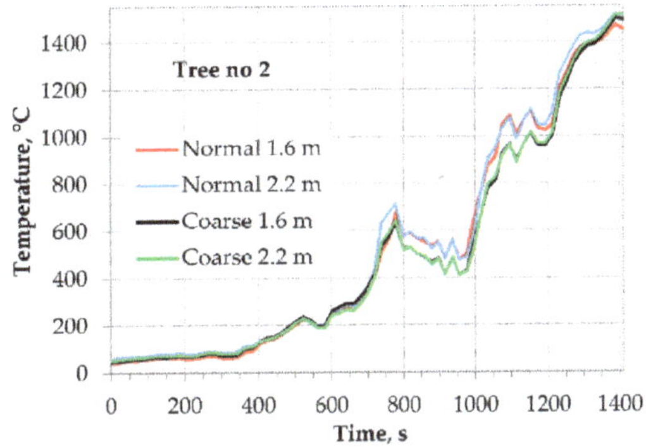

Figure 10. Numerical results for two mesh densities.

2.4. FDS Model

The proposed approach was confronted with the well-recognized FDS software. Using this software, a model of the same furnished room was built. The model accurately reproduced the experimental setup. By default, FDS uses the large eddy simulation (LES) model for turbulence modeling. The combustion process is modeled similarly to the non-premixed model mentioned above [57]. It is an option in FDS to simulate material burnout by adopting a single-step pyrolysis (SSP) model in a way similar to that presented earlier. The other possibility is to use the ignition temperature-based pyrolysis (ITP) approach, in which the ignition temperature is set, then the value of the heat release rate per unit area (HRRPUA) is applied to calculate the burnout progress [58]. However, in both cases, there is a need to carefully adjust the control parameters by the trial and error method. The data published by Park and Kwark were very helpful here [59]. For the SSP approach, the parameters were tuned with an option of fitting the TGA curve (similarly as in the Fluent model). The parameters adopted for both material burnout approaches are listed in Table 4.

Table 4. Parameters of material burnout.

	Parameter		Value
	Heat of combustion, kJ/kg		2.54×10^4
	Heat of reaction, kJ/kg		1.57×10^3
SSP	Reference temperature, °C		100.0
	Heating rate, K/min		5.0
	Pyrolysis range, °C		80.0
	Mass Fraction Exponent (n_s)		2.0
ITP	HRRPUA, kW/m^2		600.0
	Ignition temperature, °C		300.0
	Time ramp (relative intensity vs. time)	0 s	0.0
		60 s	1.0
		120 s	0.8
		240 s	0.2

The computational domain of the model consisted of a number of meshes, which allowed for applying a dense mesh (cell edge of 0.05 m) to model the combustible items and their vicinity, and a coarse mesh (cell edge of 0.1 m) for the other parts of the compartment. These values were in accordance with the NIST recommendation.

The quality of flow modeling for buoyant plumes depends on a non-dimensional term $D^*/\delta x$, where δx denotes the cell size and D^* is the characteristic fire diameter. The latter value is given as follows (ρ_∞, c_p and T_∞ are, respectively, density, specific heat and temperature of ambient air) [57]:

$$D^* = \left(\frac{\dot{Q}}{\rho_\infty c_p T_\infty \sqrt{g}}\right)^{\frac{2}{5}} \quad (9)$$

During the addressed simulation, the heat release rate ($\dot{Q}$) varied, but assuming the maximum reasonable value of 1.2 MW [46], D^* was estimated to be approximately 1 m. There are no explicit tips on how to establish the ratio $D^*/\delta x$ in advance, but in this work this value was 10 and 20 (depending on the mesh), which was consistent with the examples provided by NIST [57]. Hence, no additional mesh sensitivity analysis is discussed.

3. Results—Model Validation and Discussion

Since the first fire test was aborted before the fire could reach its maximum intensity, data from the second fire were used to validate the numerical model. The comparison of real and modeled temperature distributions is shown in Figure 11. In this experiment, the windows broke at 1193 s of fire development, and due to a rapid cold air inflow, the temperature dropped. The numerical simulation did not take this event into account. Hence, both curves diverge at this point from each other. In such a situation, only the growth phase of the fire development should be considered when validating the model.

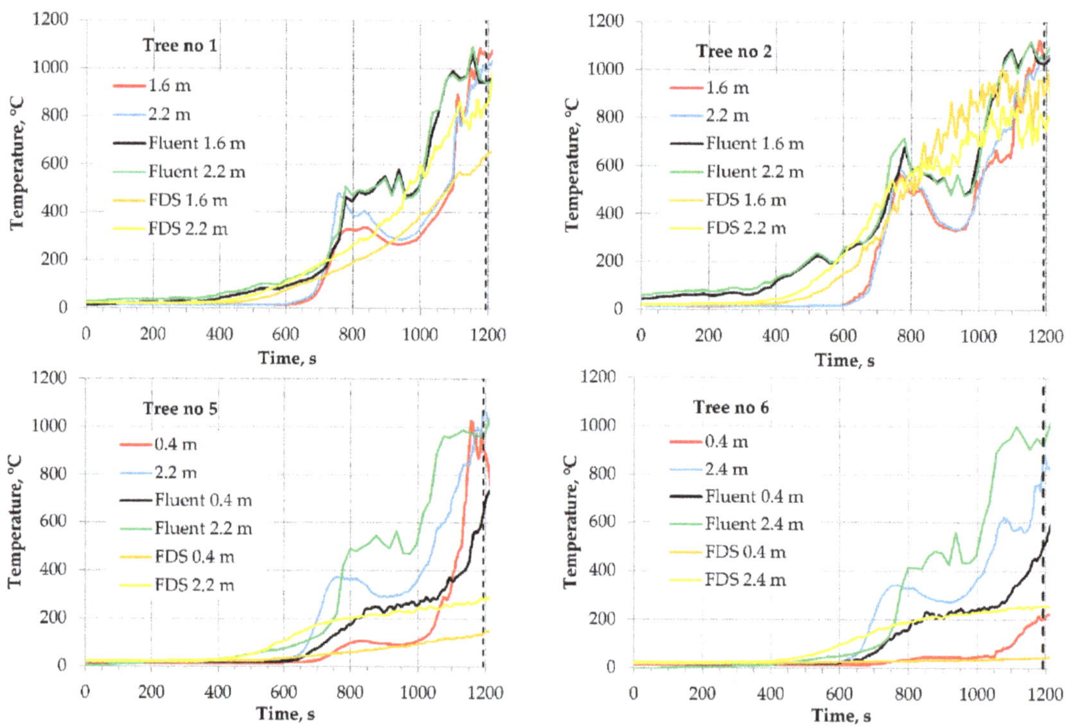

Figure 11. Comparison of the selected temperature distributions for real and modeled (Fluent and FDS—ITP approach) fire.

At first glance, one can see that the temperature distributions differ in some degree. It concerns mainly FDS results: the predicted temperatures are generally lower, but the stratification is clearer. The first local temperature maximum is not visible here in contrast to the real data and the Fluent results. Since this is the merit of the work, further analysis is focused on Fluent outcomes. Although the experimental and calculated-via-Fluent curves do not overlap exactly, their courses are very similar, and the maximum temperature values are almost the same. The differences are visible for both the temperature values (up to about 100–200 °C) and the pace the fire was developing (a shift of the forepeak up to 100 s).

The ignition process was going in a slightly different way: in the numerical model, the whole volume of the ignition source was patched to a temperature of 500 °C instead of a slow development, as in the reality. This was to speed up the calculation, but it influenced somewhat the temperature distribution in the first moments. Therefore, in the very initial stage, when only the kindling was burning (before 600 s), the calculated temperature became higher above the sofa (especially for tree no. 2).

The best match can be observed for tree no. 2, which may confirm the accuracy of the material combustion model. The moments of forepeak and the maximum temperature occurrence were in good agreement, although the calculated temperature is overestimated (sometimes up to almost 200 °C).

When analyzing data for trees no. 3 and 4, one can see that the temperature stratification observed in the experimental data is not as clear in the numerical results—the calculated curves corresponding to different heights are close to each other. In addition, for those thermocouple trees, the divergency of the calculated temperature was the highest—the pace of the temperature growth was reproduced accurately, but its values were still higher than measured (the gap sometimes exceeded even 200 °C). The same applies to other thermocouple trees placed at greater distances from the fire source, which are not shown here. This may indicate that the flows throughout the room were not reproduced in a desired level of precision.

However, considering the simplicity of the model, it can be regarded as reliable. It is important to consider that real fires can be fickle phenomena. The experiments ran in somewhat unrepeatable ways despite controlled conditions, which were kept the same as much as possible. For this reason, the principle of 'consistent crudeness' should be taken into account in fire engineering [60]. It assumes that in light of input data uncertainty and the general unpredictability of a fire, one does not need to try to achieve a very high accordance of a theoretical or a numerical model with a particular set of experimental data.

There is a clear forepeak visible in the majority of the temperature vs. time curves. It appears in both the experimental and numerical results, especially for thermocouples placed just above the burning sofa. The possible explanation is as follows:

A. At the beginning of the process, the backrest foam needs more heat to ignite, and the only burning part is the kindling. Due to the low volume of the burning material, no significant temperature rise was observed.

B. At the first stage of the fire development, the burning area slowly expands upwards and slightly on both sides, forming a u-shaped region of combustion that covers the subsequent parts of the sofa backrest, mainly due to the convectional transport of hot gases along the surface of the backrest and partially inside it because of its low porous resistance. This phase lasts to the moment where this region reaches the top of the backrest. At this phase, the volume of hot gases started to increase significantly, and a steep temperature rise was observed.

C. When the burning area reaches the top of the backrest, the fire development clearly slows down because its spread is hindered. This happens mainly horizontally by conductive heat transfer to adjacent parts of the backrest. This way of heat transfer is significantly less efficient due to the low thermal conductivity of the polyurethane foam.

D. Since the fire stops spreading quickly, the temperature above the fire source may even drop because the hot gases continue to spread along the ceiling. However, the combustion of the u-shaped area of the backrest continues and generates hot gases. Therefore, the layer of hot gases beneath the ceiling is gradually lowering.

E. When this layer reaches the top of the backrest, it causes the ignition of the upper part of the backrest, and the fire development is accelerated. A large part of the backrest is ignited, huge amounts of hot gases are generated and the hot layer lowers quickly, resulting in the fire quickly covering the sofa. The temperature rises significantly in this period.

F. In this phase, the layer of hot gases reaches far towards the room floor, and the fire may spread to other furniture items. The fire will develop to complete burnout if there is a sufficient fresh air supply or become under-ventilated in a smaller compartment.

These phases of the sofa fire development are shown in Figure 12, which presents the burning areas, and are marked in Figure 13, where a diagram of temperature distribution just above the sofa is shown (the grey dotted line marks the expected temperature rise if the windows had not fallen out). The last image in Figure 4 (timestamp 573 s) corresponds to a time moment between phases D and E; later, the large amounts of soot made the movie completely unreadable.

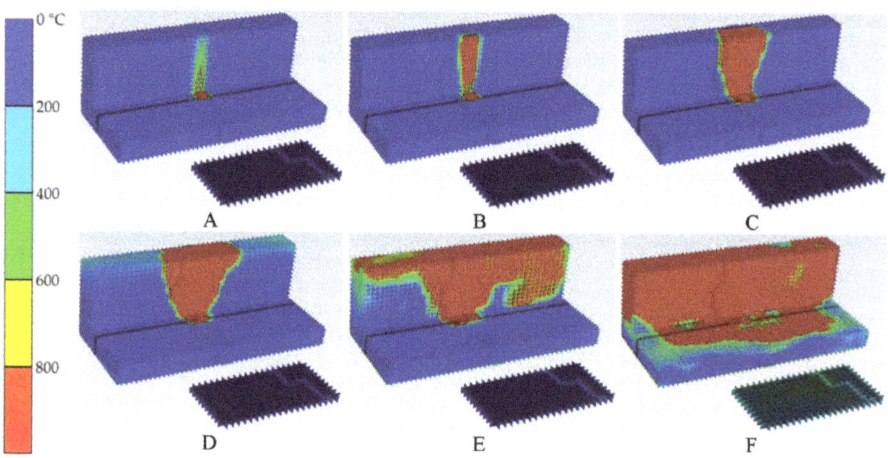

Figure 12. The subsequent phases of sofa fire development: temperature distribution on its surface (description in text).

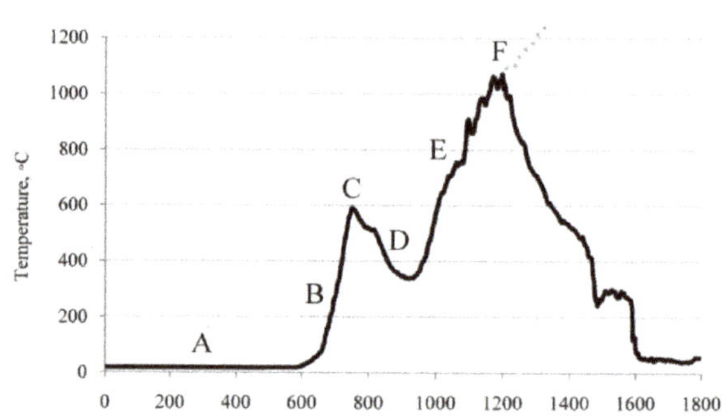

Figure 13. The phases of sofa fire marked at the temperature curve (the second fire test, tree no. 2, height 2.4 m in Figure 6) (description in text).

Figure 14 presents a snapshot taken at phase C during the first fire test. This corresponds to the frame at 493 s in Figure 4. It was almost the last moment a clear photo could be taken because, in a dozen seconds, wisps of smoke began to appear from the lowering smoke layer.

Figure 14. An image of u-shaped burning region (about 493 s of the first fire test).

The temperature distributions in the compartment at the sofa symmetry plain at these moments are shown in Figure 15. The lowering layer of hot gases is clearly visible there (C–E); the same applies to the moment when this layer reached the sofa and the fire development started to accelerate €.

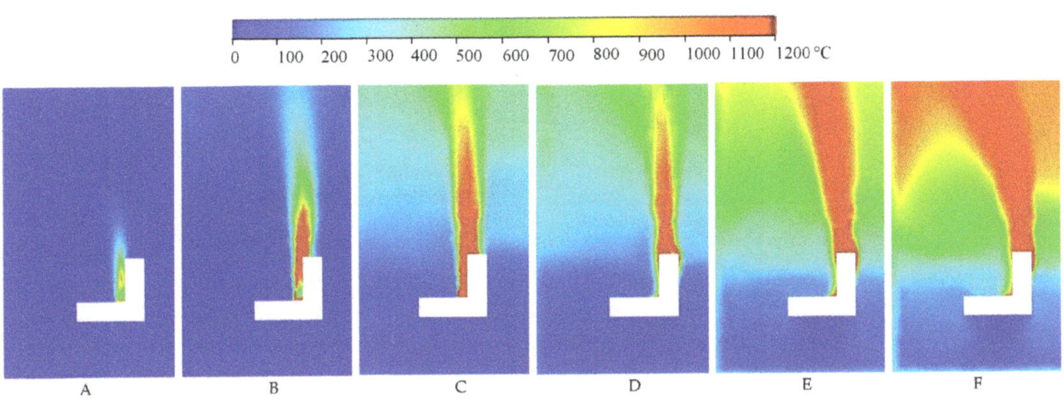

Figure 15. Temperature distributions at the sofa symmetry plain at selected moments (description in text).

Figure 16 shows the burnout of the sofa material at the same time moments (the sofa is cut by a symmetry plane, the volume corresponding to the ignition source is also shown). The observations are in line with the earlier ones. At phases B, C and D, the conical burning region was developing, but due to the low heat conductance of polyurethane foam, its horizontal spread was slow. In phases C and D, the fire was kept alive mainly due to the burnout of the material in this region rather than its spreading.

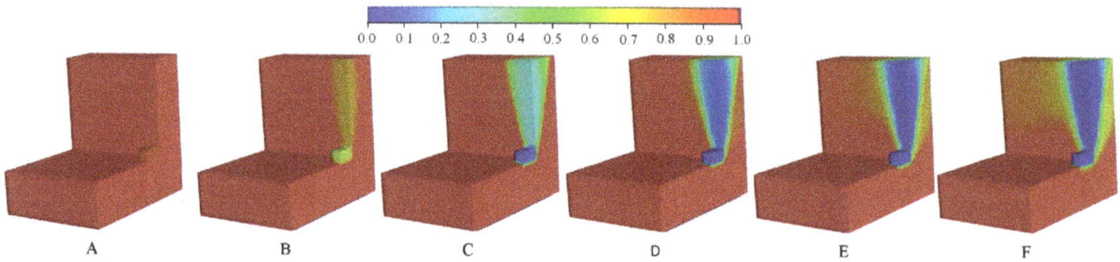

Figure 16. Progress of the material burnout (0 corresponds to the complete burnout) (description in text).

The phase marked as C is critical for fire development. This applied to both real and numerical experiments. Some real fire experiments ceased to develop at this stage. This was presumably due to slight differences in experimental conditions. It might concern the properties of polyurethane foam, which resulted from different storage conditions, despite all used sofas being exactly of the same type.

As was mentioned earlier, there was no forepeak visible in the FDS results, so the FDS simulation did not reproduce the process of fire development in such a faithful manner. However, it generally revealed a similar way of spreading the fire. As an example, the burnout progress corresponding to phase E is shown in Figure 17.

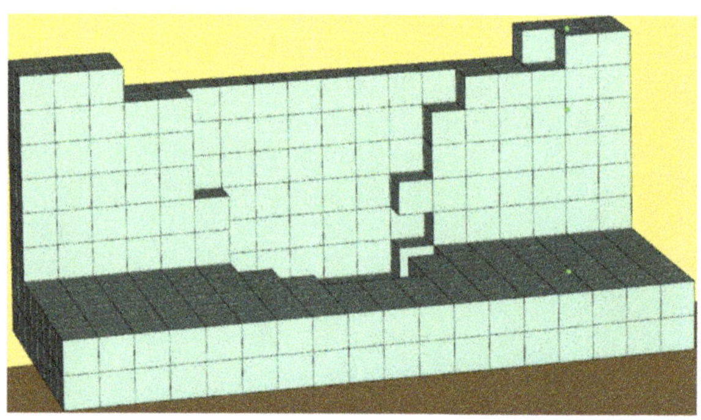

Figure 17. Material burnout in FDS simulation corresponding to phase E.

A necessary condition for the numerical model to achieve conformity with the experimental data was to adjust the parameters controlling the intensity of pyrolysis and the density of flammable material. A threshold condition was revealed here: if the pyrolysis intensity was high enough, the fire spread quite quickly over the entire item in all directions. Otherwise, the u-shaped burning region at the backrest was formed and then it spread slowly. The burning area surrounding the ignition source barely spread horizontally. Meanwhile, the height of hot layer gases in the compartment was lowering continuously and if the fuel amount was large enough to sustain the fire, it reached the sofa and caused the violent fire revival and expansion.

4. Conclusions

First, this work presents full-scale fire experiments on fire development in a furnished room in a high-rise building. Such experiments are rare due to high costs and the many organizational obstacles. Then, the second stage of the work involved numerical simulations, allowing investigators to deepen their knowledge of the processes involved during fire development in furnished spaces and explore more than the actual experimental configurations. Two software packages were used, FDS and ANSYS Fluent. The results of both numerical models were generally consistent with the real process of fire spreading.

However, those of Fluent reproduced the fire development phases in a more accurate way. In particular, since the Fluent model had to be built from scratch it allowed for accurate examining of the process of fire spreading.

Despite its relative simplicity, the proposed numerical model was able to mimic the development of the real fire. This gives hope for reaping the benefits of ANSYS Fluent capabilities in more sophisticated configurations of numerical models. However, as some clear divergences between measured and calculated temperature fields were observed, the basic model of fire spread over a single furniture item should be improved in the future. The maximum calculated temperature appeared to be overestimated, and the temperature increase rate was also higher than the real one. However, eventually, the simulated fires did not differ from the real ones to a higher degree than the real fires differed from each other. This means that there are some unknown and variable factors that can influence the course of fire development. Even though the real experiment was prepared with care to details, and thus so was the numerical model, these factors evidently were beyond researchers' control and knowledge. Anyway, despite their influence being noted, it did not alter the observed phenomena to a significant degree.

The numerical research revealed a threshold condition for fire spreading over an upholstered item. If a combustible material is less susceptible to pyrolysis, the fire spreads rather slowly and only lowering the layer of hot gases may cause the violent fire development.

Both real and numerical experiments showed non-uniform temperature distribution in the test room. This was especially visible in the fire growth phase when clear temperature stratification was observed. However, while the stratification almost ceased, the experiments did not enter the stage where the spatial temperature distribution would be more even. This could be regarded as an indicator to apply more measurement points in the future to record the results of such experiments more accurately.

Author Contributions: Conceptualization, M.K. and A.K.; methodology, M.K.; software, A.K.; validation, M.K. and A.K.; data curation, M.K. and A.K.; writing—original draft preparation, M.K. and A.K.; visualization, A.K. and M.K. All authors have read and agreed to the published version of the manuscript.

Funding: The publication of the work was supported by the Polish Ministry of Science and Higher Education, within statutory research subsidy 08/010/BK22/0060.

Data Availability Statement: Not applicable.

Acknowledgments: The authors would like to thank the Smay Company for the opportunity for participation in the research, for the provided management support and for the consent to publish the results. We would like also to thank the Headquarters of the State Fire Service in Sosnowiec for supervision, security and participation in the research.

Conflicts of Interest: The authors declare no conflict of interest.

References

1. Lange, D.; Torero, J.L.; Osorio, A.; Lobel, N.; Maluk, C.; Hidalgo, J.P.; Johnson, P.; Foley, M.; Brinson, A. Identifying the attributes of a profession in the practice and regulation of fire safety engineering. *Fire Saf. J.* **2021**, *121*, 103274. [CrossRef]
2. Chu, G.; Sun, J. Decision analysis on fire safety design based on evaluating building fire risk to life. *Saf. Sci.* **2008**, *46*, 1125–1136. [CrossRef]
3. Chu, G.Q.; Chen, T.; Sun, Z.H.; Sun, J.H. Probabilistic risk assessment for evacuees in building fires. *Build. Environ.* **2007**, *42*, 1283–1290. [CrossRef]
4. Zhao, C.M.; Lo, S.M.; Lu, J.A.; Fang, Z. A simulation approach for ranking of fire safety attributes of existing buildings. *Fire Saf. J.* **2004**, *39*, 557–579. [CrossRef]
5. Chen, Z.; Satoh, K.; Wen, J.; Huo, R.; Hu, L. Burning behavior of two adjacent pool fires behind a building in a cross-wind. *Fire Saf. J.* **2009**, *44*, 989–996. [CrossRef]
6. Khan, A.; Usmani, A.; Torero, J.L. Evolution of fire models for estimating structural fire-resistance. *Fire Saf. J.* **2021**, *124*, 103367. [CrossRef]
7. Wang, L.; Li, W.; Weimin, W.; Yang, R. Fire risk assessment for building operation and maintenance based on BIM technology. *Build. Environ.* **2021**, *205*, 108188. [CrossRef]
8. Harmathy, T.Z. A New Look at Compartment Fires. *Fire Technol.* **1972**, *8*, 196–217. [CrossRef]

9. Thomas, P.H.; Heselden, A.J.; Law, M. *Fully-Developed Compartment Fires—Two Kinds of Behavior*; H.M. Stationery Office: Richmond, UK, 1967.
10. Torero, J.L.; Majdalani, A.H.; Abecassis-Empis, C.; Cowlard, A. Revisiting the compartment fire. *Fire Saf. J.* **2014**, *11*, 28–45. [CrossRef]
11. Hua, J.; Wang, J.; Kumar, K. Development of a hybrid field and zone model for fire smoke propagation simulation in buildings. *Fire Saf. J.* **2005**, *40*, 99–119. [CrossRef]
12. Zhang, J.Y.; Lu, W.Z.; Huo, R.; Feng, R. A new model for determining neutral-plane position in shaft space of a building under fire situation. *Build. Environ.* **2008**, *43*, 1101–1108. [CrossRef]
13. Prasad, K.; Baum, H.R. Coupled fire dynamics and thermal response of complex building structures. *Proc. Combust. Inst.* **2005**, *30*, 2255–2262. [CrossRef]
14. Stern-Gottfried, J.; Rein, G.; Bisby, L.; Torero, J.L. Experimental review of the homogeneous temperature assumption in post-flashover compartment fires. *Fire Saf. J.* **2010**, *45*, 249–261. [CrossRef]
15. Jahn, W.; Rein, G.; Torero, J.L. A posteriori modelling of the growth phase of Dalmarnock Fire Test One. *Build. Environ.* **2011**, *46*, 1065–1073. [CrossRef]
16. Gupta, V.; Hidalgo, J.; Cowlard, A.; Abecassis-Empis, C.; Majdalani, H.A.; Maluk, C.; Torero, J.L. Ventilation effects on the thermal characteristics of fire spread modes in open-plan compartment fires. *Fire Saf. J.* **2021**, *120*, 103072. [CrossRef]
17. Sun, X.; Hu, L.; Zhang, X.; Yang, Y.; Ren, F.; Fang, X.; Wang, K.; Lu, H. Temperature evolution and external flame height through the opening of fire compartment: Scale effect on heat/mass transfer and revisited models. *Int. J. Therm. Sci.* **2021**, *164*, 106849. [CrossRef]
18. Lu, K.; Wang, Z.; Ding, Y.; Wang, J.; Zhang, J.; Delichatsios, M.; Hu, L. Flame behavior from an opening at different elevations on the facade wall of a fire compartment. *Proc. Combust. Inst.* **2021**, *38*, 4551–4559. [CrossRef]
19. Bonner, M.; Węgrzyński, W.; Papis, B.; Rein, G. KRESNIK: A top-down, statistical approach to understand the fire performance of building face des Rusing standard test data. *Build. Environ.* **2020**, *169*, 106540. [CrossRef]
20. Mckenna, S.; Jones, N.; Peck, G.; Dickens, K.; Pawelec, W.; Oradei, S.; Harris, S.; Stec, A.; Hull, R. Fire behavior of modern façade materials–Understanding the Grenfell Tower fire. *J. Hazard. Mater.* **2019**, *368*, 115–123. [CrossRef]
21. Sharma, A.; Mishra, K.B. Experimental investigations on the influence of 'chimney-effect' on fire response of rain screen façades in high-rise buildings. *J. Build. Eng.* **2021**, *44*, 103257. [CrossRef]
22. Chow, C.L.; Chow, W.K. Heat release rate of accidental fire in a supertall building residential flat. *Build. Environ.* **2010**, *45*, 1632–1640. [CrossRef]
23. Cheng, C.C.K. fire safety study of Hong Kong refuge floor building wall layout design. *Fire Saf. J.* **2009**, *44*, 545–558. [CrossRef]
24. Ren, F.; Hu, L.; Zhang, X.; Sun, X.; Fang, X. Temperature evolution from stratified- to well-mixed condition inside a fire compartment with an opening subjected to external wind. *Proc. Combust. Inst.* **2021**, *38*, 4495–4503. [CrossRef]
25. Lu, K.; Xu, H.; Shi, C.; Wang, Z.; Wang, J.; Ding, Y. Numerical investigation of air curtain jet effect on the upper layer temperature evolution of a compartment fire and its transition. *Appl. Therm. Eng.* **2021**, *197*, 117409. [CrossRef]
26. Chen, C.-J.; Hsieh, W.-D.; Hu, W.-C.; Lai, C.-M.; Lin, T.-H. Experimental investigation and numerical simulation of a furnished office fire. *Build. Environ.* **2010**, *45*, 2735–2742. [CrossRef]
27. Majdalani, A.H.; Cadena, J.E.; Cowlard, A.; Munoz, F.; Torero, J.L. Experimental characterization of two fully-developed enclosure fire regimes. *Fire Saf. J.* **2016**, *79*, 10–19. [CrossRef]
28. Yang, P.; Tan, X.; Xin, W. Experimental study and numerical simulation for a storehouse fire accident. *Build. Environ.* **2011**, *46*, 1445–1459. [CrossRef]
29. Byström, A.; Cheng, X.; Wickström, U.; Veljkovic, M. Full-scale experimental and numerical studies on compartment fire under low ambient temperature. *Build. Environ.* **2012**, *51*, 255–262. [CrossRef]
30. Mackay, D.; Barber, T.; Yeoh, G.H. Experimental and computational studies of compartment fire behavior training scenarios. *Build. Environ.* **2010**, *45*, 2620–2628. [CrossRef]
31. Hasib, R.; Kumar, R.; Kumar, S.; Kumar, S. Simulation of an experimental compartment fire by CFD. *Build. Environ.* **2007**, *42*, 3149–3160. [CrossRef]
32. He, Q.; Liu, N.; Xie, X.; Zhang, L.; Zhang, L.; Zhang, Y.; Yan, W. Experimental study on fire spread over discrete fuel bed—Part I: Effects of packing ratio. *Fire Saf. J.* **2021**, *126*, 103470. [CrossRef]
33. He, Q.; Liu, N.; Xie, X.; Zhang, L.; Lei, J.; Zhang, Y.; Wu, D. Experimental study on fire spread over discrete fuel bed—Part II: Combined effects of wind and packing ratio. *Fire Saf. J.* **2022**, *128*, 103520. [CrossRef]
34. Gupta, V.; Torero, J.L.; Hidalgo, J. Burning dynamics and in-depth flame spread of wood cribs in large compartment fires. *Combust. Flame* **2021**, *228*, 42–56. [CrossRef]
35. *ANSYS Fluent Theory Guide*; Release 15.0; ANSYS, Inc.: Canonsburg, PA, USA, 2013.
36. National Instruments. NI-9213 Specifications, 16-CH, ±78 mV, 24 Bit, 75 S/s Aggregate. Available online: https://www.ni.com/docs/en-US/bundle/ni-9213-specs/page/specifications.html (accessed on 21 November 2022).
37. *BS 5852-2006*; Methods of Test for Assessment of the Ignitability of Upholstered Seating by Smouldering and Flaming Ignition Sources. BSI: London, UK, 2006.
38. Tlili, O.; Mhiri, H.; Bournot, P. Empirical correlation derived by CFD simulation on heat source location and ventilation flow rate in a fire room. *Energy Build.* **2016**, *122*, 80–88. [CrossRef]

39. Król, A.; Król, M.; Krawiec, S. A Numerical Study on Fire Development in a Confined Space Leading to Backdraft Phenomenon. *Energies* **2020**, *13*, 1854. [CrossRef]
40. Modest, M.F. *Radiative Heat Transfer*, 3rd ed.; Elsevier Inc.: Amsterdam, The Netherlands, 2013; pp. 279–302.
41. Safarzadeh, M.; Heidarinejad, G.; Pasdarshahri, H. The effect of vertical and horizontal air curtain on smoke and heat control in the multi-storey building. *J. Build. Eng.* **2021**, *40*, 102347. [CrossRef]
42. Sivathanu, Y.R.; Faeth, G.M. Generalized State Relationships for Scalar Properties in Non-Premixed Hydrocarbon/Air Flames. *Combust. Flame* **1990**, *82*, 211–230. [CrossRef]
43. Jones, W.P.; Whitelaw, J.H. Calculation Methods for Reacting Turbulent Flows, A Review. *Combust. Flame* **1982**, *48*, 1–26. [CrossRef]
44. Pope, S.B. Pdf methods for turbulent reactive flows. *Prog. Energy Combust. Sci.* **1985**, *11*, 119–192. [CrossRef]
45. Glicksman, L.; Schuetx, M.; Sinofsky, M. Radiation heat transfer in foam insulation. *Int. J. Heat Mass Transf.* **1987**, *30*, 187–197. [CrossRef]
46. Eschenbacher, A.; Varghese, R.J.; Weng, J.; van Geem, K.M. Fast pyrolysis of polyurethanes and polyisocyanurate with and without flame retardant: Compounds of interest for chemical recycling. *J. Anal. Appl. Pyrolysis* **2021**, *160*, 105374. [CrossRef]
47. Font, R.; Fullana, A.; Caballero, J.A.; Candela, J.; Garcia, A. Pyrolysis study of polyurethane. *J. Anal. Appl. Pyrolysis* **2001**, *58–59*, 63–77. [CrossRef]
48. Węgrzyński, W.; Lipecki, T.; Krajewski, G. Wind and Fire Coupled Modelling–Part II: Good Practice Guidelines. *Fire Technol.* **2018**, *54*, 1443–1485. [CrossRef]
49. Hopkin, C.; Spearpoint, M.; Bittern, A. Using experimental sprinkler actuation times to assess the performance of Fire Dynamics Simulator. *Fire Saf. J.* **2018**, *36*, 342–361. [CrossRef]
50. Węgrzyński, W.; Vigne, G. Experimental and numerical evaluation of the influence of the soot yield on the visibility in smoke in CFD analysis. *Fire Saf. J.* **2017**, *91*, 389–398. [CrossRef]
51. Jomaa, G.; Goblet, P.; Coquelet, C.; Morlot, V. Kinetic modeling of polyurethane pyrolysis using non-isothermal thermogravimetric analysis. *Thermochim. Acta* **2015**, *612*, 10–18. [CrossRef]
52. Cheng, Y.; Xu, Z.; Chen, S.; Ji, Y.; Zhang, D.; Liang, J. The influence of closed pore ratio on sound absorption of plant-based polyurethane foam using control unit model. *Appl. Acoust.* **2021**, *180*, 108083. [CrossRef]
53. Garrido, M.A.; Font, R. Pyrolysis and combustion study of flexible polyurethane foam. *J. Anal. Appl. Pyrolysis* **2015**, *113*, 202–215. [CrossRef]
54. McGrattan, K.; Hostikka, S.; Mcdermott, R.; Floyd, J.; Weinschenk, C.; Overholt, K. *FDS Validation Guide*; NIST Publications: Gaithersburg, MD, USA, 2013.
55. LA Nasa, J.; Biale, G.; Ferriani, B.; Colombini, M.P.; Modugno, F. A pyrolysis approach for characterizing and assessing degradation of polyurethane foam in cultural heritage objects. *J. Anal. Appl. Pyrolysis* **2018**, *134*, 562–572. [CrossRef]
56. Król, A.; Jahn, W.; Krajewski, G.; Król, M.; Węgrzyński, W. A Study on the Reliability of Modeling of Thermocouple Response and Sprinkler Activation during Compartment Fires. *Buildings* **2022**, *12*, 12010077. [CrossRef]
57. McGrattan, K.; Hostikka, S.; Floyd, J.; McDermontt, R.; Vanella, M. *Fire Dynamics Simulator, Technical Reference Guide, Vol. 3: Validation*; NIST Publication 1019: Gaithersburg, MD, USA, 2020. [CrossRef]
58. Janardhan, R.K.; Hostikka, S. Predictive Computational Fluid Dynamics Simulation of Fire Spread on Wood Cribs. *Fire Technol.* **2019**, *55*, 2245–2268. [CrossRef]
59. Park, J.; Kwark, J. Experimental Study on Fire Sources for Full-Scale Fire Testing of Simple Sprinkler Systems Installed in Multiplexes. *Fire* **2021**, *4*, 4010008. [CrossRef]
60. Platt, D.; Elms, D.; Buchanan, A. A probabilistic model of fire spread with time effects. *Fire Saf. J.* **1994**, *22*, 367–398. [CrossRef]

buildings

MDPI

Article

Improving the Performance of Piled Raft Foundations Using Deformation Adjustors: A Case Study

Rui Zhu [1], Feng Zhou [1,*], Zhihui Wan [1], Shengjun Deng [1,2], Xin Dong [1], Zekun Zhou [3] and Wei Xing [1]

1 College of Transportation Engineering, Nanjing Tech University, Nanjing 211816, China
2 Research Center of Coastal and Urban Geotechnical Engineering, Zhejiang University, Hangzhou 310058, China
3 China Railway 19th Bureau Mining Investment Co., Ltd., Beijing 100161, China
* Correspondence: zhoufeng@njtech.edu.cn; Tel.: +86-139-1384-4622

Abstract: Complicated soil conditions are direct difficulties for high-rise building projects. A new device called a deformation adjustor, which is used to optimize the stiffness distribution in the piled raft system, has achieved good results for this challenge. This paper presents a case study on the application of deformation adjustors to improve the performance of a piled raft foundation. This case study describes the preliminary design of pile-raft foundations with deformation adjustors, followed by numerical analysis. Based on the numerical study, the potential savings are demonstrated due to the good performance of soil bearing capacity. Comparing the numerical results with the monitoring results in raft settlements, earth pressures, deformation amount of deformation adjustors, pile top reactions, and load-sharing ratios between soils and piles, the accuracy of the design schemes with an aided numerical analysis is verified. Through a long-term monitoring, soils below the raft carried 63% of the total applied loads, while the piles bear 37% of the loads. This case study proved that a piled raft foundation with deformation adjustors was an effective and economical design scheme, which can make full use of the soil bearing capacity. It is of great significance to facilitate the design and construction of piled raft foundations in complicated soil conditions.

Keywords: end-bearing piles; composite foundation; piled raft foundation; deformation adjustors; numerical analysis; differential settlement

Citation: Zhu, R.; Zhou, F.; Wan, Z.; Deng, S.; Dong, X.; Zhou, Z.; Xing, W. Improving the Performance of Piled Raft Foundations Using Deformation Adjustors: A Case Study. *Buildings* **2022**, *12*, 1903. https://doi.org/10.3390/buildings12111903

Academic Editor: Fulvio Parisi

Received: 16 September 2022
Accepted: 1 November 2022
Published: 6 November 2022

1. Introduction

The mechanism and design theory considering the utilization of soil below the raft to bear a part of the total applied load in piled raft foundation has been a research hotspot for more than half a century [1,2]. A large number of theoretical studies, experimental studies, and numerical studies have carried out in-depth explorations of the bearing characteristics of a piled raft foundation [3–5], and different optimized design methods for a composite pile foundation on soft clay subjected to vertical loads have been reported in the literature [6–9]. All of the above-mentioned works require that the pile foundation be close to its ultimate bearing capacity, thereby enforcing the pile tip piercing downward to achieve the compatible deformation between piles and soil [10].

In some cases, a cushion with a finite thickness may be laid on the top of the pile to make the pile top penetrate into the cushion so that the soil between the piles can work better [11–13]. However, it is still difficult to use for end bearing piles due to the limited role of the cushion. As the stiffness of pile and soil varies significantly, the differential deformation between pile and soil is large and cannot be easily coordinated, which is one of the major concerns in the end bearing pile foundation [14]. To solve this issue and to realize the pile-soil interaction, Zhou et al. [15] developed an optimization device called a deformation adjustor, which is placed between the top of the piles and the raft. This technique is able to accomplish the optimization of load distribution between soils and piles in complicated soil conditions. In recent years, this technique has been applied in more than twenty high-rise building projects in China [16–18].

Accordingly, this paper presents a case study on applying deformation adjustors to improve the performance of piled raft foundations in complicated soil conditions. In the original design of piled raft foundations, the majority of the total applied loads were carried by piles, while piles were prevented from being installed to the required depth due to the number of large boulders. Consequently, this technique (using deformation adjustors) was used for such a challenging project. The design scheme of a piled raft foundation with deformation adjustors in this case was presented briefly. Then, the numerical results were compared with the monitoring results for raft settlements, earth pressures, deformation amount of deformation adjustors, pile top reactions, and load-sharing ratios between soils and piles during and after construction, which proved that a piled raft foundation with deformation adjustors was an effective and economical design scheme in complicated soil conditions. Through this study, we expect to provide a reference for improving the performance of piled raft foundations.

2. Project Information and Site Conditions

The project is located in Xiamen, China, and was for the construction of two high-rise residential buildings (building A and building B) with a total construction area of 38,000 m^2. The case described herein focuses on the piled raft system of building A with a height of 120 m, as shown in Figure 1.

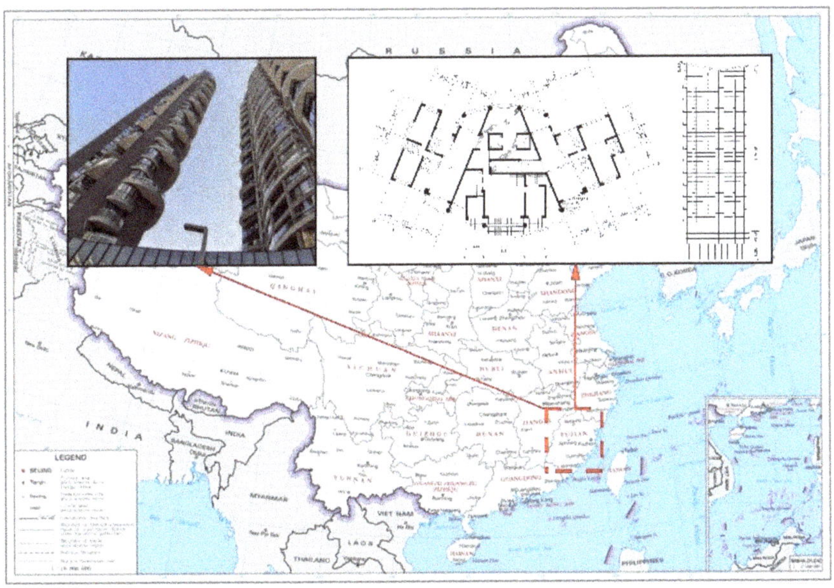

Figure 1. Project Information.

The site exploration was conducted in 2006. It was found that the soil conditions were complicated, mainly consisting of residual soil and weathered granite. Specifically, the unit weight, cohesion, internal friction angle, and deformation modulus of residual soil were 18.6 kN/m^3, 18 kPa, 28°, and 35 MPa, while those of weathered granite were 19.5~23.5 kN/m^3, 40~45 kPa, 27~30°, and 60~120 MPa, respectively. Similarly, characteristic values of the subgrade bearing capacity, skin resistance, and tip resistance for residual soil and weathered granite were 280, 100, and 1200 kPa and 400~3000, 150~800, and 1500~10,000 kPa, respectively. Soils of the site experienced highly uneven weathering and spheroidal weathering, leading to the formation of various sizes of boulders, which mainly appeared in completely and strongly weathered granite layers, as indicated in

Figure 2. Additionally, the distribution of the soil layer is highly uneven with regards to depth, resulting in the different soil characteristics in the same depth. Accordingly, the complicated soil conditions bring about severe challenge to the design and construction of piled raft foundations.

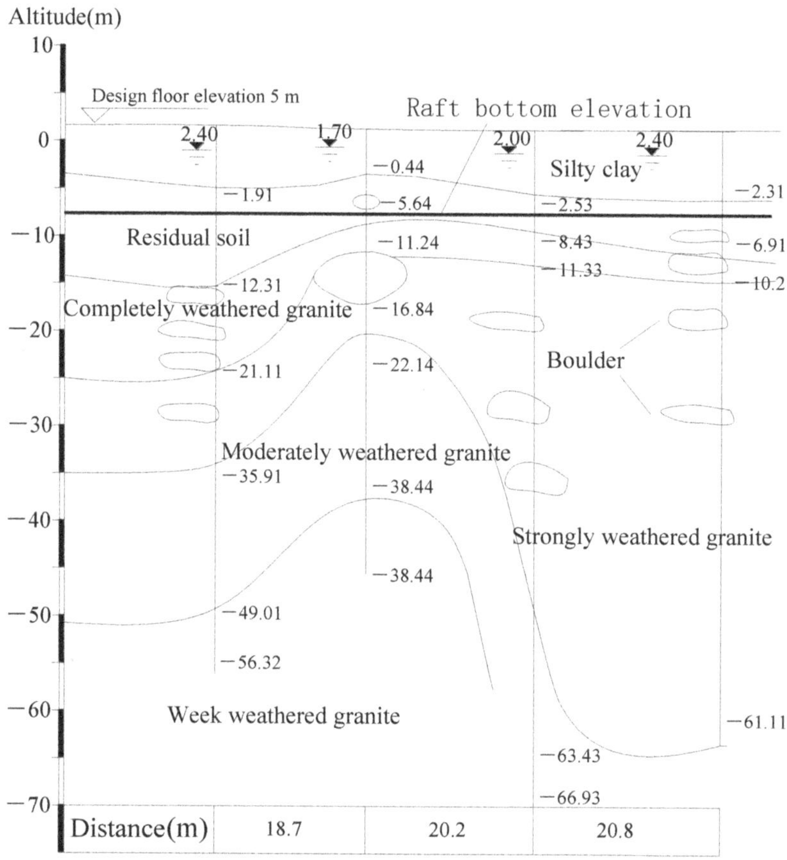

Figure 2. Typical geological section.

3. Preliminary Design Scheme

3.1. Design Concept

A deformation adjustor is a device used to adjust the bearing stiffness between two contacting objects. The overall stiffness of piled raft foundations is adjusted through the large deformation of deformation adjustors placed between the top of the piles and rafts. The schematic diagram of its mechanical model is shown in Figure 3. It can be seen that the placement of deformation adjustors is equivalent to installing springs with a specific stiffness on the top of the piles. The stiffness of the deformation adjustor is generally much lower than the stiffness of the pile. Accordingly, the applied load is partly transferred to the soils below the raft with the deformation of deformation adjustors, which significantly reduces the requirement for the bearing capacity of piles. It would save a lot of money and time for the project.

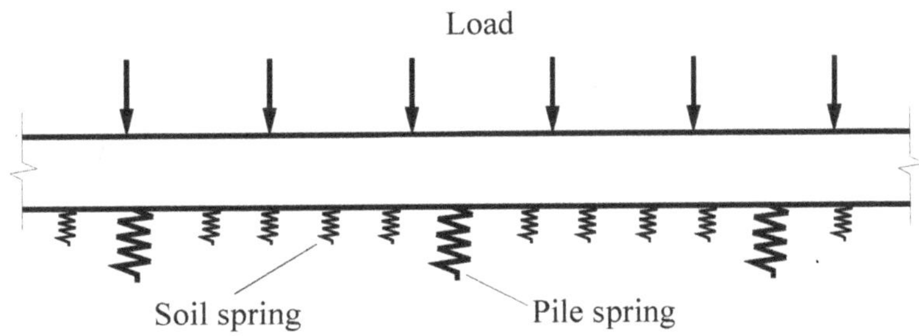

Figure 3. Mechanical model of piled raft foundations with deformation adjustors.

3.2. Parameters of the Building

The raft of building A has a width of 20.8 m, a length of 44.5 m, and an L-shaped plane close to the rectangle. The diameter (R_0) of the equivalent circular raft is 35 m. The superstructure load was approximately 524,700 kN. The area of building A was 960 m^2, and the corresponding earth pressure was 546 kPa.

3.3. Soil Bearing Capacity

The soil ultimate bearing capacity provided by the geotechnical report was 280 kPa, which was derived from laboratory tests. However, the residual soil and weathered granite are easily disturbed. The soil ultimate bearing capacity will be reduced after the disturbance. Thus, the value provided by the geotechnical report is underestimated. According to local engineering specifications, the bearing capacity was approximately 480 kPa. To confirm this value, a static plate load test, a standard penetration test, a pressure meter test, and a deep screw plate test were conducted. The test results indicated that the allowable bearing capacity was above 480 kPa, with a safety factor of 2.

3.4. Preliminary Design Scheme

The design scheme adopted a 2.5 m thick raft supported on hand-dug piles. The piles were 1 m in diameter, 12 m long, and the average center-to-center spacing of piles was four times the pile diameter. The pile tip was located at −18 m YSL, indicating that most of the piles were borne on the highly weathered granite layer. As such, the foundation should be designed as a piled raft with end-bearing piles. The design procedures are described in detail in [18] for piled raft foundations with deformation adjustors. The main parameters for this project were calculated as per [19,20], as listed in Table 1.

Table 1. Design parameters.

Quantity	Meaning	Value
P_{pr} (kN)	Loads applied to the piled raft	660,000
q_t (kN)	Allowable axial bearing capacity of a single pile	4200
R_r (kPa)	Allowable raft bearing capacity	450
A (m^2)	Gross raft area	960
η	Group stiffness efficiency factor	1.0
k_r (kN/mm)	Stiffness of the raft alone	8912
ξ	Load-sharing ratio of soils	0.62
n_a	Number of deformation adjustors per pile	1
k_{Pa} (kN/mm)	Overall stiffness of a single pile with deformation adjustors	7268
k_a (kN/mm)	Stiffness of a deformation adjustor	120
n	Pile number	60
S_t (mm)	Calculated total settlement of the building	64
S_d (mm)	Calculated differential settlement of the building	16

Besides, the calculation value of the pile number is equal to 53. Considering some special needs of the superstructure configuration, the pile number, $n = 60$, was used for the design scheme. The pile arrangement is presented in Figure 4.

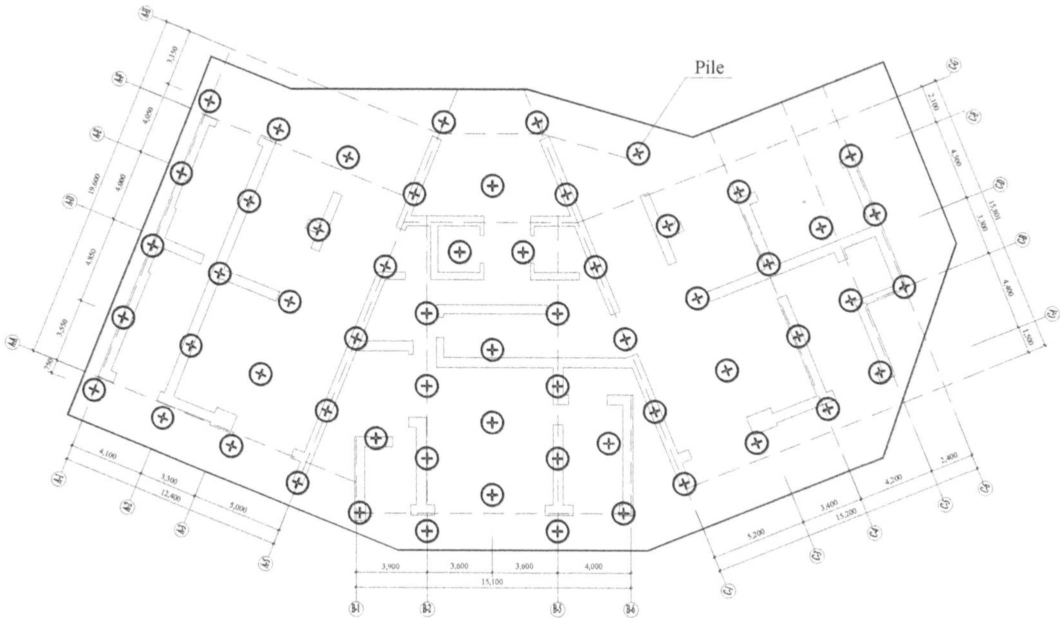

Figure 4. Plan of pile arrangement.

4. Description of the Numerical Study

4.1. Geometric Configuration of the Model

In this paper, the numerical analysis was carried out using the *ABAQUS* [21]. To be in correspondence with the construction, the three-dimensional numerical model of the piled raft foundation was simulated as it had been in the construction in both its configuration and dimensions. Figure 5 presents the basic configuration of the model. The size of model is 120 m in the x direction, 60 m in the y direction, and 40 m in the z direction. The piles are 1 m in radius and 12 m in depth. Respectively, a raft which had a thickness of 2.5 m was arranged on the pile-soil system to enable the system to support the applied load. To simulate the deformation adjusters on the pile heads in the construction, an analysis was performed on them with a simulation where two rigid plates are the top and bottom of the deformation adjustor and a spring with a length of 30 mm was between the two plates. For simplicity, the rigid plate had zero thickness. The cross-sectional geometry of the numerical deformation adjustor was identical to that of the pile (2R = 1 m). The boundary was given according to different conditions: x = 0, y = 0, and z = 0 on the upper and lower surfaces of the model, while x = 0 and y = 0 on the surface around the model. The whole model has 54,820 units and 111,728 nodes. In this paper, the model was simulated without taking advantage of the symmetry due to the irregular geometry of the piled raft foundation in the project. Additionally, the soil and the piled raft foundation are modeled with finite elements. The pile mass, raft mass and soil mass were discretized using continuum solid elements, mainly 8-noded hexahedral elements with reduced integration.

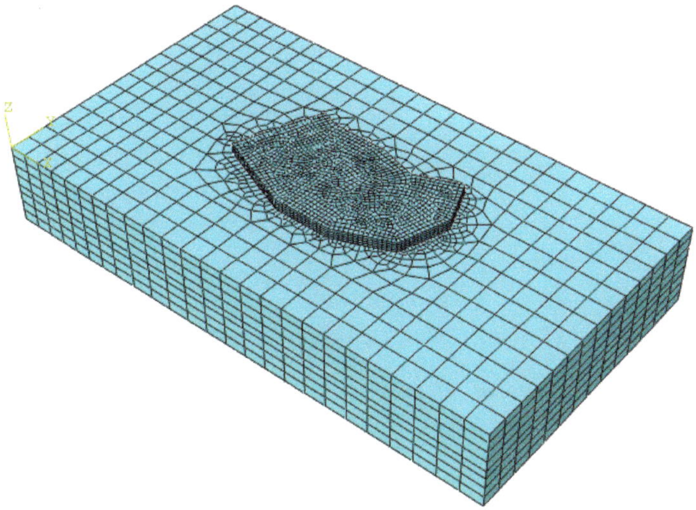

Figure 5. Typical finite element mesh.

4.2. Material Property

Actually, the soil is a multiphase medium. For simplicity, it was simulated as a one-phase media in this paper. The Mohr–Coulomb plasticity model was used to model the soil plasticity [22]. The pile and raft were simulated as elastic materials, and the properties of them were in accordance with the parameters in the construction. All of the model parameters are shown in Table 2. The validity of the model parameters in the program based on the analysis of the two static load tests, and the comparison of *P-S* curve between numerical results and static load test results was shown in Figure 6. As we can see, all of the three curves matched well. Hence, the conclusion was reached that the model parameters in the program are reasonable.

In addition, the contact between the structure and the soil was simulated as perfectly rough and no relative motion would appear between the finite elements nodes. Duplicated nodes were used to form a zero-thickness interface between the pile and the soil, the pile and the plate, the soil and the raft, and the plate and the raft.

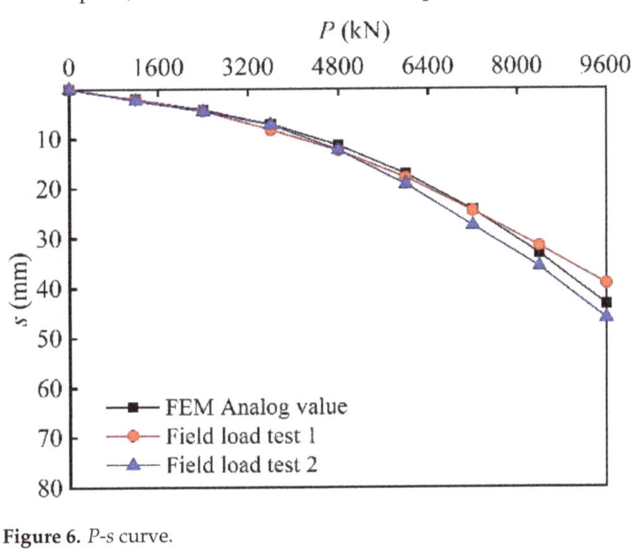

Figure 6. *P-s* curve.

Table 2. Model parameters.

Material	Model	Elastic Modulus (MPa)	Cohesion (kPa)	Internal Friction Angle	Poisson's Ratio	Unit Weight (kN/m^3)
Pile	Linear elasticity	30,000	-	-	0.2	25
Raft	Linear elasticity	30,000	-	-	0.2	25
⑥Residual soil	Mohr–Coulomb plasticity	35	18	28	0.3	18.6
⑦Completely weathered granite		60	40	27	0.3	19.5
⑧Strongly weathered granite		500	45	30	0.25	20.5

4.3. Numerical Analysis Scheme

In order to further analyze the influence of the stiffness of the deformation adjustors on the load-sharing ratio, and whether the stiffness of 12,000 kN/m in the preliminary design scheme is reasonable, numerical analysis was conducted for studying the bearing characteristics of the piled raft foundations with different stiffnesses of the deformation adjustors, which is equal to spring stiffness, as shown in Table 3.

Table 3. Different stiffness of deformation adjustors.

Adjuster Number	DA-0	DA-5	DA-12	DA-20	DA-30	DA-mix
Stiffness(kN/m)	0	50,000	120,000	200,000	300,000	Table 4

Table 4. The mixed stiffness of deformation adjustor.

Aera	1	2	3	4
Stiffness (kN/m)	50,000	100,000	200,000	300,000

The above schemes were only for the case where the deformation adjustors were a single stiffness. Actually, deformation adjustors with different stiffness could be placed on different areas to make full use of the soil bearing capacity, as shown in Figure 7a and Table 4. To limit the length of the article, the author selected path 1-1′ and path 2-2′ for the analysis, as shown in Figure 7b.

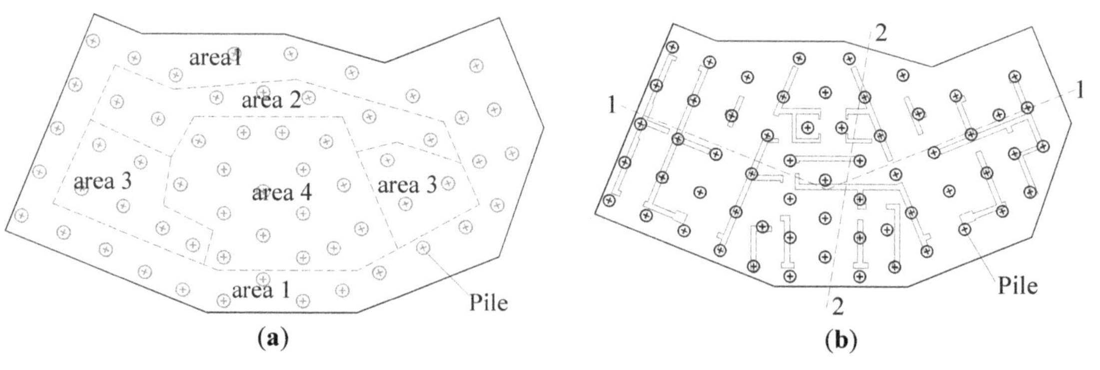

Figure 7. Numerical analysis scheme: deformation adjustors in different regions (**a**), direction of the path (**b**).

5. Numerical Results

5.1. Analysis of Settlement

The raft settlement distribution variation along path 1 and path 2 is shown in Figures 8 and 9, respectively. The stiffness of the deformation adjustors had a great influence on the raft settlement. The overall settlement and differential settlement significantly increased as the stiffness of the deformation adjustors was reduced. When the stiffness of the deformation adjustors was 120,000 kN/m, the maximum settlement and differential settlement were 41 mm and 17 mm, respectively, which was consistent with the range proposed by [23]. It indicated that 120,000 kN/m was the reasonable stiffness of the deformation adjustors. Besides, the differential settlement could be controlled nearly to zero as the deformation adjustors of different stiffness were placed in different areas.

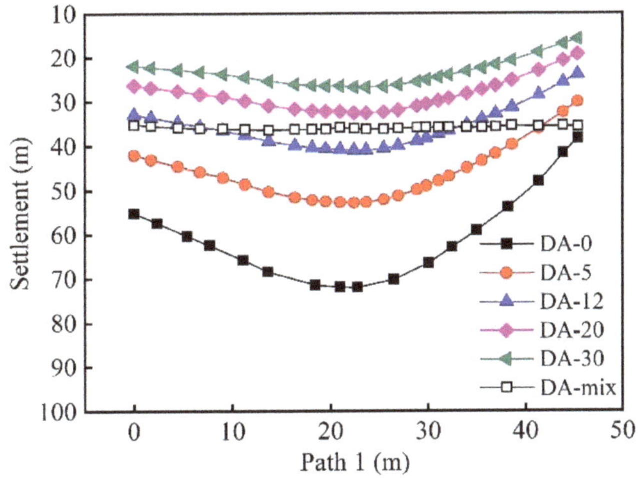

Figure 8. Curve of raft settlement along path 1.

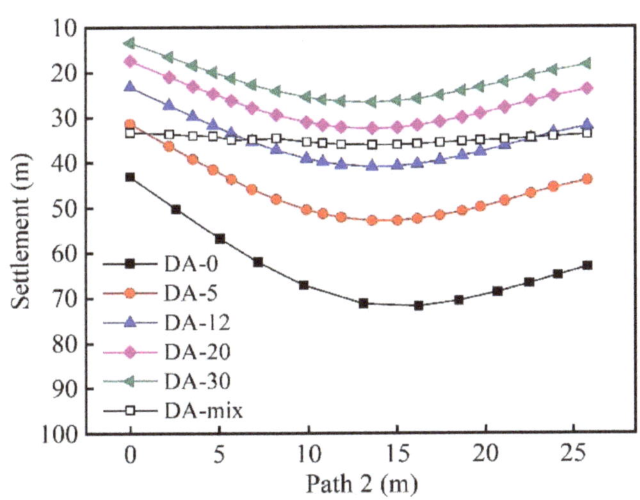

Figure 9. Curve of raft settlement along path 2.

5.2. Analysis of Earth Pressure

The earth pressure distribution below the raft along path 1 and path 2 are presented in Figures 10 and 11, respectively. The distributed load on soil gradually increased as the stiffness of the deformation adjustors reduced. When the stiffness of the deformation adjustors was zero, the average earth pressure below the raft was 602 kPa, which was greater than the actual soil bearing capacity. If 12,000 kN/m was chosen as the stiffness of the deformation adjustors, the average earth pressure below the raft was only 377 kPa, indicating that the deformation adjustors have a significant effect on promoting the pile-soil interaction. Additionally, the earth pressure distribution below the raft was more uniform in the scheme of deformation adjustors with different stiffness placed in different areas than that with a single stiffness, which is the main reason that the differential settlement was close to zero. Comparing the earth pressure distribution in Figures 10 and 11, it is found that when the earth pressure is small, the earth pressure below the raft showed characteristics of partial stiffness along the direction of the short side and somewhat showed flexibility along the direction of the long side due to the uneven distribution of the upper load. This distinction should be taken into account in the design of the raft.

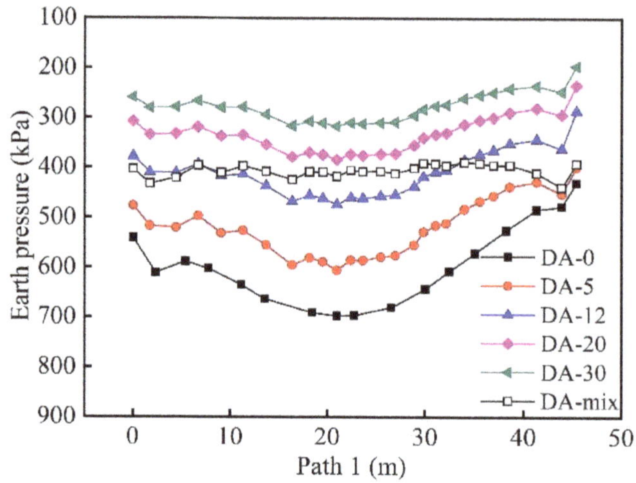

Figure 10. Curve of earth pressure along path 1.

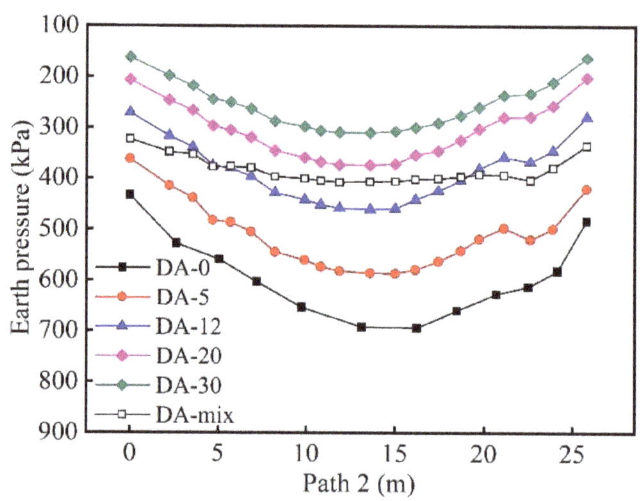

Figure 11. Curve of earth pressure along path 2.

5.3. Analysis of Deformation Amount of Deformation Adjustors

Figure 12 shows the curve of deformation amount of deformation adjustors variation with the percentage of load applied. As shown in Figure 12, the deformation amount of deformation adjustors was smaller as the stiffness of deformation adjustors was larger. When the stiffness of deformation adjustors was 12,000 kN/m, the deformation amount of the deformation adjustors was 32.52 mm after the load was applied, whereas it was 33.66 mm in the scheme of deformation adjustors with different stiffness placed in different areas.

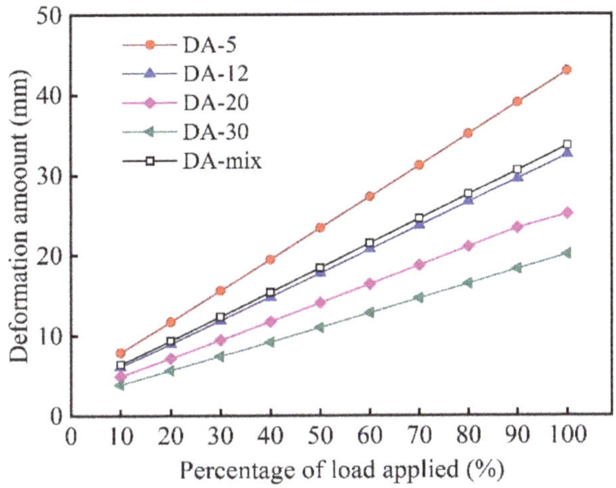

Figure 12. Curve of deformation amount of deformation adjustors variation with percentage of load applied.

5.4. Analysis of Pile Top Reaction

Figure 13 presented the curve of the pile top reaction variation with the percentage of the load applied. As shown in Figure 13, the pile top reactions increased significantly as the stiffness of deformation adjustors became larger. The average pile top reaction was 4057 kN, as the stiffness of the deformation adjustors was 120,000 kN/m, which was slightly smaller than the pile allowance bearing capacity. Additionally, in the scheme of deformation adjustors with different stiffness placed in different areas, the pile top reaction could be calculated by the deformation amount of deformation adjustors and the stiffness of deformation adjustors. Accordingly, the maximum and minimum pile top reactions were 10,098 kN and 1683 kN, respectively, which were higher than the pile allowance bearing capacity. When the deformation adjustors were adopted, the pile top reactions would be significantly different. The stiffness 120,000 kN/m was the relatively reasonable value on the basis of satisfying the design condition.

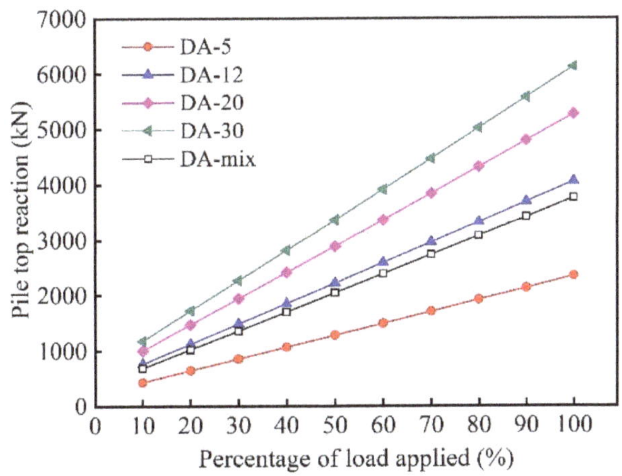

Figure 13. Curve of pile top reaction variation with percentage of load applied.

5.5. Analysis of Load-Sharing Ratio between Soil and Pile

Figure 14 is the curve of the load-sharing ratio of soil variation with the percentage of load applied. As shown in Figure 14, the load-sharing ratio of soil decreased and tended to be stable as the percentage of the load applied increased. This is because the raft, soil, and deformation adjustors were deformed and coordinated in the process of the load applied. When the stiffness of the deformation adjustors was 120,000 kN/m, the load-sharing ratio of soil was up to 59.76%. The conclusion that was reached was that the deformation adjustors could effectively change the load-sharing ratio between soil and piles, and that they made full use of the soil bearing capacity.

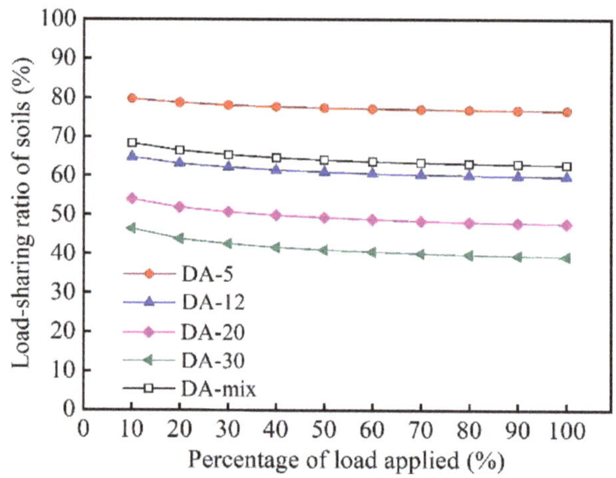

Figure 14. Curve of load-sharing ratio of soils variation with percentage of load applied.

Figure 15 presented the curve of the load-sharing ratio of soil variation with stiffness of deformation adjustors. When the stiffness of the deformation adjustors is zero, the foundation is equivalent to the natural ground, and the superstructure load was all carried by soil. With the stiffness of the deformation adjustor increasing, the decreasing trend of the load-sharing ratio of soil gradually slows down and tends to be stable. The foundation will show the characteristics of a conventional pile foundation, as the stiffness of the deformation adjustors was close to the stiffness of the piles.

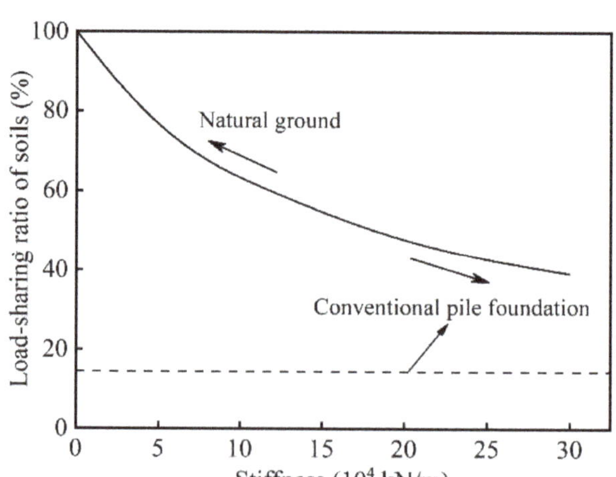

Figure 15. Curve of load-sharing ratio of soils variation with stiffness of deformation adjustors.

The essence of the deformation adjustor placed on the top of piles was an optimization of practical engineering. The deformation adjustors with different stiffness determine the difference in the load-sharing ratio of soil, resulting in different total settlements and differential settlements.

5.6. Determination of Design Scheme

Based on the above analysis, the average settlements and differential settlements of the building were all in the reasonable range when the piled raft foundation with deformation adjustors (120,000 kN/m) was adopted. The average earth pressure is 377.2 kPa and the average pile top reaction is 4057 kN. This scheme not only took full advantage of the soil bearing capacity, but also greatly reduced the project cost. The numerical analysis was also proven to be an effective method for assisting in the design of the piled raft foundation with deformation adjustors.

Additionally, deformation adjustors with different stiffness in four areas had a better effect on achieving pile-soil interaction in the numerical analysis. The differential settlement would be smaller and the distribution of the earth pressure under the raft would be more uniform. However, this scheme was more complex in the process of design and construction and had not ever been used in practical engineering. Thus, it is not adopted for the time being.

6. Monitoring Measurement

The project completed the construction of the two-story cushion and the main structure at the end of December 2010 and December 2011, respectively. During and after construction, a comprehensive monitoring of settlements, earth pressures, deformation amount of deformation adjustors, pile top reactions, steel bar stress, and load-sharing ratios between soils and piles was carried out. The monitoring measurement lasted for seventeen months from January 2011 to May 2012 and the monitoring points were arranged, as shown in Figure 16.

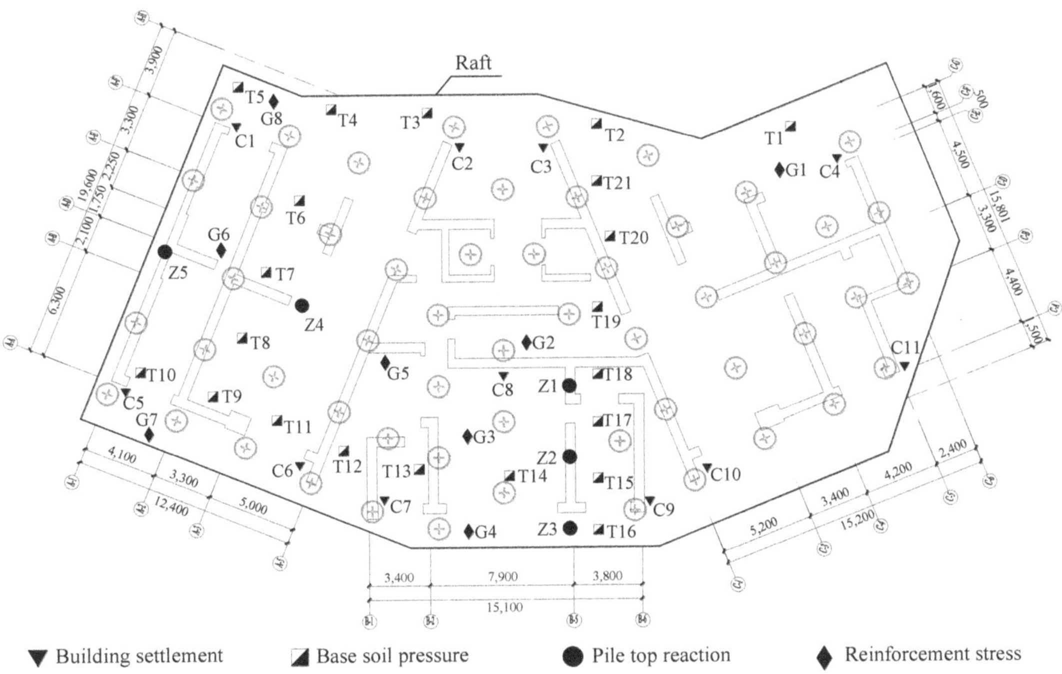

Figure 16. Arrangement of monitoring points.

6.1. *Analysis of Settlements*

Eleven monitoring points were arranged on the raft during construction and seven monitoring points were still active after construction. Figure 17 presents the curve of the building settlements variation with time. The maximum settlements were measured at C8 and C3, while the minimum settlement was at C11. Such a difference of settlement distribution may be attributed to the fact that the raft was irregular and that some monitoring points might be located at the elevator shaft or the raft corner. The average settlement and differential settlement were 24.5 mm and 4.2 mm after construction, and then, half a year later, the average settlement and differential settlement were 25.4 mm and 5.1 mm, respectively. The measured values were significantly lower than the calculated values listed in Table 2 and the numbers in [20], which were less than 200 mm for the total settlement and less than 0·001 R_0 for the differential settlement.

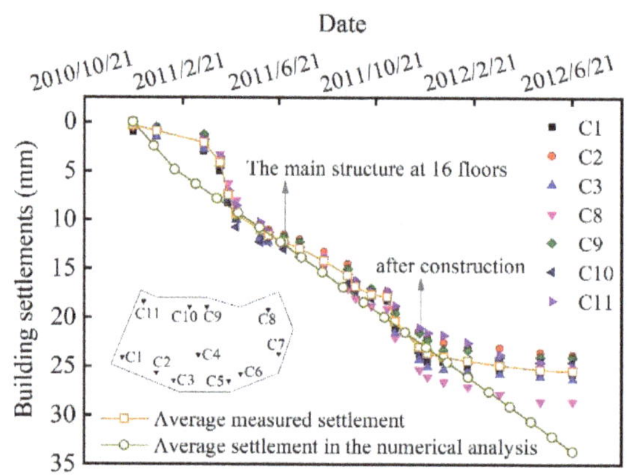

Figure 17. Curve of building settlements variation with time.

6.2. *Analysis of Earth Pressure*

During construction, 21 earth pressure cells were installed between the raft and the soil, 11 of which remained at work after construction. Figure 18 shows the curve of earth pressure variation with time. It was seen that the earth pressure increased significantly over time during construction. The maximum and minimum earth pressure were up to 433 kPa and 152 kPa, respectively. After construction, the average earth pressure tended to be stable and had a value of about 315 kPa, which was less than the value in the numerical analysis. Considering that the load applied was about 80% of the design value, the above results are relatively reasonable.

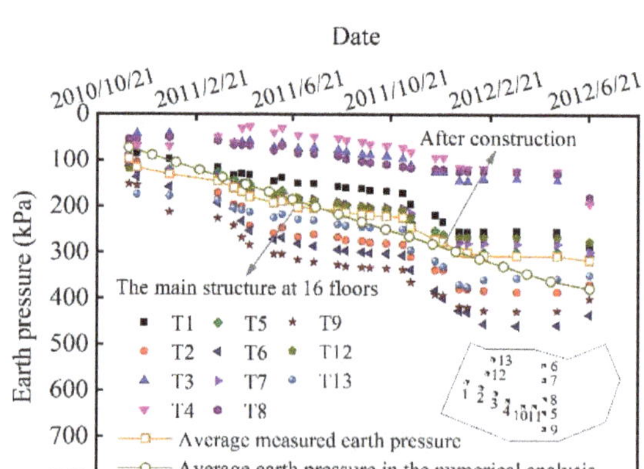

Figure 18. Curve of earth pressure variation with time.

6.3. Analysis of Deformation Amount of Deformation Adjustors

In this project, 60 deformation adjustors with monitoring sensors (1#–60#) were placed on the top of the piles, and 54 monitoring sensors were still active after construction. The deformation amount of the deformation adjustors was obtained by subtracting the pile top settlement measured from the building or raft settlement. Figure 19 shows the curve of deformation amount of deformation adjustor variation with time. As shown in Figure 19, the deformation amount increased relatively linearly with the load applied during construction, which was consistent with numerical results. After construction, the deformation amount gradually became stable. The maximum deformation amount was up to 26.1 mm and the average deformation amount was approximately 25.8 mm.

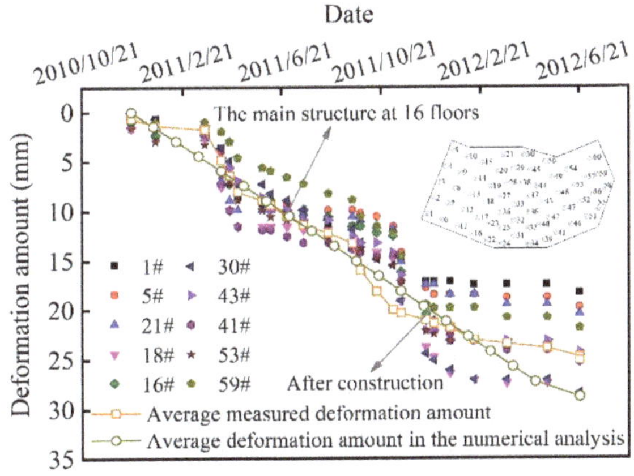

Figure 19. Curve of deformation amount of deformation adjustor variation with time.

Figure 20 is the comparison curve between the average deformation amount and average building settlement. As shown in Figure 20, the average deformation amount of the deformation adjustors was slightly smaller than the average building settlement, which is in agreement with numerical results. It indicated that the end bearing soil and pile body still have a small amount of deformation.

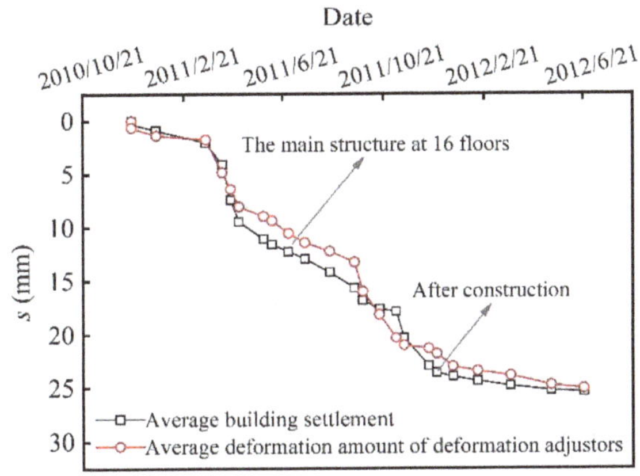

Figure 20. Comparison curve between average deformation amount and average building settlement.

6.4. Analysis of Pile Top Reaction

The sensors were installed on the top of the piles to measure the pile top reaction during and after construction. Figure 21 presented the curve of pile top reaction variation with time. As shown in Figure 21, the average pile top reaction increased significantly during construction and was up to 3246 kN after construction, which was consistent with the numerical results. However, the pile top reactions were unequal in different monitoring points. The possible reason for such a phenomenon was that each monitoring point had a different soil sharing area, leading to various measured pile top reactions. Additionally, the pile top reaction became stable and decreased slightly at 3120 kN until the termination of the monitoring in June 2012, while the pile top reaction still increased in linear in the numerical analysis. This is because the decoration load and live load had not been applied, leading to the applied load being smaller than the design value. In the numerical analysis, the load, which was consistent with the design value, was applied linearly. Consequently, the measurements were reasonable, which also indicated that the load distribution between soils and piles was further adjusted over time.

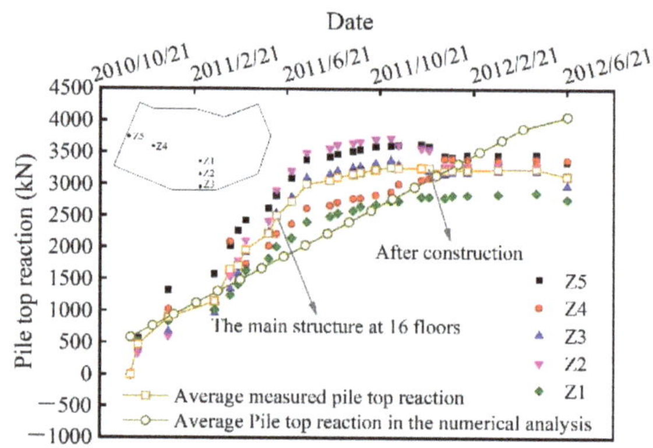

Figure 21. Curve of pile top reaction variation with time.

The average pile top reaction can be approximately calculated according to its stiffness and deformation amount. Figure 22 presented the curve of measured and calculated average pile top reaction variation with time. Since the average deformation amount of the deformation adjustor was obtained from 54 piles and the measured average pile top reaction was only obtained by five piles, the calculated average pile top reaction is significantly smaller than the measured value. However, it still indicated that the trends of measured average pile top reaction variation with time were basically in line with the calculated average pile top reaction.

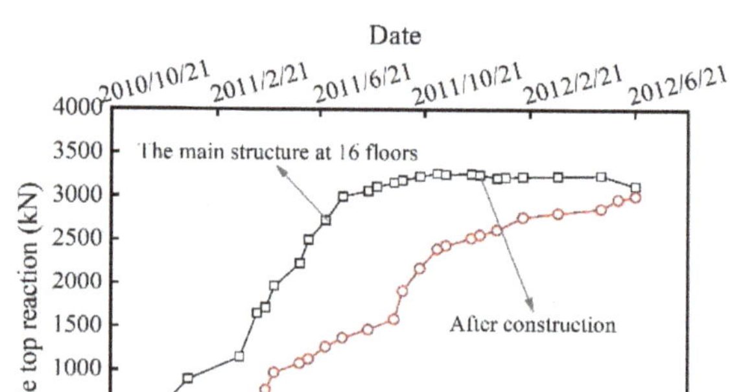

Figure 22. Curve of measured and calculated average pile top reaction variation with time.

6.5. Analysis of Steel Bar Stress

In this project, eight bar stress meters were installed within the raft in order to measure the stress distribution in the raft. Figure 23 presented the curve of steel bar stress variation with time. As shown in Figure 23, the steel bar stress significantly improved with the increase of load applied prior to the 16th floor of the main structure. Since then, the steel bar stress tended to be stable until the completion of the construction. The maximum steel bar stress was at G2, followed by G5, and then G7. This is because G2 was located in the core tube, where the structure was more rigid, and the load was concentrated. G5 and G7 were located on the edge of a shear wall and in the edge area of the raft with a small load, respectively. After construction, the maximum steel bar stress was 108.5 MPa, which did not reach the 1/3 of the design value. It was consistent with the facts written in the literature stating that the measured stress of most steel bars within the raft was low. After the field observation, there was no cracking or water seepage on the raft surface, indicating that the design concept of piled raft foundations with deformation adjustors was successful.

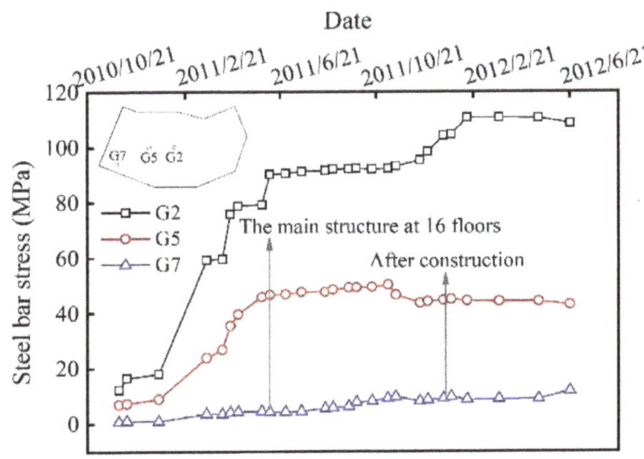

Figure 23. Curve of steel bar stress variation with time.

6.6. Analysis of Load-Sharing Ratio between Soils and Piles

In this project, the soil bearing capacity has significant utilization potential. Thus, the soil was designed to have an allowable bearing capacity of 353 kPa for supporting 62% of the superstructure load, whereas the piles were designed to bear 38% of the applied loads. The load-sharing ratio between soils and piles can be derived from the measured average earth pressure and the measured average pile top reaction directly. Figure 24 presented the curve of the load-sharing ratio of soil variation with time. As shown in Figure 24, the load-sharing ratio of soils varied with time but fluctuated around the design value. After construction, the load-sharing ratio gradually became stable. The measured value indicated that the load-sharing ratio of soils was 63%, which is slightly higher than the design value and numerical results. The application of deformation adjustors put the end-bearing piles and soils into play synchronously, which was essentially different from friction piles.

In addition, the numerical results showed that the load-sharing ratio of soils was high in the beginning of construction, and then gradually reduced to the design value with load applied, which was not exactly the same with measured results. This is because the raft and soil were in contact first, and soils deformed earlier than the deformation adjustors, leading to soil carrying more load in the beginning. As the deformation adjustors worked, the load-sharing ratio between soil and piles gradually stabilized.

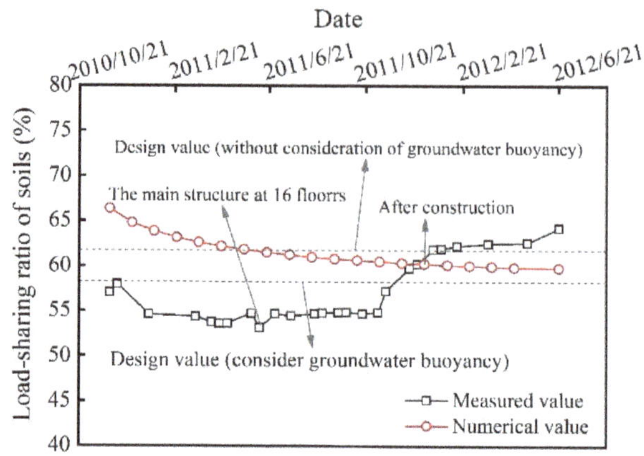

Figure 24. Curve of load-sharing ratio of soil variation with time.

7. Conclusions

(a) The pile-raft foundation with deformation adjustors was used successfully in complicated soil conditions. The deformation adjustors solved the challenges caused by abundant boulders and made the piled raft foundation into a coordinated deformation.

(b) As an application of deformation adjustors with different stiffness, the differential settlement of the building could be well controlled. The earth pressure distribution is also more uniform. However, the difference of pile top reactions was large, and the design scheme was more complex. It still needs to be tested and studied later.

(c) Deformation adjustors helped to adjust the stiffness difference so that soils and piles would jointly bear superstructure loads. In this case study, the soil was capable of supporting 63% of superstructure loads, whereas piles carried 37% of the loads.

(d) The piled raft foundations with deformation adjustors have been successfully applied to more than 20 projects, showing a better performance than conventional piled raft foundations in challenging conditions. However, the stiffness of current deformation adjustors cannot be changed after installation. Further studies are encouraged to carry out investigations on the active control technology of stiffness during the working period of deformation adjustors.

Author Contributions: Conceptualization, R.Z. and F.Z.; methodology, R.Z.; validation, Z.W., S.D. and X.D.; formal analysis, R.Z.; investigation, F.Z.; data curation, W.X.; writing—original draft preparation, R.Z. and Z.Z.; project administration, R.Z. and F.Z.; funding acquisition, R.Z., F.Z., S.D. and Z.W. All authors have read and agreed to the published version of the manuscript.

Funding: This research was funded by the financial support from Natural Science Foundation of Jiangsu Province (Nos. BK20220356, BK20200400 and BK20200707) and National Natural Science Foundation of China (Nos. 51778287, 52008100 and 52008208).

Institutional Review Board Statement: Not applicable.

Informed Consent Statement: Not applicable.

Data Availability Statement: The data used to support the findings of this study are available from the authors upon request.

Conflicts of Interest: The authors declare that they have no conflict of interest to report regarding the present study.

References

1. Wu, J.T.; El Naggar, M.H.; Zhao, S.; Wen, M.J.; Wang, K.H. Beam-unequal length piles-soil coupled vibrating system considering pile-soil-pile interaction. *J. Bridge Eng.* **2021**, *26*, 04021086. [CrossRef]
2. Zhang, Y.P.; Wu, W.B.; Zhang, H.K.; El Naggar, M.H.; Wang, K.H.; Jiang, G.H.; Mei, G.X. A novel soil-pile interaction model for vertical pile settlement prediction. *Appl. Math. Model.* **2021**, *99*, 478–496. [CrossRef]
3. Ko, J.Y.; Cho, J.Y.; Jeong, S.S. Closure: Analysis of load sharing characteristics for a piled raft foundation. *Geotech. and Eng.* **2018**, *16*, 99–102.
4. Mayne, P.W.; Poulos, H.G. Approximate displacement influence factors for elastic shallow foundations. *J. Geotech. Geoenvironmental Eng.* **1999**, *125*, 453–460. [CrossRef]
5. Randolph, M.F. Design methods for pile groups and piled rafts. In Proceedings of the 13th International Conference of Soil Mechanics and Foundation Engineering, New Delhi, India, 1 January 1994; Taylor & Francis: Leiden, The Netherlands, 1994; Volume 5, pp. 61–82.
6. Randolph, M.F.; Wroth, C.P. Analysis of deformation of vertically loaded piles. *J. Geotech. Eng.* **1978**, *104*, 1465–1488. [CrossRef]
7. Reul, O.; Randolph, M.F. Design strategies for piled raft subjected to nonuniform vertical loading. *J. Geotech. Geoenvironmental Eng.* **2004**, *130*, 1–13. [CrossRef]
8. Mandolini, A.; Russo, G.; Viggiani, C. Pile foundations: Experimental investigations, analysis and design. In Proceedings of the 16th International Conference on Soil Mechanics and Geotechnical Engineering, Osaka, Japan, 12–16 September 2005; Balkema: New York, NY, USA, 2005; Volume 1, pp. 177–213.
9. Lee, S.H.; Chung, C.K. An experimental study of the interaction of vertically loaded pile groups in sand. *Can. Geotech. J.* **2005**, *42*, 1485–1493. [CrossRef]
10. Fleming, K.; Weltman, A.; Randolph, M.; Elson, K. *Piling Engineering*; Taylor & Francis: New York, NY, USA, 2009.
11. Liang, F.Y.; Chen, L.Z.; Shi, X.G. Numerical analysis of composite piled raft with cushion subjected to vertical load. *Comput. Geotech.* **2003**, *30*, 443–453. [CrossRef]
12. Sharma, V.J.; Vasanvala, S.A.; Solanki, C.H. Behaviour of cushioned composite piled raft foundation under lateral forces. *Indian Geotech. J.* **2015**, *45*, 89–97. [CrossRef]
13. Liu, H.K.; Xia, T.D. Pile-soil stress analysis of geogrid-reinforced pile-supported embankments with piles arranged in triangle. *J. Cent. South Univ.* **2014**, *45*, 3957–3963.
14. Poulos, H.; Mattes, N. Settlement and load distribution analysis of pile groups. *Aust. Geomech. J.* **1971**, *1*, 18–28.
15. Zhou, F.; Lin, C.; Wang, X.D.; Cheng, J.Y. Application of deformation adjustors in piled raft foundations. *Proc. Inst. Civ. Eng. Geotech. Eng.* **2016**, *169*, 527–540. [CrossRef]
16. Zai, J.M. *Theory and Practice of Composite Pile Foundation*; Intellectual Property Publishing House: Beijing, China, 2004.
17. Zhou, F.; Qu, W.; Guo, T.X.; Lin, S.Z. Engineering practice of composite end-bearing pile foundation based on settlement control. *Chin. J. Rock Mech. Eng.* **2015**, *5*, 1071–1079.
18. Zhou, F.; Lin, C.; Zhang, F.; Lin, S.Z.; Wang, X.D. Design and field monitoring of pile-raft foundations with deformation adjustors. *J. Perform. Constr. Facil.* **2016**, *30*, 04016057. [CrossRef]
19. CDC. *Technical Code for Design of Building Foundation (GB 50007)*; China Building Industry Press: Beijing, China, 2002.
20. CDC. *Technical Code for Building Pile Foundations (JGJ 94)*; China Building Industry Press: Beijing, China, 2008.
21. Hibbitt, H.; Karlsson, B.; Sorensen, P. *ABAQUS. Theory Manual, Verion. 6.12*; SIMULIA: Providence, RI, USA, 2012.
22. Tradigo, F.; Pisanò, F.; Prisco, C.D. On the use of embedded pile elements for the numerical analysis of disconnected piled rafts. *Comput. Geotech.* **2016**, *72*, 89–99. [CrossRef]
23. Bjerrum, L. Allowable settlement of structures. In Proceedings of the 3rd European Conference on Soil Mechanics and Foundation Engineering, Weisbaden, Germany, 1963; Volume 2, pp. 135–137.

buildings

MDPI

Article

Structural Response Analysis of FPSO under Pool Fire

Dapeng Zhang [1], Bowen Zhao [2,*] and Keqiang Zhu [3]

1 Ship and Maritime College, Guangdong Ocean University, Zhanjiang 524088, China
2 Ocean College, Zhejiang University, Zhoushan 316000, China
3 Faculty of Maritime and Transportation, Ningbo University, Ningbo 315211, China
* Correspondence: zju_zhbw@zju.edu.cn

Abstract: There is a large proportion of pool fire occurrence on the upper part of offshore platforms. In order to reduce the occurrence of fire disasters, the fire risk assessment of FPSOs should be carried out. According to the temperature characteristics of offshore platform fires based on computational fluid dynamics, the temperature field of the superstructure of the offshore platform under pool fire has been analyzed, the regularities of the distribution of the wall temperature of the platform of FPSO under different wind speeds are studied, and research on the distribution of heat radiation flux of different fire is made. Based on the finite element method, the structural response of the platform structure in different fire scenarios has been made. In consideration of the pool fires caused by liquid leakage of the upper part of the platform structure, with the basis of the changes of temperature field and radiation field being obtained by CFD, a structural response analysis of the offshore platform structure using the finite element method and a risk assessment method based on quantitative analysis for pool fires caused by liquid leakage is proposed.

Keywords: pool fire; risk assessment; structural response; finite element method

Citation: Zhang, D.; Zhao, B.; Zhu, K. Structural Response Analysis of FPSO under Pool Fire. *Buildings* **2022**, *12*, 1665. https://doi.org/10.3390/buildings12101665

Academic Editor: Elena Lucchi

Received: 7 September 2022
Accepted: 9 October 2022
Published: 12 October 2022

1. Introduction

With the rapid development of science and the economy, the demand for oil and natural gas is increasing year by year, and human beings are not satisfied with the development of land energy. As an important facility of marine resources, the number of offshore oil platforms, especially FPSOs, is increasing. An FPSO, floating production storage and offloading unit, also known as a floating production storage offloading tanker, can be initially used to store crude oil [1], and have been widely used in deep sea areas. The processing equipment of the FPSO is built into the modules and located on the vessel deck. The fluid transferred from underwater is processed into oil, gas, and water. The FPSO stores the processed oil. The processed gas is exported through an export riser or injected again into the subsea reservoir. Among all floating devices, the FPSO has functions and advantages such as strong wind wave resistance, wide water depth range, large oil storage, unloading capacity, and reuse [2]. In addition, its vessel-shaped features make it more popular than other offshore platforms such as spars and jack-ups [3].

Although this platform has many advantages, the high level of congestion caused by pipe networks and upper crude oil processing equipment poses a higher risk to the FPSO [4–6]. As the offshore platform or FPSO are far from the land, once fire and explosion accidents occur, it is difficult to conduct a timely rescue, which results in disastrous loss [7]. The UK health and safety research report shows that fire accidents are the most common of all marine accident reports and blowouts [8]. Pipeline leakage, ship collision, structural failure, and other accidents can lead to a fire. A pool fire is a kind of fire that is common in marine fires. It has a large scale of combustion and is difficult to extinguish. When a fire occurs, the platform structure of the FPSO is completely exposed to fire, the temperature of the structure rises, the elastic modulus and yield strength of steel decrease, and the bearing capacity of the platform will be reduced. All of these factors can lead to the collapse of the platform and make the entire FPSO invalid.

The fire load was first defined in the structural fire protection, which laid the foundation for the quantitative assessment of fire. Anderberg [9] defined the fire load of the steel structure through the risk assessment method and then conducted a limit analysis and collapse analysis of the whole platform and gave the fire analysis program. Ruert A. and Schaumann P. [10] conducted the fire resistance test of different steel structure frames; through the analysis of a large number of data fitting, the temperature distribution of various forms of frame structure was calculated and the failure temperature of different structures was obtained which provided reference data for the analysis of fire resistance of steel structures for later scholars. Walker et al. [11] evaluated marine structures for fire safety and being explosion-proof, the overall structure and components were analyzed, the structural dynamic response analysis under fire and explosion was obtained, and the deformation analysis and structural dynamics under the temperature and pressure time curves were conducted. Tolloczko [12] proposed some general situations for fire research which should be used for marine structures and studied the dynamic response of structures under fire loads. Wingerden et al. [13] conducted analysis through the principle and method of computational fluid dynamics (CFD) and developed special analysis software for the risk of fire and explosion of platform structures. Soares and Shetty [14] firstly applied the joint probability method to the fire risk assessment of offshore platforms and applied it to the passive fire protection design of the superstructure. From 2008 to 2010, Pusan National University, Nowatec AS, and other research institutions studied fires and explosions of upper deck modules of FPSOs based on the CFD principle and the nonlinear finite element theory and established an integrated analysis and evaluation system [15].

In the past 30 years, offshore oil engineering accidents caused by fire and explosion have occurred frequently, which not only led to casualties and economic losses but also caused damage to the marine environment. Given the serious consequences, it is necessary to model accident scenarios and their consequences in order to assess and manage fire risks in such situations. In recent years, considerable efforts have been made in maritime fire modeling and risk assessment. Kim et al. [16], for example, evaluated the load characteristics of steel and concrete tubular members under jet fire. To obtain reliable load values, jet fire tests were carried out in parallel with a numerical study. Paik et al. [17] described a few procedures for the quantitative assessment and management of fire and gas explosion risks in offshore installations. Kim and Paik [18,19] studied the feasibility of applying the computational fluid dynamics (CFD) method for the analysis of fire heat flow in an FPSO upper module, taking into account wind effects. The CFD results agreed well with the experimental results. Luketa [20] and Vasanth et al. [21] studied the characteristics of two fire pools with fuel surfaces at different elevations using CFD. Sun et al. [22] studied the load characteristics in process modules of offshore platforms under jet fire conditions. The effects of different parameters on the load characteristics of the considered scenarios were discussed. Betteridge and Steven [23] described an empirical model to account for the reduced pool fire size and discussed the effect of water on combustion. Wu et al. [24] derived the available safe evacuation time estimation equation by simulating the fire development process using a fire dynamics simulator field model. Yi et al. [25] and Ahmadi et al. [26] established a CFD model to accurately predict the incident radiation of large LPG pool fires and the consequence of large-scale pool fires in storage terminals, respectively. Xu et al. [27] introduced a quantitative risk assessment method for offshore platforms exposed to gas explosions. Bhardwaj et al. [28] proposed a methodology for risk assessment and probabilistic modeling of fire and explosion accidents in FPSO. Li et al. [29] focused on the use of CFD to simulate the fire risk of offshore facilities caused by subsea gas release.

According to the literature, past studies mainly concentrated on the characteristics of fire occurrence and the assessment of the damage caused by fire; the structural response of an FPSO caused by a fire load and explosion was seldom reported in the literature. In fact, the high temperature of a fire will cause the platform steel structure to lose its structural capacity and result in large deformation, and the blast wave of an explosion can cause devastating damage to life and equipment. Therefore, a detailed study of fire risk assessment and structural response analysis of FPSO is urgent for developing safety measures.

In this paper, based on the finite element analysis software ABAQUS, a finite element analysis of the response of a platform structure in different fire scenarios has been made. In consideration of the pool fires caused by liquid leakage of the upper part of the platform structure, with the basis of the changes of temperature field and radiation field being obtained by CFD, the structural response analysis of the offshore platform structure is made by ABAQUS using the finite element method. The novelty of this paper is to present some results and analysis to reveal the impact of pool fires on offshore platform structures. The wall temperature of the FPSO under different wind speeds and the distribution of heat radiation flux are studied. The structural response of the offshore platform caused by a pool fire is analyzed. The rest of this paper is organized as follows: Section 2 presents some principles of fire risk assessment for FPSO. Section 3 presents the simulation modeling of pool fires and analyzes the flame shapes, temperature field, and heat radiation of a pool fire under different wind speeds. Section 4 illustrates the structural response and mechanism of FPSP under a pool fire. Besides, the validity of the numerical method is verified in this section. Finally, Section 5 summarizes the conclusions drawn from this work.

2. Principle of Fire Risk Assessment for Platform Structure of FPSO

2.1. Fire Types

Quantitative risk analysis of marine structure fire assessment includes deck fire and fire in the sea. The most serious structural unit of the offshore platform fire is the deck structure, so a platform fire risk assessment usually takes the fire on deck into consideration. Types of fire include pool fire, jet fire, ball fire, and flash fire. As the operating conditions of the offshore platforms are different, the fire types are different. When the leakage is liquid, a pool fire usually occurs.

Offshore platform fire is mostly hydrocarbon fire, the heat of the fire is large and the release is fast, so the temperature rise of the steel structure is fast. As a result, the strength of the steel structure decreases rapidly; when the temperature is above 600 °C the strength of the structure almost disappears. Therefore, when a fire happens to large steel structures, it is prone to distortion, deformation, and collapse. Although there are four types of platform fire, pool fire and jet fire are the main form. As oil and gas leakage occurs mostly on the platform deck, the deck structure is a typical unit affected by fire, so it is of practical significance to study the deck fire.

2.2. Pool Fire Risk Assessment Method Based on Quantitative Analysis

Risk assessment includes risk identification, probability analysis, and consequence analysis, as shown in Figure 1. When carrying out a quantitative risk assessment for offshore platforms, it makes no sense to analyze the safety of the entire offshore platform directly. First, it is necessary to identify the hazard sources of liquid leakage, which can be used to determine the fuel information. After the hazard identification, we need to determine the geometry of the structure, the pool size, the pool location, wind speed, and wind direction, which make a leak and fire scene clear. The main cause of the risk is the failure of some equipment and parts. Failure of equipment and parts will cause fuel leakage. In order to determine the probability of fire, we must first determine the probability of leakage, and the leakage probability of an offshore platform can be determined according to the failure probability of equipment and parts, as shown in Table 1 [30,31].

The probability of leakage is calculated according to the quantity of the equipment:

$$P = np \tag{1}$$

The assumption for the equation that the events are independent is necessary. Among them, P is the probability for some components of the leakage, p is the leakage probability for a single component, and n is the number of a certain part. If the leakage probability of

1 flange is 2.34×10^{-9}, the leakage probability of the 10 flanges is 2.34×10^{-8}, the leakage probability of 1 m pipeline is 1.71×10^{-8}, and the leakage probability of 100 m is 1.71×10^{-6}.

In this paper, a method for the quantitative analysis of marine fire is put forward according to some specifications and related literature. It needs to be pointed out that this method is a pool fire scenario that is caused by liquid leakage on an offshore platform.

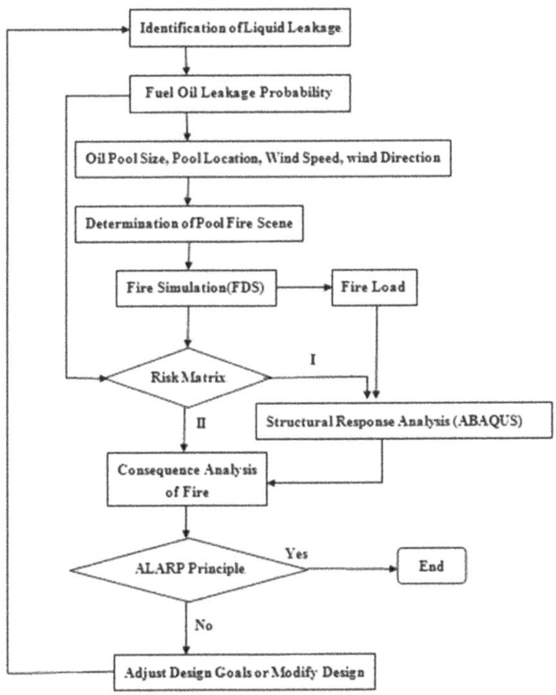

Figure 1. Quantitative risk assessment method for fire risk.

Table 1. Probability of different components.

Device Name	Manual Valve	Emergency Shutoff Valve	Flange	Pipeline	Pressure Vessel	Heat Exchanger
Failure Probability	3.5×10^{-9}	4.39×10^{-8}	2.34×10^{-9}	1.71×10^{-8}	1.16×10^{-9}	8.43×10^{-8} [1]

[1] The data in the table indicate the failure probability of the unit. The unit of the pipeline is m.

2.3. Risk Matrix and ALARP Principle

To determine the level of the risk matrix, the probability of the occurrence and the consequences of fire are put into a matrix, which is the risk matrix, as shown in Figure 2. The risk matrix is divided into three levels and the consequences are expressed in terms of temperature. Risk consequences are expressed in terms of temperature and the risk matrix is used to screen the simulated fire scenarios. I represents an unacceptable risk as a high risk, when it is at this level, further evaluation will be needed. II represents critical risk also known as medium risk. III represents an acceptable risk which means it is a low risk. This method is a rough analysis of the probability of the size and the consequences of the accident; the general calculation will ignore the effect of low risk. The analysis focuses on the medium and high risk. When it is a medium risk, the direct assessment of the consequences will be made without structural analysis.

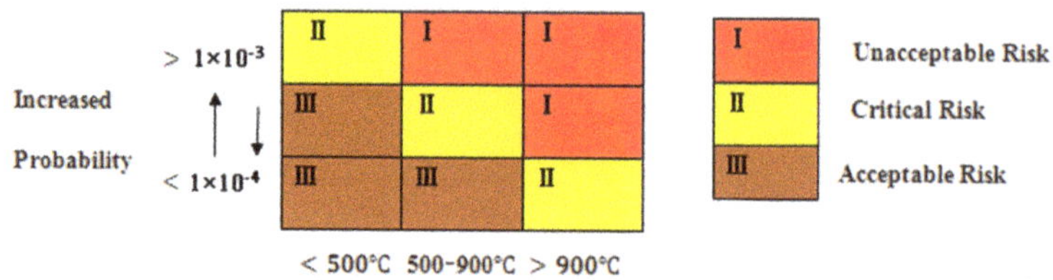

Figure 2. Risk matrix.

The ALARP principle, as shown in Figure 3, (as low as reasonably practicable) is the principle of risk evaluation at present, which is a high degree of risk evaluation at home and abroad. It is a kind of minimum feasibility principle. If the risk is in the ALARP region, the risk can be accepted.

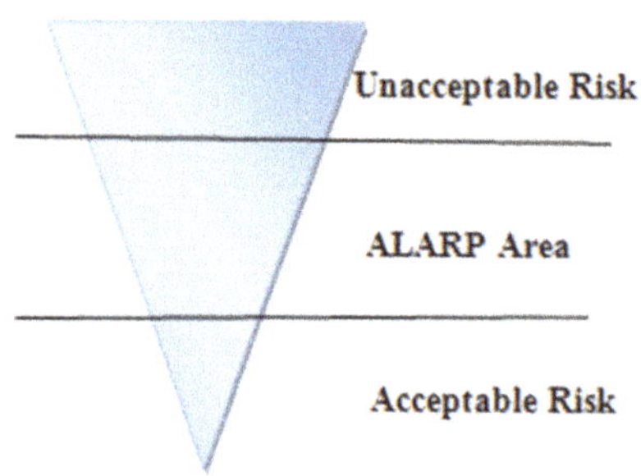

Figure 3. ALARP principle.

2.4. Risk Measurement

The general measure includes three aspects: personnel risk, property damage, and environmental risk of fire accident consequences. In the actual calculation of risk, the probability of the occurrence of different events should be calculated separately, and then the probability of all events will be accumulated.

Although individual risk can be used to measure the risk of death, the largest concern is the risk of the entire group of the accident, so the consequences of group risk are generally calculated by an f-N curve.

The f-N curve, as shown in Figure 4, represents the relationship between the probability of an accident and the number of casualties, the value of which is cumulative, indicating an acceptable level of risk.

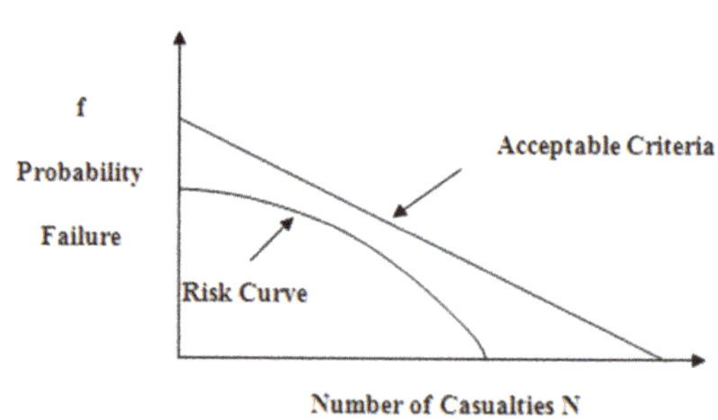

Figure 4. f-N curve.

For the risk assessment method proposed in this paper, due to limited time and energy, in addition to the risk probability structural analysis of fire and temperature, the risk assessments of other aspects are not the focus of this paper, the results of which can be found in related documents and specifications. Therefore, this paper focuses on the study of fire characteristics and the structural response analysis of a platform structure under the action of a pool fire; the characteristics of the temperature of a platform structure of a pool fire under different wind speeds are obtained by the CFD principle. The characteristics of fire heat flux are studied under different wind speeds, and the ABAQUS software is used to conduct the thermal–mechanical coupling analysis of the platform.

There are numerous possibilities for simulating pool fires of offshore platforms. Because of the numerous possibilities, the choice of a fire scenario is more complicated in the fire simulation. In order to simulate a fire, the equipment on the platform is simplified to the uniform load, which simplifies the deck model.

3. Analysis of the Temperature Field of the Platform under Fire

The traditional temperature field for the steel structure usually adopts the ISO834 standard heating curve, but the actual fire environment is different from the temperature curve. Offshore platform fires occur outdoors, where oxygen and combustion are more complete, the fire generated by heat radiation is relatively strong, and harm is relatively large [32]. The study of the characteristics of fire is the premise of the fire resistance for the structure; the temperature response of the offshore platform and the temperature changing with time seem to be the known conditions for studying the structural response of the platform. It is very difficult to calculate the temperature distribution and temperature field of the platform with the theoretical model. Therefore, the simulation of a pool fire based on FDS and the analysis of different fire scenarios of a pool fire are the key link of fire analysis and the premise of the structural response analysis, which makes it consistent with the real ones.

In this paper, the platform of the superstructure of an FPSO on active service in the South China Sea was studied. The platform has two decks. The first deck is for the oil and gas transmission area, in which the simulation of the fire caused by a pipeline leakage is made. The range of a pool fire is generally very small compared with the whole platform, so only a part of this platform was studied. In an FDS simulation, the size of the upper deck is 20 m × 14 m × 6 m and the size of the bottom deck is 20 m × 14 m × 6 m. The FPSO model is shown in Figure 5.

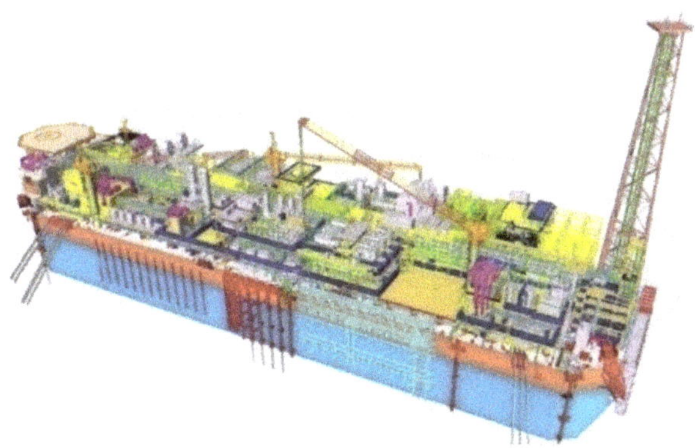

Figure 5. FPSO CAD model.

3.1. Simulation Principle of FDS

FDS (fire dynamic simulation) is a software based on the fluid dynamics model, which can be used for the numerical simulation of the thermal smoke propagation and spread of fire.

The large eddy simulation theory is used to simulate the turbulent motion of fire and the basic equations of the dynamic model in the fire process are as follows.

Mass conservation equation:

$$\frac{\partial \rho}{\partial t} + \nabla \cdot \rho \vec{u} = 0 \tag{2}$$

Momentum equation:

$$\rho\left(\frac{\partial u}{\partial t} + \left(\vec{u} \cdot \nabla \cdot\right)\vec{u}\right) + \nabla p = \rho g + f + \nabla \cdot \tau_{ij}\frac{\partial \rho}{\partial t} + \nabla \cdot \rho \vec{u} = 0 \tag{3}$$

Component equation:

$$\frac{\partial}{\partial t}(\rho Y_i) + \Delta \cdot \rho Y_i \vec{u} = \Delta \cdot \rho D_i \Delta Y_i + W_i \tag{4}$$

Energy equation:

$$\frac{\partial}{\partial t}(\rho h) + \nabla \cdot \rho h \vec{u} = \frac{Dp}{Dt} - \nabla \cdot \vec{q_r} + \Delta \cdot k \nabla T + \sum_L \nabla \cdot h_i \rho D_i \nabla Y_i + \vec{u} \Delta p \tag{5}$$

Gas state equation:

$$p = \rho T R \sum_i \frac{Y_i}{M_i} \tag{6}$$

where ρ is density (in kg/m^3); $\vec{u}$ is velocity vector, m/s; p is pressure, Pa; h is total enthalpy, J/kg; T is temperature, K; Y_i is the mass concentration of component i; D_i is the diffusion coefficient of component i, m^2/s; W_i is the chemical reaction rate of component i, kg/(m^3 s); M_i is the molecular weight of component i, kg/mol; R is the molar gas constant, J/(mol. K); $\vec{q_r}$ is fire heat radiation, W/m^2; q is the combustion reaction heat, J/mol; k is the thermal conductivity, W/(m·K); g is the gravity acceleration, m/s^2; t is the simulation time, s; τ_{ij} is the viscous stress tensor, Pa. The contribution of those existing equations and numerical procedures comes from the commercial software FDS, the same as in Section 4.1.

The flame of liquid hydrocarbons in the pool on the ocean platform can quickly spread to the entire oil pool, so the ignition process is very short. Thus, it isn't necessary to consider the ignition process, only to study the characteristics of the formation of flame.

3.2. FDS Model and Results of Pool Fire

Fire simulation is based on a real situation to simulate a virtual fire scene. The heat of combustion for liquefied petroleum gas transfers mainly in the form of heat radiation which produces a large number of CO and CO_2, resulting in a certain area of casualties and equipment damage and a large number of people's death by suffocation.

Figure 6 depicts the instantaneous temperature contours of a mid-longitudinal section of a pool fire. It was obtained through the post-processing function of the FDS software. The pool fire creates a thermal vortex at the bottom of the combustion and the small eddy draws air into the center of the flame and continues to rise as the flame rises. At the height of a small diameter of the oil pool, the small eddy begins flow separation with the fire plume and gradually disappears at a turbulent fire plume. At the height of the half diameter of the oil pool, it can be clearly seen that the flame has shrunk into a neck, which is known as the pool fire flame neck-in phenomenon. Most of the combustion will be completed in the area between where the flame neck-in phenomenon occurs and in front of a certain separation point at the height of 1–2 pool oil diameters. At the same time, a new heat vortex will be formed on the combustion surface. Due to the pulsation characteristics of the fire, it will cause the periodic generation and shedding of a thermal vortex. It is important to note that the FPSO has coamings limiting the size of pools of leaked oil. The problem with gas is not a pool fire, but a jet fire, due to the pressure of the gas inside the pipes.

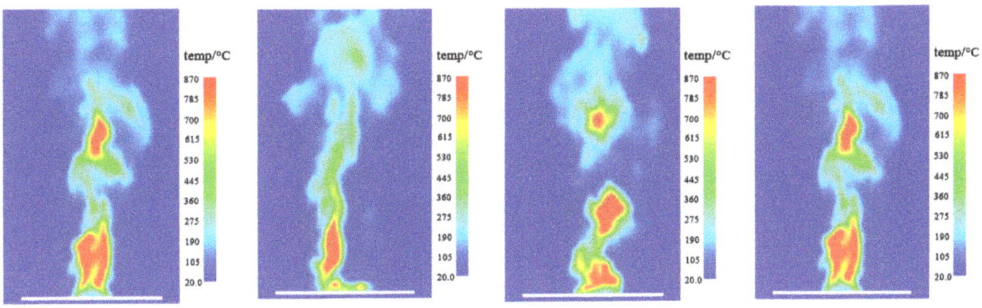

Figure 6. Instantaneous temperature contours of pool fire.

3.2.1. Flame Shapes of Pool Fire

The CAD model of the entire FPSO offshore platform structure is shown in Figure 5. The simulation of a pool fire for 600 s of combustion is created by FDS. Different parameters according to the characteristics of a pool fire have been set to obtain the influence degree of different parameters on the pool fire. The fuel of the oil pool is assumed to be heptane. The pressure is assumed to be the atmospheric pressure, the ambient air density is 1.258 kg/m^3, and the leakage is continuous leakage. The distribution of the whole temperature field and the flow field of the fire is unstable, but this instability is limited. The beating of the flame is going up and down at a certain amplitude, and when the fire falls into a pattern, we can choose the fire at a certain moment as the research object. The open environment condition: the temperature is 20 °C; the air pressure is 101 kPa; the wind speeds are 0 m/s, 1.5 m/s, and 3 m/s, respectively. In this paper, we only study a part of this platform and simplify it into an FDS model as shown in Figure 7.

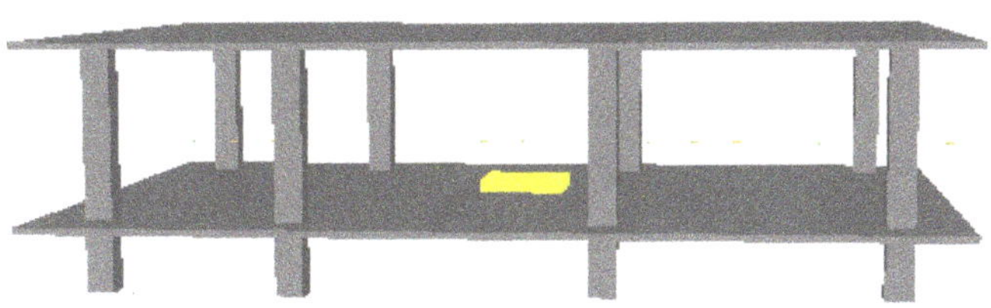

Figure 7. The simplified model in FDS.

Figures 8–10 depict the flame shapes under different wind speeds. The flame shapes are obtained through the software Smokeview. Under the action of each single wind speed, the four instantaneous shapes of the fire development process have been selected and compared at the moment of 2 s and 10 s, and 100 s and 400 s during the fire development process. It can be found from Figures 8–10 that the flame will tilt; the greater the wind speed, the greater the tilt angle of the flame. When the wind blows, the flame is subjected to the wind force W_f along the wind direction, as the flame has buoyancy F at the same time, so the combined action of the two forces causes the flame to tilt. It can also be found that the fire spreads quickly, reaching a prosperous period of fire development at 10 s, which shows that the burning speed of a pool fire is very quick. Thus, at the beginning of the fire, personnel should be evacuated without delay. When the fuel is leaking while it is burning, the burning area will become larger and significant heat will be released. As the flame spreading direction is unknown, it can also cause more damage, especially in the vicinity of the oil pool which hasn't burnt up. Once the fire spreads, it is more difficult to keep the fire under control. It will cause inestimable damage to the surrounding equipment and buildings.

Under the action of the wind, the flame deviates from the center of the flame and then returns to the center; this is due to the return force of the flame itself which can cause the flame to become out-of-center and then return back to the central position. The interaction of wind force and return force causes the flame to tilt significantly, destroying the symmetry of the pool fire performance, and increasing the complexity of the research.

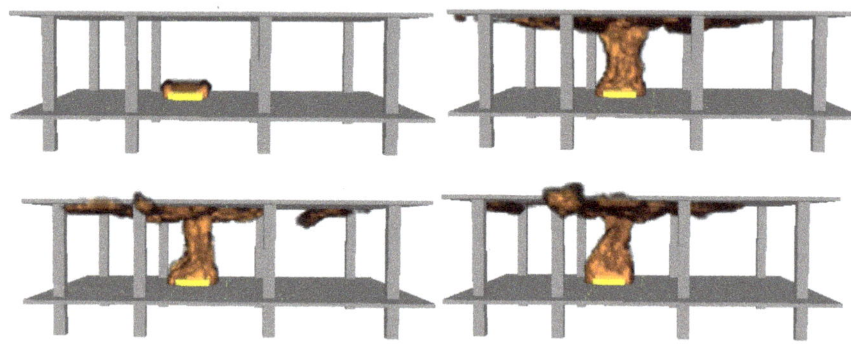

Figure 8. Flame shapes under the wind speed of 0 m/s.

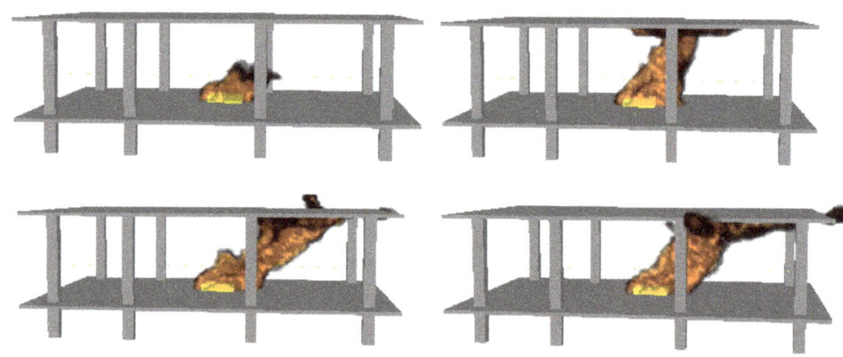

Figure 9. Flame shapes under the wind speed of 1.5 m/s.

Figure 10. Flame shapes under the wind speed of 3 m/s.

3.2.2. Temperature Field of Pool Fire

The temperature field of the flame under different wind speeds is shown in Figure 11. Observing Figure 11, it can be seen that with the increase in wind speed, there is an inclination angle of the flame when it has been in a stable state. In addition, the distribution of temperature is tilted with the inclination of the flame; the wind can change the temperature distribution of the flame area and will also change the temperature distribution of the deck area so that the high-temperature distribution of the deck is mainly concentrated in the area downwind.

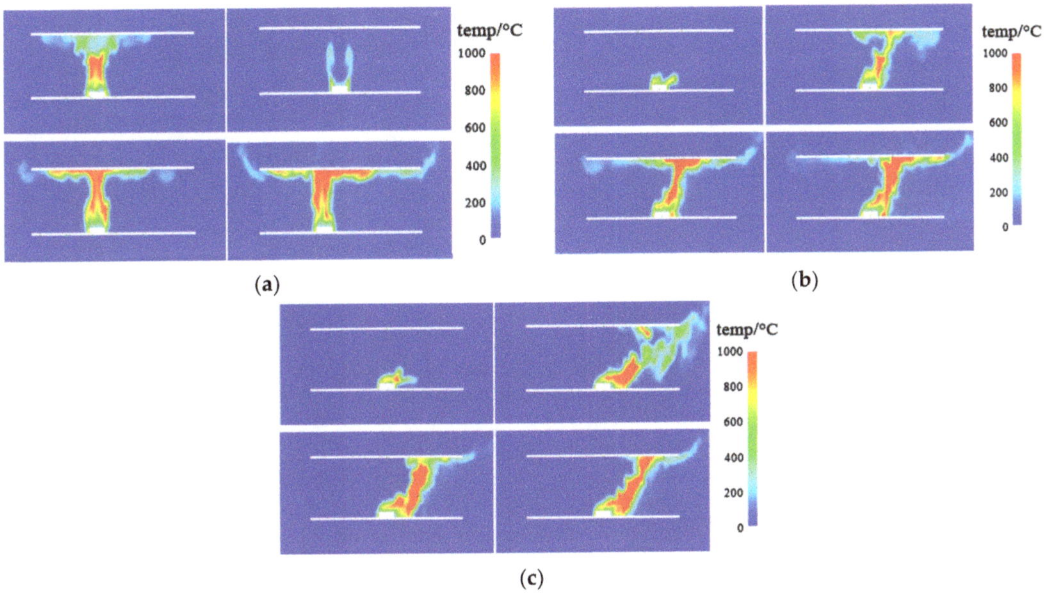

Figure 11. Temperature field of XZ plan, Y = 8 m: (**a**) wind speed = 0 m/s; (**b**) wind speed = 1.5 m/s; (**c**) wind speed = 3 m/s.

Under different wind speeds, when the wind speed reaches 3 m/s, the temperature of the flame can be reduced. It can be found from the temperature contours that the maximum temperature of 0 m/s and 1.5 m/s is 1000 °C, and when the wind speed reaches 3 m/s, the maximum temperature is 870 °C. Excessive wind speed will accelerate the heat exchange between the flame and air and as the speed of cold air blowing in from the outside increases, the flame temperature will be dispersed in time; therefore, when the wind speed is too high, sometimes it will not increase the temperature, but will reduce the flame temperature. The wind can bring oxygen to the flame and cause the fuel to burn more intensely and cause a more complete release of heat. The wind can also promote the heat exchange between the flame and the air, which causes the thermal vaporization of the fuel gas diffusion to become faster and the combustion to release more heat. However, with the increase in wind speed, the flame temperature will decline. This is because when the wind is too powerful it brings in cold air more quickly, causing the exchange rate of hot air and cold air to accelerate, and as a result, the temperature loss of the oil pool is also more significant. With the increase in wind speed, the combustion of the oil pool is intensified and the heat convection is also increased.

3.2.3. Heat Radiation of Pool Fire

Many scholars have conducted significant research on heat conduction and heat convection in fire research, however, during the development and spread of fire, heat radiation can cause other no-fire areas to catch flame and expand the fire. Offshore platform fires belong to open environment fires; the pool fire in the open environment causes the most serious casualties and equipment damage. The overturning of the platform is mainly reflected in the intensity of the heat radiation flux of the fire; the radiation heat flux is the heat radiation energy per unit area per unit time. The offshore platform is the base of marine production and operation and is also the infrastructure of offshore oil exploitation. The study on the radiation field of the pool fire has instructive significance for the measures of marine fire and the means of escape.

To obtain the size of radiation heat flux, different monitoring points are chosen for detection, the oil pool center (X = 7 m, Y = 7 m) is set up as the monitoring center, the 18 points along the X axis with an equal distance (1 m) of the oil pool are chosen as the detecting points, and the measured height is 1.65 m (assuming that the height of the person is 1.65 m). The locations of different points are shown in Figure 12.

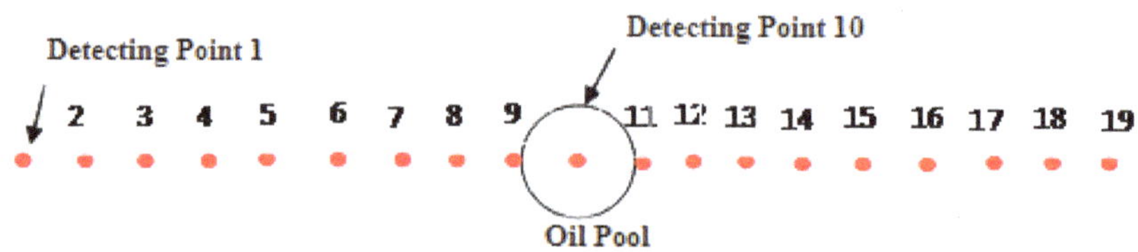

Figure 12. Top view of detecting points.

As there is usually wind at sea, we discuss the radiation heat flux under different wind speeds. Because of the large fluctuation of the radiation heat flux, the average values of different wind speeds have been compared. The average radiation heat flux of different detecting points is shown in the figure below: the 0 point is the center of the oil pool, the right side of the pool is positive, and the left side of the pool is negative.

As shown in Figure 13, with the observation of the changes in the radiation heat flux of each detecting point, it can be found that when the wind speed is 0, the maximum heat flux is in the oil pool center. When the wind speed is 1.5 m/s, the maximum radiation heat flux is under the downwind direction with a distance of 3 m from the oil pool center. When wind speed is 3 m/s, the maximum radiation heat flux is under the downwind direction with a distance of 5 m from the oil pool center. This is because under the action of wind a large amount of heat moves in downwind, resulting in the radiation heat flux of the downwind area being higher than that of the upwind area, which also leads to the increase in the risk of the downwind direction with the increase from 0 m/s to 3 m/s of the wind speed. The value of radiation heat flux also increases. It is also found that when the wind speed is 0, the radiation heat flux curve is symmetric with respect to the center of the pool.

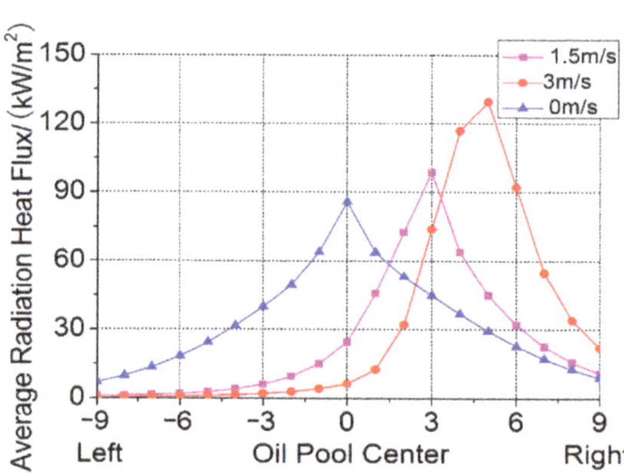

Figure 13. The average radiation heat flux of different detecting points under different wind speeds.

4. Structural Response and Mechanism of Platform under Pool Fire

4.1. Simulation Principle of ABAQUS

The main component of the offshore platform is steel, and in the study of the fire risk of the offshore platform, the research of the thermal performance and mechanical properties of the steel structure at high temperatures is also an important part.

In order to better reflect the influence of the pool fire on the structural response of the platform structure, based on the FDS software to calculate the various conditions of the structure temperature field and the environmental temperature field, the structural response analysis of offshore platforms is made using ABAQUS software, which obtains the structural deformation distribution and stress distribution of specific pool fire conditions for offshore platform structures.

ABAQUS can be used for the analysis of non-coupling heat transfer, sequential coupled thermal stress analysis, fully coupled thermal stress analysis, and thermal analysis. The stress and strain field in the process of sequential coupling thermal analysis is determined by the temperature field, but the temperature field is not affected by the stress and strain field; compared with the sequential coupling analysis, the heat transfer analysis and stress analysis are carried out at the same time in fully coupled thermal stress analysis. The study of this paper is the structural response of offshore platform structures under a pool fire, and the temperature field is obtained by FDS software, so we adopt the sequential coupled thermal stress analysis method for the simulation of the structural response and mechanism of the offshore platform structure on the FPSO. Under the action of fire, the elastic modulus and yield strength of the steel will decrease sharply at high temperatures and the bearing capacity of the platform structure will change. Based on the ABAQUS software, this paper analyzes the structural response under high temperature; firstly, the simply supported beam under high temperature is analyzed and compared with the experiment to prove the validity of the research method. Then, the structural response of the two decks of the offshore platform is analyzed.

When ABAQUS is used to create the sequential coupled thermal stress analysis, the heat transfer parameters must be set reasonably to ensure the accuracy of the temperature distribution in each position and time step. The basic theory of ABAQUS thermal stress analysis is based on the Fourier heat conduction law and the law of conservation of energy.

The energy conservation equation and the specific temperature distribution are solved by the following equations in ABAQUS.

$$\rho_s c\dot{\theta} = W - I_q \tag{7}$$

where ρ_s is the density of the steel structure, c is the specific heat of the steel structure, θ is the temperature rate of change, W is the external heat, and I_q is the internal heat.

The unit area heat balance equation based on the Fourier heat conduction law in ABAQUS is shown in Equation (8).

$$Q = -\lambda \cdot gradT \cdot F \tag{8}$$

where Q is the heat flux per unit area, λ is the thermal conductivity, *gradT* is the temperature gradient vector, and F is the heating area of the steel structure. Equation (8) can also be expressed as a heat conduction differential equation just as the following.

$$\rho_s c\frac{\partial T}{\partial t} - \frac{\partial}{\partial x}\left(\lambda_x\frac{\partial T}{\partial x}\right) - \frac{\partial}{\partial y}\left(\lambda_y\frac{\partial T}{\partial y}\right) - \frac{\partial}{\partial z}\left(\lambda_z\frac{\partial T}{\partial z}\right) - q_v = 0 \tag{9}$$

T is the temperature of the steel at t time and is the coefficient of thermal conductivity of steel, x, y, z are the coordinates of the structure, and q_v is the heat generated by the heat source inside the steel structure.

The steel of the platform used in this paper is Q235 steel, a homogeneous steel, meaning $\lambda_x = \lambda_y = \lambda_z = \lambda$, q_v=0, so the Equation (9) can be simplified to the following equation.

$$\rho c\frac{\partial T}{\partial t} - \lambda\left(\frac{\partial^2 T}{\partial x^2} + \frac{\partial^2 T}{\partial y^2} + \frac{\partial^2 T}{\partial z^2}\right) = 0 \tag{10}$$

After the pool fire, the temperature change of the offshore platform structure is related to the fire time. Therefore, the transient heat conduction analysis step in ABAQUS is used to simulate the heat transfer process. In addition, the ambient air temperature change curve of the structure is obtained by FDS; therefore, the heat transfer analysis can be carried out directly in ABAQUS by the heat flow between the air and the surface of the structure, so as to obtain the temperature field in ABAQUS.

When heat conduction occurs, the structure itself has the temperature difference between $T(x, y, z)$. The temperature difference can cause the expansion of the structure, the expansion coefficient is α_T. The physical equation of the structure due to thermal expansion is as below.

$$\begin{cases} \varepsilon_{xx} = \frac{1}{E}\left[\sigma_{xx} - u\left(\sigma_{yy} + \sigma_{zz}\right)\right] + \alpha_T \Delta T \\ \varepsilon_{yy} = \frac{1}{E}\left[\sigma_{yy} - u\left(\sigma_{xx} + \sigma_{zz}\right)\right] + \alpha_T \Delta T \\ \varepsilon_{zz} = \frac{1}{E}\left[\sigma_{zz} - u\left(\sigma_{xx} + \sigma_{yy}\right)\right] + \alpha_T \Delta T \\ \gamma_{xy} = \frac{1}{G}\tau_{xy}, \gamma_{yz} = \frac{1}{G}\tau_{yz}, \gamma_{xz} = \frac{1}{G}\tau_{xz} \end{cases} \quad (11)$$

4.2. ABAQUS Model and Results of Pool Fire

As it is impossible to simulate the real ocean environment by computer, the following assumptions are made. (1) The temperature of the marine environment is 20 °C and the temperature of all the structures is also at a temperature of 20 °C. (2) The strength of the joints of all the components is the same as that of the welded parts. (3) For the shell element, the temperature gradient along the shell thickness is not considered. (4) The shell element should be used for a thin structure, the temperature of the upper surface of the shell is equal to that of the lower surface of the shell, and the temperature on the surface of the shell is to be considered, but not in the direction of thickness.

4.2.1. Experimental Verification

Before making the structural response analysis, the validity of the ABAQUS method should be verified. Because of the great fire experiment cost, therefore, we can verify the correctness of the ABAQUS method with the experiment results of Southeast University by Shuping Cong et al. In the experiment, Q235 I steel beam was mainly used (it is assumed to be the ideal type, one end is sliding hinged and the other end is hinged–fixed). The beam is placed in the oil-fired furnace to be heated by the constant load, whose value is 10 kN/m. The span of the beam is 4.2 m, the height of the web is 250 mm, and the thickness is 6 mm; the width of the flange plate is 125 mm, and the thickness of the flange plate is 9 mm. The beam loading and the loading temperature curve of the beam are shown in Figures 14 and 15, respectively.

According to the experimental temperature curve, a steel beam model was built with ABAQUS. In order to maintain the consistency between the experimental model and the finite element model, the dimensions of the model are the same as that of the I-beam used in the experiment, as is the fixed form. The finite element model is established with the shell element. The first step is to create the temperature load on the beam, the second step is to apply the uniform load on the upper panel of the I-beam, and the third step is to submit the calculation.

As shown in Figure 16, it can be seen from the simulation that the displacement of the middle position of the beam is larger, which is similar to the deformation of the experimental component, which also has the largest displacement change in the middle position.

According to the finite element model and the experimental data, we can make a comparison of the maximum displacement values of the beam center position of the two methods. The results obtained by the two methods are in excellent agreement.

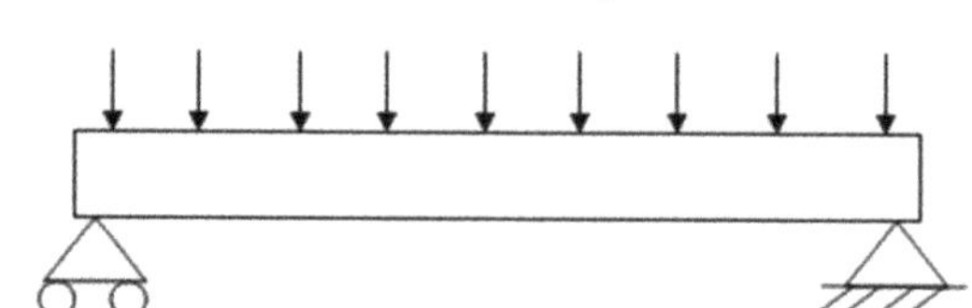

Figure 14. Schematic diagram of beam loading.

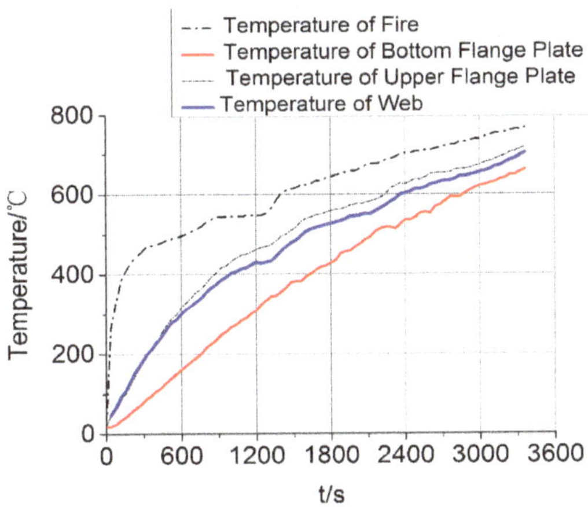

Figure 15. Loading temperature curves versus time.

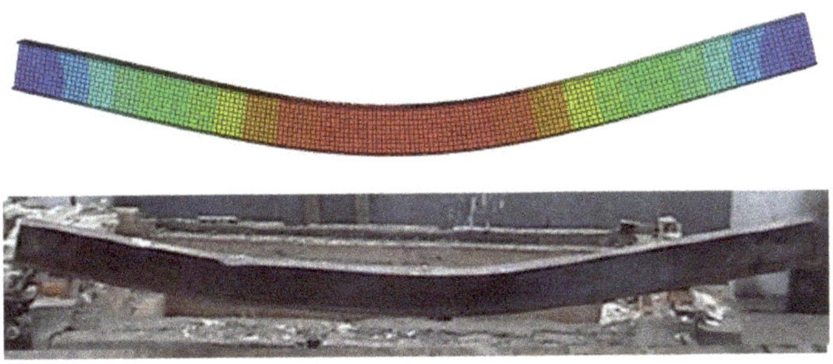

Figure 16. The comparison between the simulation and the experiment.

Figure 17 shows that the maximum displacement of the middle position of the steel structure increases linearly with the increase in temperature when the temperature is less than 600 °C. However, when the temperature is higher than 600 °C, the displacement changes rapidly, and then the displacement variable changes exponentially with the increase in temperature. Figure 18 shows that the displacement of steel structure members is small and smooth when the time is less than 2700 s; with time, the mutation of the displacement variation occurs at 2700 s and the displacement increases sharply, which indicates that the fire resistance time of the steel structure is 2700 s.

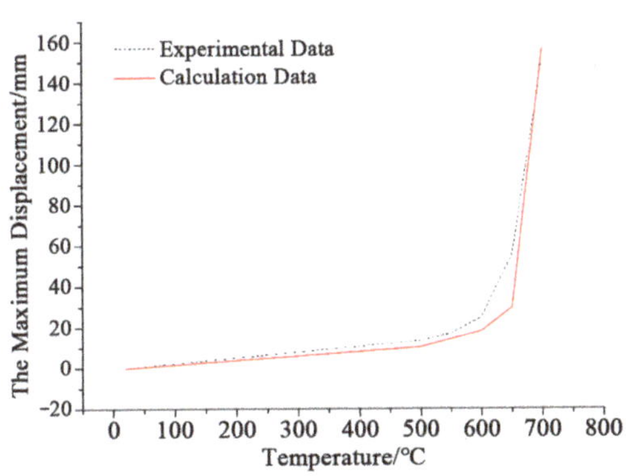

Figure 17. Displacement–temperature curve.

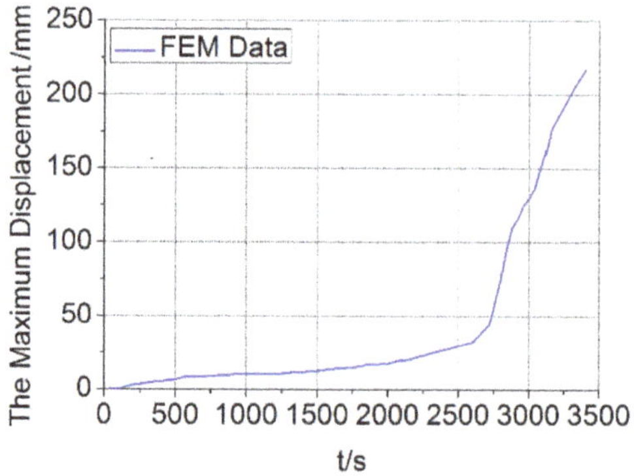

Figure 18. The maximum displacement of the middle position of the beam versus time.

It can be concluded that the finite element method can be applied to the coupling analysis of a fire temperature structure by comparison with the experimental results.

4.2.2. Finite Element Model of Offshore Platform Frame Structure

The superstructure model of offshore platform size is 20 m × 14 m × 6 m, including two deck structures. In order to maintain the continuity and effectiveness of numerical simulation, this simplified model is consistent with that of the fire simulation. Although the model is simplified, the most basic structural framework is retained. The vertical and horizontal support beams of each deck are Q235 steel I-beam shape, as shown in Figure 19. The height is 6 m, the material support column for Q235 steel. In order to reduce the calculation time, considering that the secondary beam is far less than the support effect of the main girder, we only set up the main girder of the deck without modeling the secondary beam.

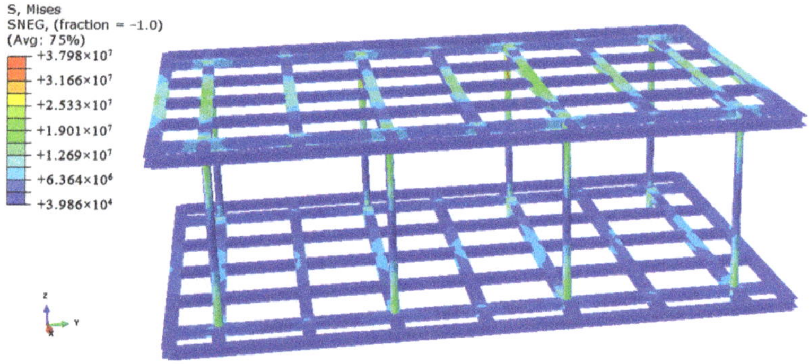

Figure 19. Model in ABAQUS.

The analytical parameters of heat transfer (surface radiation) and interaction (surface film condition) in relation to fire heat transfer are defined in the interaction module. Before setting the conditions of thermal convection and heat radiation, the rising temperature curve of the steel frame structure must be obtained by FDS software. First of all, the upper and lower two decks should be numbered, as shown in Figure 20; then, the temperature variation curves of the corresponding observation nodes are obtained by FDS software; in order to make the results as accurate as possible, the temperature of each node in the target area is selected and then the average value is obtained, which is used for the temperature change curves of the two layers of the steel frame.

According to the observed values, 48 kinds of temperature amplitude curves are established in ABAQUS, and then the parameters of heat convection and heat radiation are defined in the Interaction module. The surface film condition is chosen to define the convection condition in the constraint condition and the convective heat transfer coefficient is 1500. It should be noted that the temperature curve must correspond to the corresponding heating area. Then the surface radiation is chosen to define the radiation condition in the constraint condition. The parameter selection method is similar to the convection condition, and the thermal radiation coefficient is 0.5.

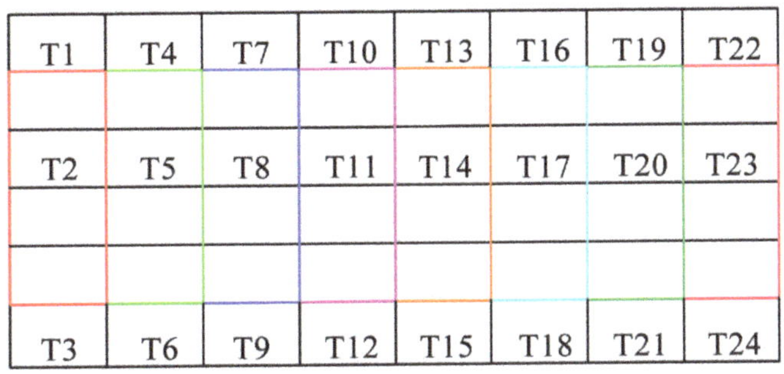

Figure 20. The observation area of the deck.

To create and define the boundary conditions in the load module, the boundary conditions of the platform column are defined as fully fixed, and the initial temperature is set as 20 °C. When the heat transfer analysis is completed, it is necessary to add the mechanical boundary conditions, and the weight of the mechanical equipment on the deck is applied to the upper deck in an equivalent gravity load. After referring to the specific platform weight distribution and calculation, 8 times the equivalent gravity load is applied to the weight of the equipment, the temperature of the ABAQUS is then used as the predefined field of structural analysis, and the ABAQUS model is used to calculate the stress and displacement of the steel frame structure.

Due to using the shell element, the grid partition is quadrilateral mesh; as it is a sequential coupling analysis, the element unit type of heat transfer analysis is not the same as that of the structural analysis. Among them, the element unit type used in heat transfer analysis is Heat Transfer and the element unit type used in structural analysis is Shell.

4.2.3. Structural Response Analysis of Steel Frame Structures

Here, the structural response of the structural frame is studied when the oil pool size is 2 m $\times$ 2 m. In order to more accurately reflect the integrity of the pool fire combustion process for frame structure strength of offshore platform steel, the total simulation time is set to be 20,000 s so that it can make sure the transfer process is thorough and closer to the actual situation.

Figure 21 shows the stress and deformation of the structure at different times. As can be seen from Figure 21, the stress and deformation of the structure increase with the time of the fire; the region with the largest deformation after heating occurs in the middle T8 and T9 area, which is the first coverage area of the pool fire combustion. The maximum deformation occurs at the bottom of the third beam. In addition, the steel frame structure begins to suffer an obvious collapse phenomenon after the fire has been burning for 8000 s; the steel in the middle part of the upper deck is broken and the other transverse members extend to both sides. During the fire, the mechanical behavior of the steel is nonlinear. Although the steel itself will not burn, its properties will change greatly at high temperatures. When the temperature is 400, the yield strength will drop to the average at room temperature. When the temperature is 600 °C, the yield strength and elastic modulus drop to one-fifth of the room temperature, and the bearing capacity has been basically lost. According to Figure 11, the temperature of the platform in the fire can reach 600–1000 °C, and the bearing capacity of steel greatly changes. During the whole combustion, the stress of steel changes and redistributes all the time. The joint position is the weak link of the steel frame in fire resistance, especially the stress concentration at the root of the beam flange at the junction of the beam and column, which has a great impact on the fire resistance bearing capacity of the beam.

The maximum deformation can be found in the middle parts of the upper deck, which are the weakest ones in the structure. The results show that the deformation at different times after heating can be increased with the increase in time, and the maximum deformation is the area covered by the pool fire. Other parts will have different levels of deformation, just as shown in Figure 22.

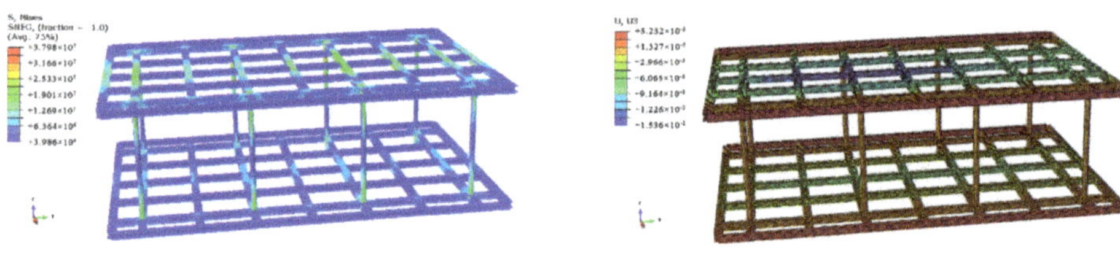

(a)

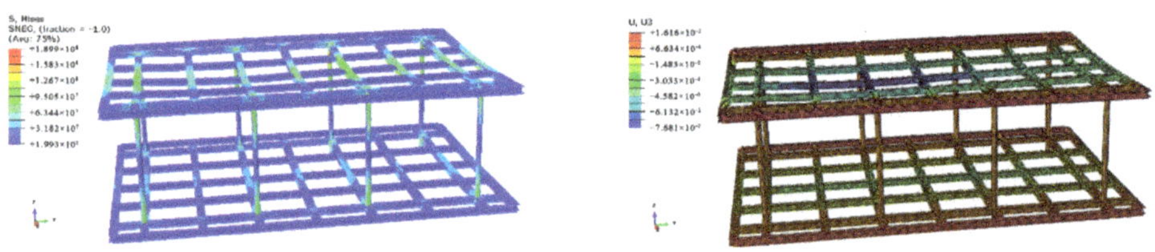

(b)

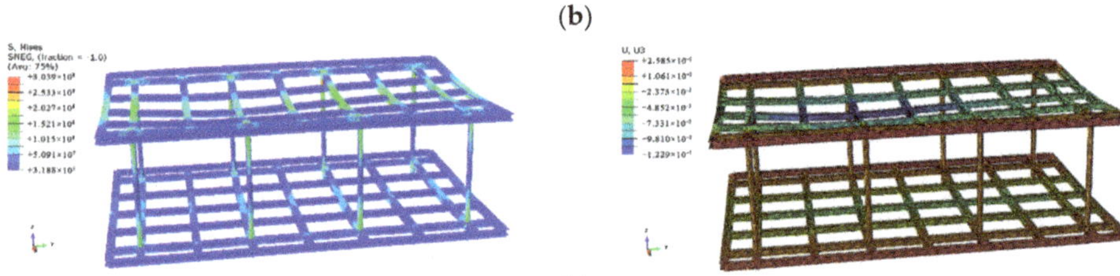

(c)

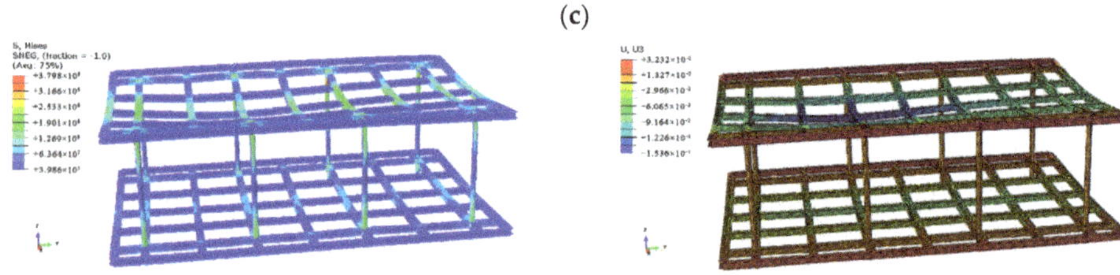

(d)

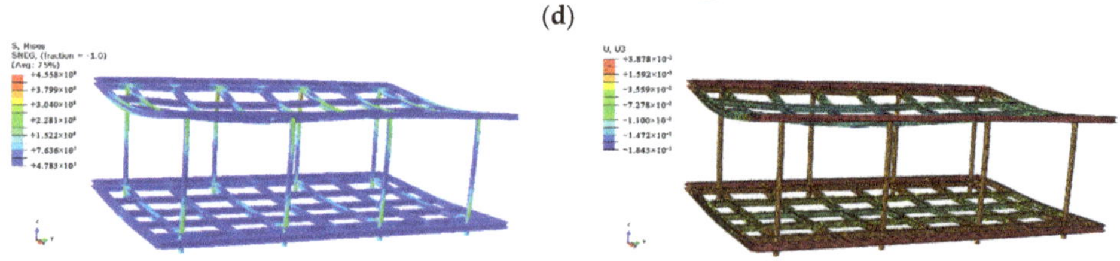

(e)

Figure 21. *Cont.*

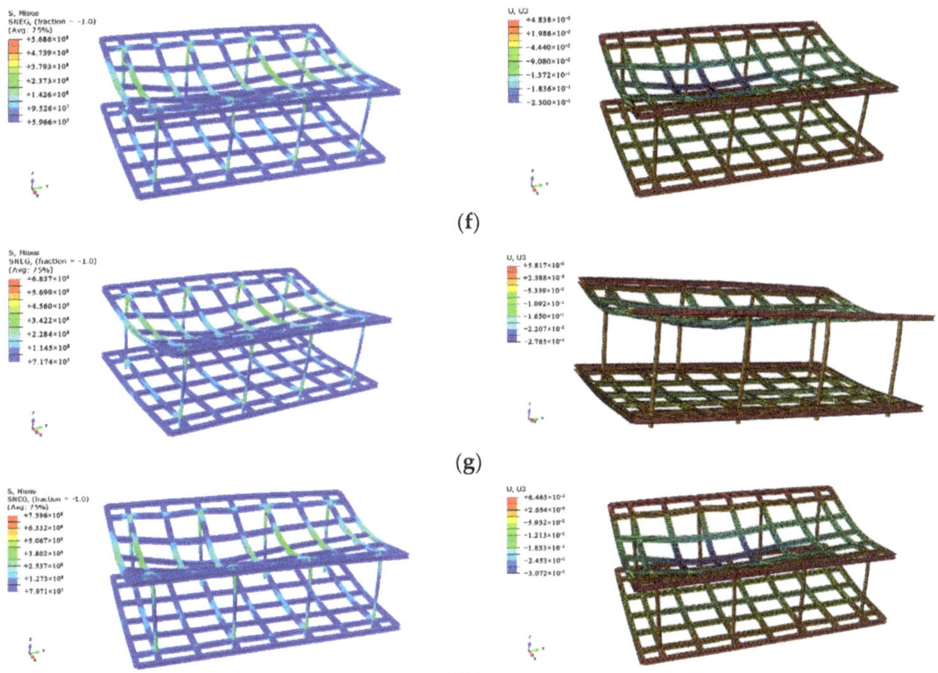

Figure 21. Stress and deformation at different time: (**a**) 1000 s; (**b**) 5000 s; (**c**) 8000 s; (**d**) 10,000 s; (**e**) 12,000 s; (**f**) 15,000 s; (**g**) 18,000 s; (**h**) 20,000 s.

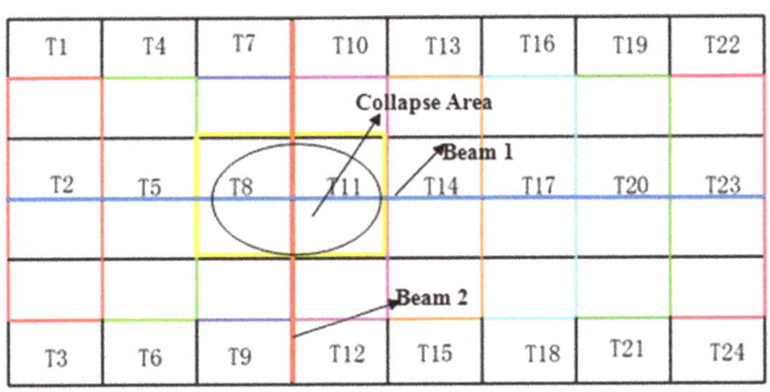

Figure 22. Model in ABAQUS.

5. Conclusions

In this paper, the structural response analysis of the offshore platform structure is made using the finite element method, and a risk assessment method based on the quantitative analysis of a pool fire caused by liquid leakage is proposed. This study demonstrates the following:

(1) A pool fire is a common type of offshore platform fire and the consequences are incalculable. Under different wind speeds, when the wind speed is lower than 1.5 m/s, with the increase in wind speed steel structure temperature increases; but when the wind speed is 3 m/s, the temperature of the steel structure is lower than that of 1.5 m/s and 0 m/s. Although different wind speeds can lead to different temperatures of the same part and the difference of the highest temperature of the whole steel structure, the temperature change trend of a steel structure caused by a pool fire is basically the same. Under the

action of the outside wind, the amount of heat moves along with the wind direction, which leads to higher thermal radiation in the downwind direction and an increase in the risk for the downwind area.

(2) When the pool fire occurs, the temperature of the steel members of the platform gradually increases under the action of the fire. When it has been under the action of the fire for 1000 s, although the steel structure has serious buckling, the integrity of the platform is maintained well. When it is 5000 s, the steel members at the top of the pool fire flame will have a large deformation; with the increase in the fire time, the deformation of the platform becomes more serious, causing platform structure failure and overturning. The FPSO has complete fire protection systems, such as a fire water system, foam system, water sprinkler system, and lifeboat. The aim of these safety systems and the protection of structural elements subject to large fires ensures enough time to evacuate people before any critical structural collapse. According to the calculation results, unless the structure collapses significantly in a short time, 1000 s is enough time to ensure the evacuation of personnel to a safe area.

(3) FDS is based on the solution of the linear grid. The accuracy of the simulation calculation of rectangular objects is very high, so in the establishment of the rules of the offshore platform model, the calculation accuracy will be high. However, the FDS software is not a universal simulation software of fire, the modeling and meshing are not flexible, and the solution is simple. In this study, the mechanical calculation data of the structural response were obtained after the completion of the fire simulation. The fire simulation and structural response simulation were conducted separately. This caused some delay. In future steps, we hope to couple a fire simulation and structural response simulation, so that the process of structural failure and deformation occurs at the same time as a fire occurs. This requires some linkage of the software.

Author Contributions: Conceptualization, D.Z. and B.Z.; methodology, K.Z.; software, D.Z.; validation, D.Z. and K.Z.; investigation, D.Z.; resources, D.Z.; data curation, D.Z.; writing—original draft preparation, D.Z.; writing—review and editing, B.Z.; visualization, B.Z. All authors have read and agreed to the published version of the manuscript.

Funding: This research was funded by the Program for Scientific Research Start-up Funds of Guangdong Ocean University, grant number 060302072101, and the Zhanjiang Marine Youth Talent Project-Comparative Study and Optimization of Horizontal Lifting of Subsea Pipeline, grant number 2021E5011.

Data Availability Statement: Not applicable.

Conflicts of Interest: The authors declare no conflict of interest.

References

1. Mather, A. *FPSO Handbook*; Witherby Seamanship International Ltd.: Livingston, UK, 2009.
2. Duggal, A.; Minnebo, J. The Floating Production, Storage and Offloading system–past, present and future. In Proceedings of the Offshore Technology Conference, Houston, TX, USA, 4 May 2020.
3. Shimamura, Y. FPSO/FSO: State of the art. *J. Mar. Sci. Tech.-Jpn.* **2002**, *7*, 59–70. [CrossRef]
4. Khakzad, N.; Reniers, G. Safety of Offshore Topside Processing Facilities: The Era of FPSOs and FLNGs. In *Methods in Chemical Process Safety*; Elsevier: Amsterdam, The Netherlands, 2018; Volume 2, pp. 269–287.
5. Vinnem, J.; Røed, W. *Offshore Risk Assessment Vol. 2 Principles, Modelling and Applications of QRA Studies*; Springer: Berlin, Germany, 2014; pp. 154–196.
6. Buaprommart, K.; Mahgerefteh, H.; Martynov, S.; Striolo, A. Shale gas well blowout fire and explosion modelling. *Appl. Therm. Eng.* **2019**, *149*, 1061–1068. [CrossRef]
7. Wilhoit, L.; Supan, C. 2010 worldwide survey of floating production, storage and offloading (FPSO) units. *Offshore* **2010**, *8*, 1–16.
8. HSE, O.A. *Incident Statistics Reports, OTO96*; 954; Health and Safety Executive: London, UK, 1996.
9. Anderberg, Y. Computer Methods for Optimising Passive Fire Protection on Offshore Platforms. In Proceedings of the International Conference on Offshore Structural Design-Hazards, Safety and Engineering on November, Jakarta, Indonesia, 16–20 January 1994; pp. 15–16.
10. Rubert, A.; Schaumann, P. Structural steel and plane frame assemblies under fire action. *Fire Saf. J.* **1986**, *10*, 173–184. [CrossRef]
11. Walker, S.; Hu, C.; Williams, J.R. Assessment of the Blast Resistance of Offshore Topsides Structures. In Proceedings of the Offshore Technology Conference, Houston, TX, USA, 6–9 May 1991.

12. Tolloczko, J. Blast and Fire: The Engineering Background. In Proceedings of the Offshore Technology Conference, Houston, TX, USA, 6–9 May 1991.
13. van den Berg, A.C.; van Wingerden, C. Numerical simulation of offshore gas explosions. In Proceedings of the Offshore Technology Conference, Houston, TX, USA, 6–9 May 1991.
14. Shetty, N.K.; Soares, C.G.; Thoft-Christensen, P.; Jensen, F.M. Fire safety assessment and optimal design of passive fire protection for offshore structures. *Reliab. Eng. Syst. Safe.* **1998**, *61*, 139–149. [CrossRef]
15. Suardin, J.A.; McPhate, A.J., Jr.; Sipkema, A.; Childs, M.; Mannan, M.S. Fire and explosion assessment on oil and gas floating production storage offloading (FPSO): An effective screening and comparison tool. *Process. Saf. Environ.* **2009**, *87*, 147–160. [CrossRef]
16. Kim, B.J.; Kwan Seo, J.; Hyo Park, J.; Sung Jeong, J.; Keun Oh, B.; Hoon Kim, S.; Hee Park, C.; Kee Paik, J. Load characteristics of steel and concrete tubular members under jet fire: An experimental and numerical study. *Ocean. Eng.* **2010**, *37*, 1159–1168. [CrossRef]
17. Paik, J.K.; Czujko, J.; Kim, B.J.; Seo, J.K.; Ryu, H.S.; Ha, Y.C.; Janiszewski, P.; Musial, B. Quantitative assessment of hydrocarbon explosion and fire risks in offshore installations. *Mar. Struct.* **2011**, *24*, 73–96. [CrossRef]
18. Kim, B.J.; Yoon, J.Y.; Yu, G.C.; Ryu, H.S.; Ha, Y.C.; Paik, J.K. Heat flow analysis of an FPSO topside model with wind effect taken into account: A wind-tunnel test and CFD simulation. *Ocean Eng.* **2011**, *38*, 1130–1140. [CrossRef]
19. Paik, J.K.; Kim, B.J.; Jeong, J.S.; Kim, S.H.; Jang, Y.S.; Kim, G.S.; Woo, J.H.; Kim, Y.S.; Chun, M.J.; Shin, Y.S.; et al. CFD simulations of gas explosion and fire actions. *Ships Offshore Struc.* **2010**, *5*, 3–12. [CrossRef]
20. Luketa, A.; Blanchat, T. The phoenix series large-scale methane gas burner experiments and liquid methane pool fires experiments on water. *Combust. Flame* **2015**, *162*, 4497–4513. [CrossRef]
21. Vasanth, S.; Tauseef, S.M.; Abbasi, T.; Abbasi, S.A. CFD simulation of pool fires situated at differing elevation. *Process. Saf. Environ.* **2015**, *94*, 89–95. [CrossRef]
22. Sun, L.; Yan, H.; Liu, S.; Bai, Y. Load characteristics in process modules of offshore platforms under jet fire: The numerical study. *J. Loss Prevent. Proc.* **2017**, *47*, 29–40. [CrossRef]
23. Betteridge, S. Modelling large LNG pool fires on water. *J. Loss Prevent. Proc.* **2018**, *56*, 46–56. [CrossRef]
24. Wu, B.; Zong, L.; Yip, T.L.; Wang, Y. A probabilistic model for fatality estimation of ship fire accidents. *Ocean Eng.* **2018**, *170*, 266–275. [CrossRef]
25. Yi, H.; Feng, Y.; Wang, Q. Computational fluid dynamics (CFD) study of heat radiation from large liquefied petroleum gas (LPG) pool fires. *J. Loss Prevent. Proc.* **2019**, *61*, 262–274. [CrossRef]
26. Ahmadi, O.; Mortazavi, S.B.; Pasdarshahri, H.; Mohabadi, H.A. Consequence analysis of large-scale pool fire in oil storage terminal based on computational fluid dynamic (CFD). *Process. Saf. Environ.* **2019**, *123*, 379–389. [CrossRef]
27. Xu, Y.; Bai, Y.; Paik, J.K.; Dai, W. An improved method for quantitative risk assessment of unconfined offshore installations subjected to gas explosions. *Structures* **2020**, *25*, 566–577. [CrossRef]
28. Bhardwaj, U.; Teixeira, A.P.; Guedes Soares, C.; Ariffin, A.K.; Singh, S.S. Evidence based risk analysis of fire and explosion accident scenarios in FPSOs. *Reliab. Eng. Syst. Safe.* **2021**, *215*, 107904. [CrossRef]
29. Li, X.; Khan, F.; Yang, M.; Chen, C.; Chen, G. Risk assessment of offshore fire accidents caused by subsea gas release. *Appl. Ocean Res.* **2021**, *115*, 102828. [CrossRef]
30. Meng, X.; Chen, G.; Zhu, G.; Zhu, Y. Dynamic quantitative risk assessment of accidents induced by leakage on offshore platforms using DEMATEL-BN. *Int. J. Nav. Arch. Ocean.* **2019**, *11*, 22–32. [CrossRef]
31. Wang, Y.; Wang, K.; Wang, T.; Li, X.Y.; Khan, F.; Yang, Z.; Wang, J. Reliabilities analysis of evacuation on offshore platforms: A dynamic Bayesian Network model. *Process. Saf. Environ.* **2021**, *150*, 179–193. [CrossRef]
32. Shebeko, Y.N.; Smolin, I.M.; Korolchenko, A.Y.; Shevchuk, A.P.; Borodkin, A.N.; Malkin, V.L.; Simonov, O.A.; Gurinovich, L.V.; Popov, S.A.; Kolosov, V.A. Some aspects of fire and explosion hazards of large LPG storage vessels. *J. Loss Prevent. Proc.* **1995**, *8*, 163–168. [CrossRef]

buildings

MDPI

Article

The Effect of Relative Humidity Dependent Thermal Conductivity on Building Insulation Layer Thickness Optimization

Wen Yang [1,2,*], Guanjie Zhang [2], Wenfang He [1,2] and Jiaping Liu [1,2]

1 State Key Laboratory of Green Building in Western China, Xi'an University of Architecture and Technology, No. 13 Yanta Road, Xi'an 710055, China
2 School of Architecture, Xi'an University of Architecture and Technology, No. 13 Yanta Road, Xi'an 710055, China
* Correspondence: yangwen@xauat.edu.cn; Tel.: +86-187-10-93-5688

Abstract: Optimization of insulation layer thickness is a significant factor in energy-efficient building design. Accurate determination of the thickness of the insulation layer will contribute to building energy conservation. In this study, ten typical cities in five thermal zones were selected, and the external thermal insulation of a typical residential building was taken as the research object. Using the degree day method and the economic model of full life cycle cost analysis, the optimal thickness of seven kinds of building insulation materials under absolute dry conditions, the lowest humidity and the highest humidity of the monthly average of the annual daily average were obtained. In addition, the carbon emission, energy saving and recovery period of materials under different working conditions were further obtained through numerical calculation. The results show that the optimum thickness of seven building insulation materials in ten typical cities under three working conditions is 18.21–346.05 mm. Their carbon emission change rate is between −2.7% and 38.6%, energy saving change rate is between −0.4% and 18.4%, and the payback period growth is within 1.5 years. Among them, polystyrene foam is the material least affected by humidity. It is recommended to be the main building insulation material in high humidity areas.

Keywords: building insulation materials; thermal conductivity; relative humidity; optimum thickness; carbon emissions

Citation: Yang, W.; Zhang, G.; He, W.; Liu, J. The Effect of Relative Humidity Dependent Thermal Conductivity on Building Insulation Layer Thickness Optimization. *Buildings* **2022**, *12*, 1864. https://doi.org/10.3390/buildings12111864

Academic Editor: Baojie He

Received: 1 September 2022
Accepted: 31 October 2022
Published: 3 November 2022

1. Introduction

Choosing the right insulation materials for the building envelope and determining the right thickness of insulation are effective methods to achieve energy conservation. Determining the thickness of insulation layers is vital for understanding heat transfers in which thermal conductivity is a crucial factor [1–3]. ASTM standard C168–97 [4] defines the thermal conductivity (W/m·K) as the time rate of steady-state heat flow through the unit area of uniform material, which is caused by the unit temperature gradient perpendicular to the direction of unit area. However, a constant thermal conductivity value is usually substituted in building energy evaluation. The existing specifications in China consider the influence of temperature on the thermal conductivity of building insulation materials but do not mention the influence of humidity on building insulation materials. Lack of specifications can lead to poor thermal performance, thereby increasing building energy consumption and other problems [5].

In the laboratory, in the process of measuring the thermal conductivity of building insulation materials, the moisture content of the materials, as well as the temperature and humidity of the laboratory, will affect the measurement. Thermal insulation materials in buildings differ from their performance under laboratory conditions because they are subjected to a real-world climate during use. See Table 1 for details.

Table 1. Research status of building thermal insulation materials affected by ambient temperature and humidity.

Researcher	Research Contents	Research Results/Innovation
I. Budaiwi, A. Abdou. [6]	The impact of the k-value change of fibrous insulation materials (i.e., fiberglass) in a typical wall–roof system due to moisture content levels on the thermal and energy performance of a typical residential building under hot–humid climatic conditions is investigated.	Moisture performance is investigated utilizing theoretical longterm hygrothermal performance modeling and simulation techniques. Layer-and time-averaged levels of moisture content in the fibrous insulation are determined, and the corresponding k-value change is evaluated from measured relationships. The impact of the k-value change due to moisture on the building thermal load and cooling energy performance of a residential building is then assessed utilizing detailed building energy simulation software.
K.J. Kontoleon, C. Giarma. [7]	This paper investigates the impact of moisture content on the thermal inertia parameters of building material layers. Their consideration is essential to enhance the design of building elements, from a thermal point of view, when exposed to varying moisture content conditions.	Moisture content and relative humidity variations of each analysed layer, as defined by specific moisture storage functions, are shown to interrelate non-linearly with the layer resistor–capacitor circuit section parameters (thermal conductivity and volumetric heat capacity) showing notable consequences on the thermal inertia parameters.

Although the thermal characteristics of building thermal insulation materials in the real environment are different from those under laboratory conditions, the influence of temperature and humidity on the thermal conductivity measured under laboratory conditions is still the basis of this research. This paper conducts a literature survey on the influence of temperature and humidity on the thermal conductivity, as shown in Table 2.

Table 2. Research status of building thermal insulation materials affected by temperature and humidity.

Researcher	Research Contents	Research Results/Innovation
I. Budaiwi, A. Abdou, M. Al-Homoud. [8]	They revealed the relationship between the temperature and thermal conductivity of various locally produced insulating materials.	The impact of thermal conductivity variation with temperature on the envelope-induced cooling load for a theoretically modeled building is quantified and discussed.
M. Khoukhi, M. Tahat. [9]	This study is to investigate the relationship between the temperature and thermal conductivity of various densities of polystyrene, which is widely used as building insulation material in Oman.	The impact of thermal conductivity variation with temperature on the envelope-induced cooling load for a simple building model is discussed.
Hoseini, Atiyeh, Majid Bahrami. [10]	This work presents a comprehensive investigation of aerogel blankets thermal conductivity (k-value) in humid conditions at transient and steady state regimes. Transient plane source (TPS) tests revealed that the k-value of aerogel blankets can increase by up to approximately 15% as the ambient relative humidity (RH) increases from 0% to 90% at 25 °C.	This paper mechanisms affecting the k-value of aerogel blankets as a function of RH and T are investigated.
Alvey, Jedediah B, Jignesh Patel, Larry D, Stephenson. [11]	In this paper, the thermal conductivity of several commercially available insulating materials (three kinds of aerogel composite blankets, two kinds of extruded polystyrene foam (XPS) and one kind of foamed polyurethane foam (PUR)) is evaluated as affected by ambient temperature and humidity.	Results indicate that humidity levels play a significant role in PUR performance but not a significant role in XPS performance. The three aerogel composites have mixed results: one has little relationship between moisture content and thermal performance, one is strongly affected by moisture and the remaining is moderately affected by moisture.

Table 2. *Cont.*

Researcher	Research Contents	Research Results/Innovation
Nosrati, Roya Hamideh, Umberto Berardi. [12]	In this paper, the change of thermal conductivity of aerogel reinforced materials in the temperature range of −20 °C to +60 °C, and the relative humidity (RH) range of 0% to 95% is studied.	This study shows that compared to the standard testing condition, the maximum increase in the thermal conductivity was 100% under 95% RH, while the greatest temperature-driven increase in the thermal conductivity was 12% at the maximum tested temperature. The humidity-driven changes in the thermal conductivity of aerogel-based products are significantly greater than temperature-driven changes.

In addition, researchers try to better specify the optimal thickness of insulation for different types of materials when studying factors affecting thermal conductivity. See Table 3 for the research status of optimal thickness. Based on the above research on the status quo, almost all the current research on the optimal thickness ignore the influence of environmental humidity on the thermal conductivity. Thermal conductivity is based on the measured thermal conductivity of the material in the absolute dry state. This will cause the design value of thermal conductivity to be lower than the actual value, resulting in insufficient design of insulation thickness, affecting users' thermal comfort and increasing building energy consumption. Research on the influence of temperature and humidity on building thermal insulation materials shows that the humidity-driven change of thermal conductivity is significantly greater than the temperature-driven change. Therefore, in practical engineering applications, we should first consider the influence of humidity when selecting the thickness of building insulation materials, especially when building walls are exposed to extreme humidity for a long time, so as to better realize building energy conservation and reduce greenhouse gas emissions. Although in some areas with high humidity, the construction of buildings may consider the installation of moisture-proof and moisture-reducing components inside the envelope, in the production, transportation and installation of building materials, the humidity inside the building insulation materials will also reach a dynamic balance with the surrounding environment, so it is important to consider the impact of humidity on the building.

Therefore, the effects of humidity on the thermal conductivity of seven building insulation materials, including polystyrene foam (EPS), extruded polystyrene foam (XPS), polyurethane (PUR), rock wool, centrifugal wool, aerogel-enhanced hollow glass microspheres (HGM) and foamed cement, were obtained through experiments in this paper. Secondly, the heat transfer load of typical insulation structure is analyzed by the degree day method, and then the optimal insulation thickness is calculated using the full life cycle cost analysis (LCCA) economic model under different humidity conditions. Finally, the input–output method is used to accurately evaluate the carbon reduction after adopting the optimal thickness under the condition of environmental humidity. See Figure 1 for details. By analyzing the research results of this paper, building engineers will be able to rationally design and choose the thickness of the building insulation materials which is more accurate and fits the actual operation.

Table 3. Research status of optimum thickness.

Researcher	Research Contents	Research Results/Innovation
T.M.I. Mahlia, B.N. Taufiq, Ismail, H.H. Masjuki. [13]	This paper analyzes that the relationship between the thickness of building wall insulation materials.	The thermal conductivity is nonlinear and follows the polynomial function $x_{opt} = a + bk + ck^2$.

Table 3. *Cont.*

Researcher	Research Contents	Research Results/Innovation
K. Comakli, B.Yuksel [14]	They studied the optimal insulation thickness of foamed polystyrene in the coldest city in Turkey and concluded that when the optimal thickness is used, obvious energy-saving effect can be achieved for cities with large number of humid days.	They proved the energy-saving effect of the optimal thickness and demonstrated the influence of the number of days on the optimal thickness and the payback period of investment.
A. Ucar, F. Balo [15]	The optimum insulation thickness of the external wall for four various cities from four climate zones of Turkey, energy savings over a lifetime of 10 years and payback periods are calculated for the five different energy types and four different insulation materials.	They proposed the optimal thickness corresponding relationship for different types of energy and different types of building insulation materials, and adopted P1-P2 method as the calculation method of the optimal thickness.
Huakun Huang, Yijun Zhou, Renda Huang, Huijun Wu, Yongjun Sun, Gongsheng Huang, Tao Xu. [16]	Taking the typical subtropical humid climate office building as the model, they established a full life cycle assessment model to calculate the optimal economic thickness of the new aerogel super insulation material d and further evaluated the energy saving rate, economic benefits, greenhouse gas emissions, etc.	They compared the building energy-saving effect caused by the optimal thickness of the new super thermal insulation material and the traditional building thermal insulation material and compared the energy saving rate, economic benefits, greenhouse gas emissions, etc., when using different building thermal insulation materials.

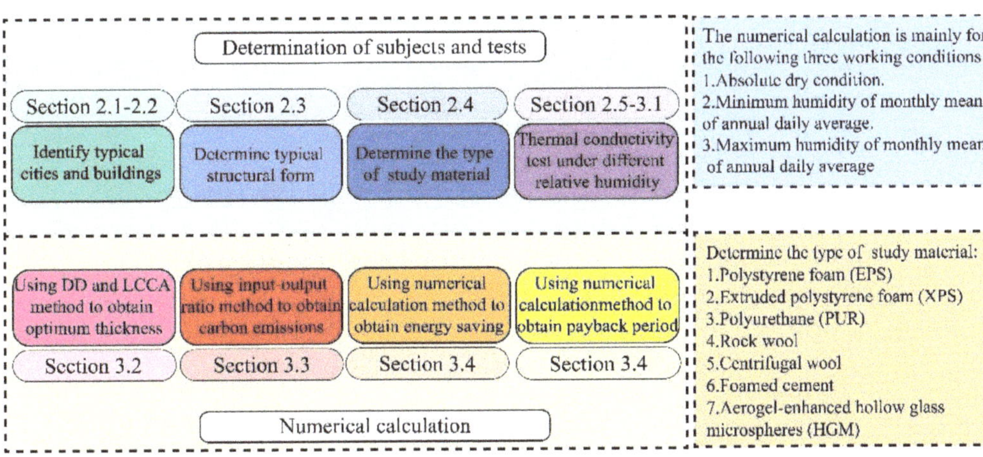

Figure 1. The research mind map.

2. Materials and Methodology

2.1. Typical Cities and Weather Data

In terms of land area, China covers about 1/15 of the world's landmass with a 9.73 million square kilometer territory. The vast land has resulted in the diversity of natural conditions, and various climates have been formed under the specific topographic and geomorphic characteristics. According to the climate characteristics, we divide China into severe cold area, cold area, hot summer and cold winter area, hot summer and warm winter area and mild area. In the study, according to the indicators in the code for thermal design of civil buildings [17], we selected ten typical cities with high relative humidity in five thermal zones as the research objects to study the optimal thickness change of thermal insulation materials under the influence of relative humidity and its environmental impact. Using the Xi'an University of Architecture and Technology's measured meteorological data from 1988 to 2017, we determined the cold and warm seasons of the different cities and calculated the equivalent temperature difference and other basic data. See Table 4 for specific data of the cities.

Table 4. Basic data of ten typical cities.

Thermal Zoning	Province	City	Maximum Humidity of Monthly Mean of Annual Daily Mean	Minimum Humidity of Monthly Mean of Annual Daily Mean	HDD (°C·d)	CDD (°C·d)	DD (°C·d)
Severe cold area	Heilongjiang	Tonghe	80.65%	59.65%	5701	2	3001.22
	Jilin	Erdao	81.10%	55.77%	5390	0	2836.84
Cold area	Shandong	Chengshantou	90.16%	62.84%	2698	2	1420.69
	Henan	Yucheng	82.26%	63.71%	2306	99	1247.82
Hot summer and cold winter area	Shaanxi	Hanzhong	83.16%	69.45%	1920	68	1033.97
	Hubei	Enshi	82.03%	72.84%	1541	98	844.85
Hot summer and warm winter area	Fujian	Xiamen	82%	65.32%	516	199	340.20
	Hainan	Haikou	83.81%	76.97%	95	403	188.97
Mild area	Guizhou	Zhijin	81.90%	73.87%	1762	1	927.71
	Yunnan	Longling	89.42%	72.87%	1285	0	676.32

2.2. The Typical Building Model

As the main body of climate, architecture is an important intermediate link in the process of analyzing the impact of climate on building materials. Therefore, a typical slab high-rise residential building in China was chosen and is modeled as a typical building in Table 5.

Table 5. Building model information.

Model	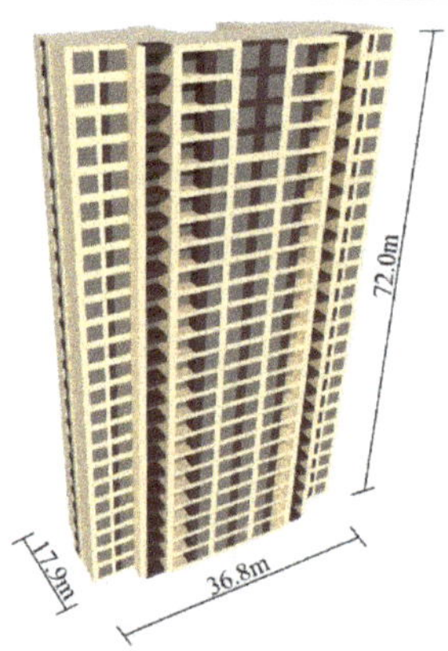		
Model information	Typical group	Control group A	Control group B
Number of layers	24th floor	16th floor	24th floor
Window wall ratio	0.7	0.5	0.7
External wall area of each floor	320 m^2	320 m^2	385 m^2

The typical building is a 24-story residential building with a length of 36.8 m, a width of 17.9 m and a height of 72 m. Each floor is composed of four apartment types and two elevators. The floor height of each floor is 3 m, and the building area is 530 m^2. The window wall ratio is 0.7, and the external wall area of each floor is 320 m^2. External walls are insulated using an external insulation structure. The roof adopts an inverted insulation form, and the floor between each floor does not have an insulation layer. The built-in partition wall is also separated by a light partition wall. In this study, another two control groups are set according to typical buildings to assess the environmental impact caused by the optimal thickness of the number of floors and the window wall ratio. The number of floors of control group A is set as 16, the window wall ratio of control group B is 0.5, and the external wall of each floor is 385 m^2.

2.3. Construction

In the whole building envelope, the area of the external wall accounts for more than half, and the wall heat transfer energy consumption can reach about one-third of the total building energy consumption [18]. Building envelopes use heat transfer energy depending on the insulation material they are constructed of and the thickness of their insulation layers. According to different components of thermal insulation materials, the external wall insulation system is generally divided into a self-insulation system and a composite wall insulation system in the existing wall insulation system. The wall structure composed of a kind of building material with low thermal conductivity is called a self-insulation system. It can realize energy saving through the characteristics of building materials. Depending on where the insulation layer is located, there are three types of composite wall insulation: external wall external insulation system, external wall internal insulation system and sandwich insulation [19]. Compared with the four insulation systems, the exterior wall external insulation has strong applicability, which can effectively solve the heat loss problem of the weak links such as the thermal bridge in the external enclosure components and achieve the building energy-saving effect without affecting the indoor decoration and living conditions. Therefore, we chose the exterior wall external insulation system as the research object.

According to the national standard atlas external wall external insulation building structure (10J121) [20], A-type pasted insulation board external insulation system was selected as a typical structure. A-type pasted insulation board external insulation system is only mentioned in the standard drawings as applicable to EPS, XPS and PUR. However, it was learned from reading other documents [21,22] that A-type adhesive external insulation system is also applicable to adhesive external insulation materials such as rock wool, centrifugal cotton, aerogel-enhanced HGM and foamed cement. The structure of type A-pasted insulation board external insulation system in the standard drawings has a facing layer and a bonding layer, but it does not play a role in thermal insulation in the actual project. For the convenience of calculation, we simplified the structure and determined the specific thickness of each layer. The specific structure and thickness are shown in Figure 2 and Table 6.

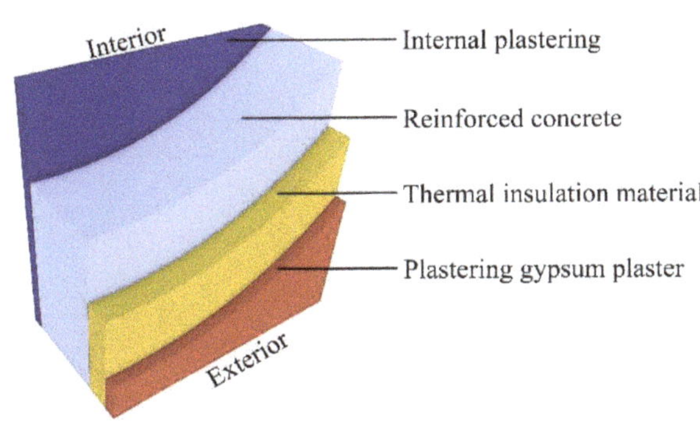

Figure 2. Simplified wall structure of A-type pasted insulation board external insulation system.

2.4. Test Materials

Building insulation materials can achieve low thermal conductivity in different forms, including loose filling, cotton wool or blanket, rigid form, in situ foaming or reflection form. Choosing the appropriate form and type of building insulation materials depends on the type of application and the physical, thermal and other properties of the materials required. However, traditional thermal insulation materials have become the preferred materials in many building and thermal energy storage applications because of their low thermal conductivity and low cost [23]. However, with the continuous research on building thermal insulation materials, researchers have recently proposed a new type of building super thermal insulation material-aerogel-enhanced HGM, which is also taken as the research object in this study [24].

Table 6. Basic data of each layer of materials after simplification of A-type external insulation system.

Material	Thickness (mm)	Thermal Conductivity (W/m·K)	Dry Density (kg/m^3)
Plastering gypsum plaster	4	9.44	1500
Thermal insulation material	δ_{im}	k_{im}	ρ_{im}
Reinforced concrete	300	1.74	2500
Internal plastering	25	0.81	1600

Thermal conductivity can be affected by temperature, water content and humidity according to existing literature and published studies [25]. However, few people have further demonstrated that the change of thermal conductivity caused by the change of relative humidity will have an impact on the optimal thickness of insulation layer and the environment caused by the change of optimal thickness. As a result, we chose XPS, EPS, PUR, rock wool, centrifugal cotton, aerogel-enhanced HGM and foamed cement as the research objects for analyzing.

2.5. Instrument

Thermal conductivity of building insulation materials must be measured through experiments in order to be accurate. Due to the different applicable materials, test accuracy and test range of the measurement method, it is very important to select the appropriate test method for accurately obtaining the required measurement data. We selected the test method of constant temperature and humidity chamber and transient plane source method. The constant temperature and humidity chamber was used to provide stable working conditions for the experiment. Transient plane source is responsible for accurate determination of thermal conductivity. According to the transient plane source method, we selected TPS2200, and its basic test parameters are shown in Table 7. The TPS test principle is based on the thermal field generated by the plane circular heat source in the theoretically infinite medium to measure the instantaneous temperature rise of the tested piece. The plane probe designed by this technology is a metal nickel sheet with double helix structure after etching, and is wrapped with a Kapton or mica insulation layer about 0.025 mm thick to measure the change of thermal field. During the test, the heating time and heating power were set by the monitoring software and executed by the test equipment, as shown in Figure 3.

Table 7. TPS2200 Basic test parameters.

Apparatus	Model	Test Temp Rage	Test Sample Requirements	Manufacturer	Range	Sample Type
Hot Disk Thermal constant analyzer	TPS2200	RT °C	The min thickness is 2 mm, the min diameter is 10 mm, and the max size is not limited	Hot Disk AB	Thermal Conductivity: 0.01–500 W/m·K Thermal Diffusivity: 0.01~300 mm^2/s Specific Heat: 0.01~5 MJ/m^3·K	Solid, liquid, powder, paste, foam etc.

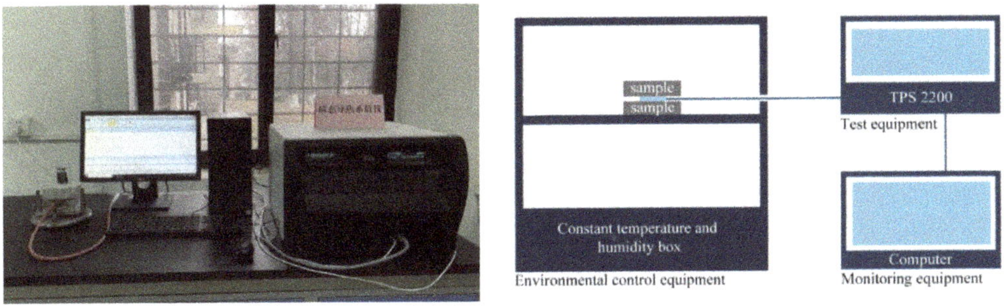

(a) (b)

Figure 3. TPS2200 test instrument: (**a**) physical drawing; (**b**) schematic diagram.

3. Results and Discussion

3.1. Results of the Effect on Thermal Conductivity by Humidity

The thermal conductivity of building insulation materials is significantly affected by environmental humidity [7,26]. A transient plane source and a constant temperature and humidity chamber were used to measure the thermal conductivity under different humidity conditions in this study. Finally, we obtain the functional relationship between the thermal conductivity and relative humidity of the seven insulation materials.

3.1.1. Experimental Process

A perfect experimental process will make the experimental results more accurate. After the research on the standards and previous papers, we determined the experimental process as follows:

1. Test piece drying stage: Place the prepared test pieces in the drying oven. For materials that may undergo chemical changes or irreversible structural damage, 70 °C is used as the drying temperature. For materials that will not undergo chemical changes or irreversible structural damage, 105 °C is used as the drying temperature [27]. When measuring the mass change of the test piece over time at an interval of 24 h, it should not exceed 1% of its own mass three times in a row. The test piece is considered to have reached the constant weight.
2. Sealing and cooling stage of test piece: Take the test piece out of the drying oven, wrap it with plastic film and place it indoors until it is cooled to room temperature.
3. Test phase: After the sample is cooled, remove its plastic film and quickly measure its actual size and mass. Set the temperature of the constant temperature and humidity box to 25 °C and set up the humidity at 0%, 30%, 50%, 70%, 85% and 98% in succession. In the case of three consecutive weight measurements taken at intervals of 24 h, if the mass change of the test piece does not exceed 1%, the test piece has reached its constant weight. At this time, as the test pieces are placed up and down in pairs, the metal probe is placed in the middle of each and fixed, and then the thermal conductivity value is measured.
4. Test repetition stage: Three groups of tests should be repeated for each insulation material a total of five times.

3.1.2. Experimental Data and Analysis

The thermal conductivity of building insulation materials was measured under the condition that the relative humidity was also changing gradually. Finally, the functional relationship between thermal conductivity and relative humidity was measured, as shown in Figure 3 and Table 8.

Table 8. Functional relationship between thermal conductivity and relative humidity of different materials.

Material	Fitting Formula
EPS	$y = 1.5 \times 10^{-8}x^3 - 1.57 \times 10^{-6}x^2 + 5.66 \times 10^{-5}x + 0.02724$
XPS	$y = 8.58 \times 10^{-9}x^3 - 1.24 \times 10^{-6}x^2 + 6.76 \times 10^{-5}x + 0.02963$
Centrifugal cotton	$y = 3.01 \times 10^{-7}x^3 - 3.42 \times 10^{-5}x^2 + 9.54 \times 10^{-4}x + 0.02971$
PUR	$y = 2.93 \times 10^{-8}x^3 - 4.43 \times 10^{-6}x^2 + 2.21 \times 10^{-8}x + 0.0445$
Rock wool	$y = 2.67 \times 10^{-7}x^3 - 3 \times 10^{-5}x^2 + 8.48 \times 10^{-4}x + 0.05062$
Aerogel-enhanced HGM	$y = 8.05 \times 10^{-8}x^3 - 1.41 \times 10^{-5}x^2 + 8.71 \times 10^{-4}x + 0.04789$
Foam cement	$y = 2.16 \times 10^{-7}x^3 - 2.59 \times 10^{-5}x^2 + 0.00108x + 0.07134$

Figure 4 and Table 8 show the change of thermal conductivity of building insulation materials with the change of relative humidity. The fitting formula in Table 4 was used to accurately determine the corresponding thermal conductivity under different humidity. There was an almost linear relationship between thermal conductivity and relative humidity for EPS, XPS, PUR and aerogel-enhanced HGM. In environments with a relative humidity over 98%, changes in thermal conductivity of EPS, XPS and PUR are 16.8%, 9.4%, 13.3% and 52.1%, respectively. As the relative humidity began to rise, when it reached 30%, the thermal conductivity of foamed cement began to increase with the increase of relative humidity. Foamed cement undergoes a 16.7% change in thermal conductivity when relative humidity is 30%. Th thermal conductivity of foamed cement increases linearly with changing relative humidity when the relative humidity is increasing. In the presence of relative humidity above 70%, the thermal conductivity of foamed cement will rise sharply. When the relative humidity reaches 98%, the change of thermal conductivity is 85.4%. Thermal conductivity varies linearly and slowly with relative humidity before it reaches 70% for centrifugal cotton and rock wool. The thermal conductivity of centrifugal cotton and rock wool begins to increase sharply when the humidity reaches 70%, compared to the standard thermal conductivity under dry conditions. At 98% relative humidity, the variation in thermal conductivity is 167.9% and 95.3% respectively.

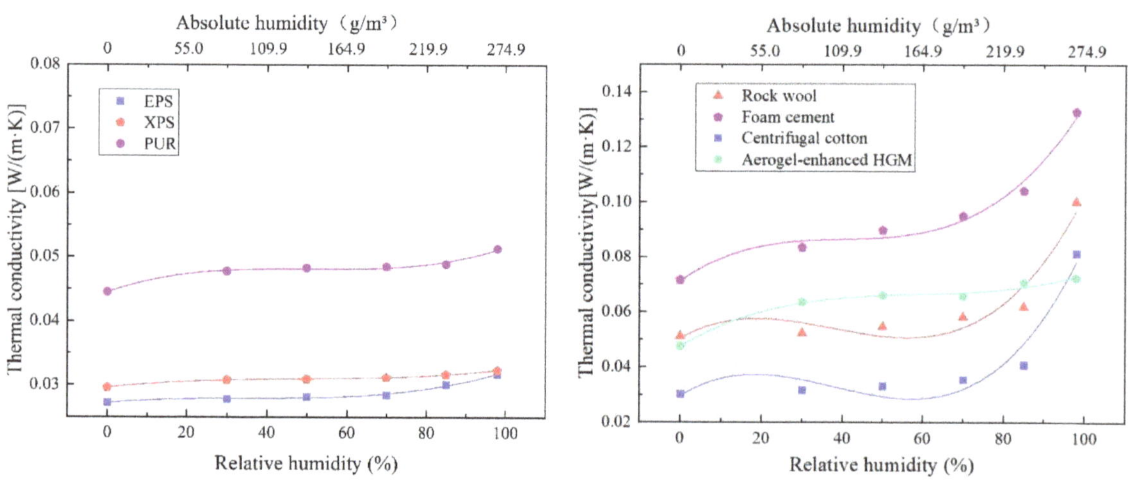

Figure 4. Functional relationship between thermal conductivity and relative humidity of different materials.

Thermal insulation materials will exhibit a great deal of difference in thermal conductivity based on the pore structure and porosity of its pores. EPS, XPS and other materials all have many small holes that are completely sealed, and XPS structures achieve closure rates of more than 99%. Such pore structure will make the wet air not shuttle well in the material but can only stay in the closed pores on the surface, so that the thermal conductivity of a substance will not increase rapidly with changes in relative humidity. In contrast, foamed cement and other cement-based materials have strong water absorption capacity, and there are interconnected holes in the structure. Because the material itself has an adsorption capacity, the thermal resistance between solids decreases as relative humidity increases. A high relative humidity leads to a saturated state and liquid water begins to appear, which makes the thermal conductivity of the material increase sharply. Aerogel-reinforced materials are concrete materials before aerogel modification, and the concrete itself has interconnected cavities, so its thermal conductivity increases rapidly when the relative humidity just starts to increase. However, with the increase of thermal conductivity, its thermal conductivity does not increase significantly. This is because the aerogel itself has a strong hydrophobicity, so that its internal moisture does not increase much, resulting in such a result.

3.2. Results of the Optimum Thickness

In this study, the degree day method was used to calculate the heat transfer load of the steady-state enclosure structure, and then the LCCA was used to calculate its economic cost to solve the optimal thickness. For six different building insulation materials, the corresponding optimal insulation thickness of ten typical cities in China under three conditions was obtained, analyzed and compared. The relevant parameters are given in Tables 9 and 10.

Table 9. Performance of thermal insulation materials.

Material	Dry Density (kg/m^3)	Price (\$/m^3)
EPS	16	56.51
XPS	4	117.74
Centrifugal cotton	23	75.35
Rock wool	200	91.05
Foam cement	374	59.65
Aerogel-enhanced HGM	260	493.88
PUR	40	204.08

Table 10. Parameters used in calculation.

Parameter	Value
CDD and HDD	Table 4
Related parameters in construction	Table 6
k_{im}	Figure 4 and Table 7
C_{im}	Table 9
φ	0.90%
Φ	3.70%
N	20 years
C_f	0.1275
μ_f	3.6×10^6
η_{hs}	0.99

3.2.1. Numerical Calculation Process

The heat loss of a building refers to the heat transfer and heat consumption through the enclosure structure and the air infiltration and air conditioning through the gaps of doors and windows. As the air infiltration through the gaps of doors and windows and the heat consumption of air conditioning are determined by the sealing performance and thermal insulation performance of the external windows, we will not consider it again. The insulation on the walls of the enclosure structure primarily determines the heat transfer and the heat consumption. Insulating a wall primarily involves choosing the right insulation material and setting the thickness of the insulation sufficiently. The focus of this study is to consider the heat loss of the wall to optimize the thickness of the insulation layer.

The heat loss of the building envelope accounts for more than half of the total energy consumption of the building. According to the report of Hasan [28], the heat loss Q (W/m^2) per unit area of the enclosure structure can be obtained by the following formula:

$$Q = U(T_i - T_{md}) \tag{1}$$

where U (W/(m^2·K)) is the total heat transfer coefficient of the external wall, T_i (°C) is the constant indoor comfortable temperature, and T_{md} (°C) is the average daily temperature. The annual heat loss per unit area of external wall Q_{an} (W/m^2) can be determined by the following formula:

$$Q_{an} = 86400UDD \tag{2}$$

where degree day DD (°C·d) is the function of heating degree day, air conditioning degree day and energy efficiency of heating and cooling system, which can be calculated by the following equation:

$$DD = \frac{CDD}{COP} + \frac{HDD}{\eta} \tag{3}$$

Annual energy demand E_{an} (J/m^2-year) can be calculated by dividing the annual heat loss by the efficiency of the heating system η_{hs} to determine, as follows:

$$E_{an} = \frac{Q_{an}}{\eta_{hs}} = \frac{86400UDD}{\eta_{hs}} \tag{4}$$

The total heat transfer coefficient U of the exterior wall structure is obtained from the reciprocal of the sum of the thermal resistances of each layer of the enclosure structure:

$$U = \frac{1}{R_{iaf} + R_{oaf} + R_w + R_{im}} \tag{5}$$

where R_{iaf} ((m^2·K)/W) and R_{oaf} ((m^2 K)/W) is the thermal resistance of inner and outer air layers, R_w ((m^2·K)/W) is the total thermal resistance of all structural layers in the external wall except the insulation layer, and R_{im} ((m^2·K)/W) is the thermal resistance of the insulation layer. As the thermal resistance of other parts of the enclosure except for the insulation layer is very small and subject to little environmental impact, the impact of environmental temperature and humidity on other parts of the enclosure is not considered. R_{iaf} and R_{oaf} have not been considered in some studies of optimal thickness [29]. In this study, these resistance terms are considered, which is closer to the actual use. R_{im} can be expressed as:

$$R_{im} = \frac{\delta_{im}}{k_{im}} \tag{6}$$

where δ_{im} (m) and k_{im} (W/(m·K)) represent the insulation thickness and thermal conductivity of insulation materials, respectively. The annual energy demand can be calculated as follows:

$$E_{an} = \frac{86400DD}{(R_{iaf} + R_{oaf} + R_W + \frac{\delta_{im}}{k_{im}})\eta_{hs}} \tag{7}$$

Annual fuel consumption m_{af} (kg/m^2) can be calculated by dividing Equation (7) by μ_f:

$$m_{af} = \frac{86400DD}{(R_{iaf} + R_{oaf} + R_W + \frac{\delta_{im}}{k_{im}})\mu_f\eta_{hs}} \tag{8}$$

where μ_f is the lower calorific value of a given fuel, usually in J/kg, J/m^3, or J/kWh, depending on the fuel type.

In the process of determining the optimal thickness through financial analysis, there are generally two methods: investment recovery method and LCCA. The investment recovery method is based on the time required to repay the initial investment and the cost of energy saving in the operation stage brought by the investment. The disadvantage of this simple analysis method is that it does not consider currency inflation, which is also a very important financial consideration [30]. However, insulation thickness of building exterior walls can also be optimized using LCCA [31]. Therefore, we used an LCCA analysis method in this study.

Annual cost per unit area C_{an} ($) can be obtained from the following formula:

$$C_{an} = \frac{86400DD}{(R_{iaf} + R_{oaf} + R_W + \frac{\delta_{im}}{k_{im}})\mu_f \eta_{hs}} C_f \tag{9}$$

where C_f is the fuel cost in $/kWh. The total cost of the whole life cycle is calculated through LCCA and converted to present value by multiplying the total cost by the present value factor (*PWF*), where the service life of the building is *N* (year). *PWF* is the inflation rate (φ) and interest rate (Φ) function of.

$$PWF = (\frac{1-\varphi}{1+\phi})\ [1 - (\frac{1+\varphi}{1+\phi})^N\ ]\ \text{If } (\phi \neq \varphi) \tag{10}$$

$$PWF = \frac{N}{1+\phi} = \frac{N}{1+\varphi}\ \text{If } (\phi = \varphi) \tag{11}$$

Building insulation material cost C_{in} ($) is the unit price of the material multiplied by the thickness as follows:

$$C_{in} = C_{im}\delta_{im} \tag{12}$$

where C_{im} ($\$/m^3$) is the unit price of insulation material; δ_{im} (m) is the thickness of the insulation material. Total cost of building heat loss C_{tot} ($) should be the heating cost plus the cooling cost as follows:

$$C_{tot} = \frac{86400C_f PWFDD}{(R_{iaf} + R_{oaf} + R_W + \frac{\delta_{im}}{k_{im}})\mu_f \eta_{hs}} + C_{im}\delta_{in} \tag{13}$$

δ_{im} is obtained by derivation of Equation (13), and then the optimal thickness of the insulation material is as follows:

$$\frac{dC_{tot}}{d\delta_{im}} = \frac{d}{d\delta_{im}}[\frac{86400C_f PWFDD}{(R_{iaf} + R_{oaf} + R_W + \frac{\delta_{im}}{k_{im}})\mu_f \eta_{hs}} + C_{im}\delta_{im}] = 0 \tag{14}$$

$$\delta_{im} = 293.938\sqrt{\frac{DDC_f PWFk_{im}}{C_{im}\mu_f \eta_{hs}}} - (R_{iaf} + R_{oaf} + R_w)k_{im} \tag{15}$$

3.2.2. Data and Analysis

In the thermal insulation design of the building's peripheral protective structure, there is no need for excessive thermal insulation, which is planned to increase the initial investment cost in exchange for the reduction of energy cost, and there can be no lack of thermal insulation, which is to exchange the increase of energy cost for the reduction of initial investment cost. Therefore, it is vital to determine the optimal insulation thickness accurately in order to minimize the economic cost [32].

By solving the optimal thickness of seven kinds of insulation materials under three conditions, the change of their optimal thickness under the influence of humidity was obtained. The calculation was carried out successively for the selected ten different cities. See Table 11 for the specific values in detail.

The optimal thickness of building thermal insulation materials for a specific city increases with increasing humidity, as shown in Table 11. This is because the heat consumption of buildings are certain, and the thermal conductivity of materials increases with the increase of humidity. If you want to achieve a comfortable indoor thermal environment, you can enhance the thermal insulation performance of the enclosure structure through the thickness of the insulation layer. Building insulation material thickness will be smaller if relative humidity is not taken into account, which will lead to a higher operating cost of the life cycle.

Table 11. Optimum insulation thickness of different insulation materials in ten typical cities under three working conditions (mm).

Material	Working Condition	City									
		Severe Cold Area		Cold Area		Hot Summer and Cold Winter Area		Hot Summer and Warm Winter Area		Mild Area	
		Tonghe	Erdao	Chengshantou	Yucheng	Hanzhong	Enshi	Xiamen	Haikou	Zhijin	Longling
EPS	Dry	192.71	187.09	129.54	120.79	109.08	97.66	58.40	41.04	102.81	86.35
	Minimum humidity	195.97	189.93	131.91	123.05	111.57	100.20	59.45	42.10	105.60	88.56
	Maximum humidity	200.03	194.34	136.66	125.54	113.88	101.38	60.46	42.45	106.73	90.77
XPS	Dry	135.63	131.57	90.00	83.68	75.22	66.98	38.62	26.08	70.69	58.81
	Minimum humidity	138.65	134.41	91.97	85.50	76.94	68.53	39.34	26.53	72.38	60.14
	Maximum humidity	139.57	135.42	93.04	86.05	77.35	68.79	39.52	26.59	72.62	60.61
PUR	Dry	120.25	116.47	77.75	71.87	63.99	56.31	29.89	18.21	59.77	48.70
	Minimum humidity	124.36	120.42	80.20	74.08	65.95	58.00	30.43	18.28	61.63	50.07
	Maximum humidity	125.19	121.18	81.31	74.58	66.35	64.05	30.54	18.29	61.86	50.53
Centrifugal cotton	Dry	174.47	169.32	116.63	108.62	97.90	87.44	51.50	35.60	92.15	77.09
	Minimum humidity	170.19	164.99	114.77	107.26	100.01	92.01	51.20	38.33	97.98	81.03
	Maximum humidity	203.22	198.57	155.84	128.61	117.09	102.58	59.31	40.92	108.11	100.02
Rock wool	Dry	200.58	194.50	132.26	122.80	110.13	97.78	55.32	36.54	103.34	85.55
	Minimum humidity	200.16	193.74	132.66	123.39	112.68	101.58	55.70	37.88	114.07	88.76
	Maximum humidity	221.47	215.60	158.51	136.50	123.02	180.03	59.84	38.88	107.99	99.98
Foam cement	Dry	294.14	285.26	194.35	180.52	162.02	143.98	81.96	54.53	152.11	126.11
	Minimum humidity	324.50	313.08	214.39	199.21	180.70	161.67	88.71	58.79	171.59	141.04
	Maximum humidity	346.05	336.21	238.22	211.65	190.07	167.28	92.46	59.60	177.00	150.92
Aerogel-enhanced HGM	Dry	73.63	71.11	45.3	41.38	36.10	31.01	13.10	5.61	33.31	25.93
	Minimum humidity	83.22	80.17	49.83	45.21	39.00	32.99	12.14	2.78	35.74	26.97
	Maximum humidity	84.02	81.04	50.54	45.50	39.20	33.07	11.99	2.60	35.83	27.02

Taking Chengshantou, the coastal city with the highest humidity, as an example, it can be observed in Figure 5. that the thickness changes of EPS, XPS and PUR are relatively small, and the change rates is between 1.8–5.5%. The thickness changes of centrifugal cotton and rock wool are significantly different from EPS, XPS and PUR at high humidity, but there is little difference with the above three materials under the minimum humidity condition. Their change rates under the minimum humidity condition, the maximum humidity condition and the absolute dry condition are −1.6% and 33.4%, 0.3% and 19.8%, respectively. Aerogel-enhanced HGM shows great changes under the condition of the lowest humidity, and the change rate is 10.0%. The change rate of the highest humidity condition is very small. Foamed cement shows great difference compared with the other five materials under either the lowest humidity condition or the highest humidity condition. This is because their thermal conductivity will increase greatly when the humidity is 0–30% and when the humidity exceeds 70%. Their change rates under the lowest humidity condition, the highest humidity condition and the absolute dry condition are 2.6% and 19.5%, respectively. The reason for this result is that when calculating the optimal insulation thickness for the same kind of thermal insulation material in the same building, the only variable is the thermal conductivity, while the minimum humidity condition and the maximum humidity condition in Chengshantou are 62.84% and 90.16%, respectively. Except for EPS, XPS, PUR, aerogel-enhanced HGM and other materials whose thermal conductivity changes almost linearly with the change of humidity, other materials will change before and after 70% of materials. This makes the optimal thickness of thermal insulation material show a strong correlation with the thermal conductivity. In Figure 6, we bring in typical buildings to discuss the impact of three materials with different moisture change laws on their economic costs under different working conditions when the thickness changes. By comparing EPS, rock wool and foamed cement, it can also be observed that EPS has the least influence under the change of humidity. Rock wool changes little in low humidity but greatly in high humidity. Foamed cement is always changing with the change of humidity, which is also in line with the previous description of these three materials.

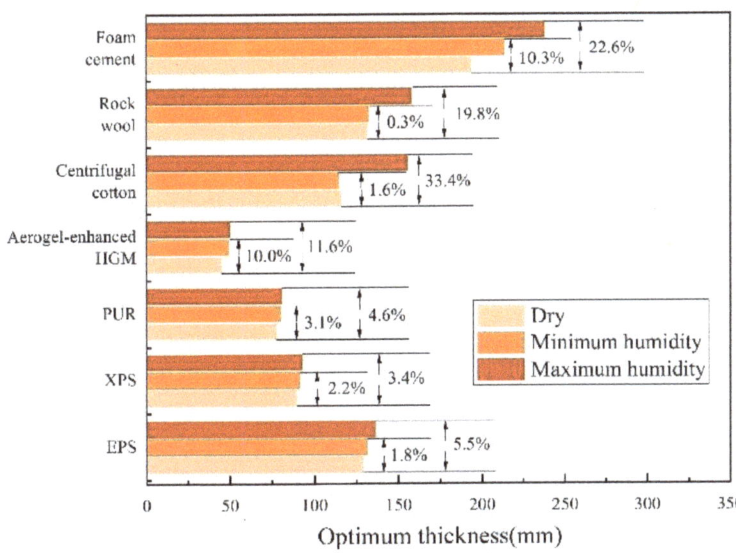

Figure 5. Optimum thickness of different thermal insulation materials under three working conditions in Chengshantou.

In Table 11, we can also observe that for different cities, the main factors affecting the optimal thickness of materials are DD value and the relative humidity of the city. Haikou and Hanzhong, which have little difference in maximum humidity but a large difference in DD value, are compared, as shown in Figure 7. The difference values of EPS and centrifugal cotton under absolute dry condition and maximum humidity condition are 165.8% and 168.2%, 175% and 186.1%, respectively. It can be seen that DD value is an indispensable and important parameter in the calculation of the optimal thickness. For Tonghe and Haikou with the largest difference in DD value, although their relative humidity is not exactly the same, under the absolute dry state, the optimal thickness difference values of EPS and centrifugal cotton are 369.6% and 344%, respectively. Therefore, the optimal thickness of each material in each city largely depends on the urban climate. On the basis of satisfying the urban climate, considering the influence of humidity change on the thermal conductivity will play a great role.

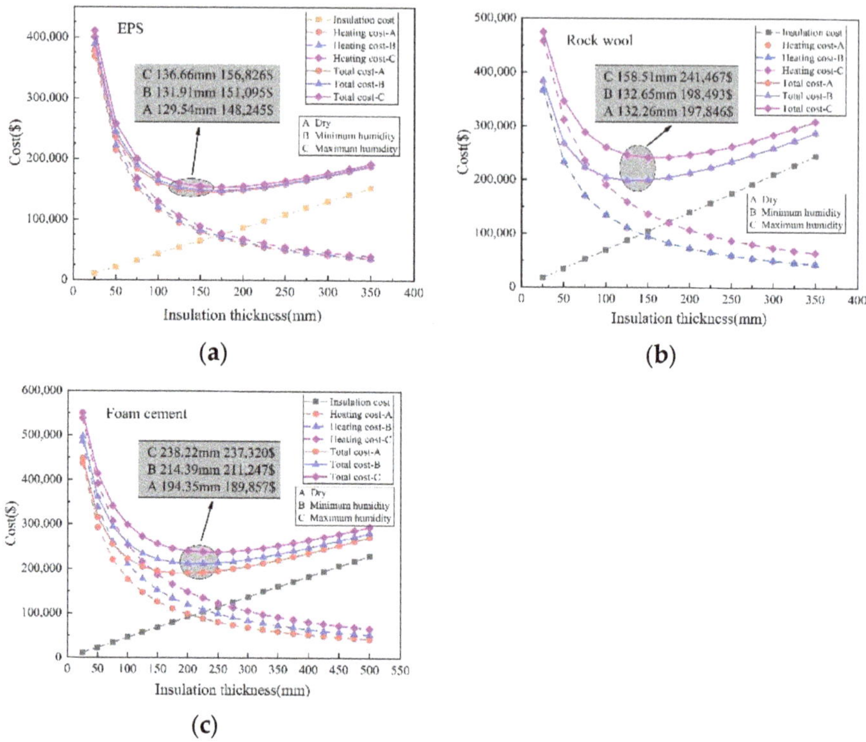

Figure 6. The influence of the thickness of different insulation materials on the economic cost of typical buildings in Chengshantou under three working conditions: (**a**) EPS; (**b**) rock wool; (**c**) foamed cement.

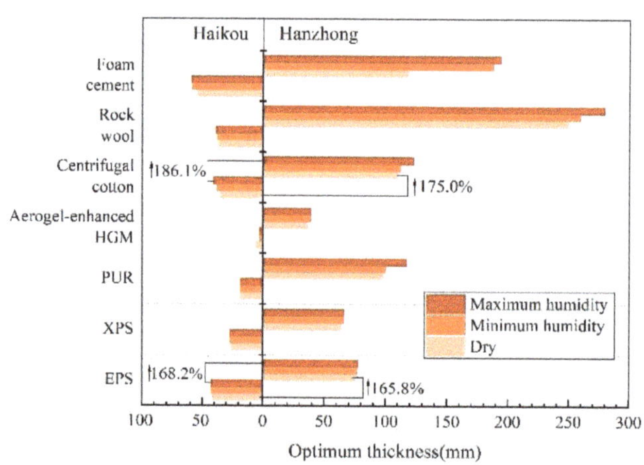

Figure 7. Comparison of the best thickness between Hanzhong and Haikou.

3.3. Results of the Greenhouse Gas Emissions

The optimal thickness of building insulation is used to reduce operating costs and greenhouse gas emissions, but changes in thermal conductivity will change the optimal thickness. However, the change of the optimal thickness will cause the change of carbon emissions [32]. In this paper, the input–output method is used to calculate the carbon emissions of the optimal insulation thickness corresponding to seven different building insulation materials under three conditions in ten typical cities in China, and a comparative analysis is made.

3.3.1. Numerical Calculation Process

Building insulation materials and thickness settings serve a vital purpose in determining the efficiency of heating and cooling. The appropriate thickness setting and material selection will directly reduce the heating and cooling energy consumption, thus reducing carbon emissions and realizing building energy conservation. Among the existing carbon footprint analysis methods, the carbon emission factor method and the input–output method are commonly used. To calculate the annual carbon emissions, we used the input–output method to capture the relationship between energy efficiency and carbon emissions. The general equation for combustion is as follows [33,34]:

$$C_gH_yO_zS_wN_t + \alpha A(O_2 + 3.76N_2) \rightarrow gCO_2 + \frac{y}{2}H_2O + wSO_2 + (\alpha - 1)AO_2 + BN_2 \quad (16)$$

The constants g, y, z, w and t in the formula are inconsistent for different fuel types. Considering China's economic situation, we take coal as the main reference, where $g = 7.078$; $y = 5.149$; $z = 0.517$; $w = 0.01$; and $t = 0.086$. Constants A and B can be determined by element equilibrium:

$$A = (g + \frac{y}{4} + w - \frac{z}{2}) \quad (17)$$

$$B = 3.76\alpha(g + \frac{y}{4} + w - \frac{z}{2}) + \frac{t}{2} \quad (18)$$

The emission rate (ER) of combustion products produced by burning 1 kg fuel is calculated by the following formula:

$$ER_{CO_2} = \frac{gM_{CO_2}}{M_f} \equiv kgCO_2/kgfuel \quad (19)$$

$$ER_{SO_2} = \frac{wM_{SO_2}}{M_f} \equiv kgSO_2/kgfuel \quad (20)$$

where M_f is the molecular weight of the fuel

$$M_f = 12g + y + 16z + 32w + 14t \quad (21)$$

CO_2 and SO_2 emissions can be calculated as follows:

$$m_{CO_2} = \frac{gM_{CO_2}}{M_f} m_{af} = \frac{44g}{M_f} \frac{86400DD}{(R_{iaf} + R_{oaf} + R_W + \frac{\delta_{im}}{k_{im}})\mu_f \eta_{hs}} \quad (22)$$

$$m_{SO_2} = \frac{wM_{SO_2}}{M_f} m_{af} = \frac{64w}{M_f} \frac{86400DD}{(R_{iaf} + R_{oaf} + R_W + \frac{\delta_{im}}{k_{im}})\mu_f \eta_{hs}} \quad (23)$$

3.3.2. Data and Analysis

The burning of fossil energy will produce greenhouse gases, such as CO_2 and SO_2, which will seriously damage the ecosystem. This has attracted widespread attention in China. Considering that most of China still uses coal-fired power generation, we bring the general equation constants of coal into the calculation. According to the observation of Formulas (22) and (23), the change rates of CO_2 and SO_2 emissions are consistent. We will not discuss them separately below.

For a specific city, the change law is obvious. Taking Chengshantou city as an example, the change rates of EPS, XPS and PUR insulation materials under the lowest and highest humidity conditions and absolute dry conditions are 2.0% and 6.0%, 2.5% and 3.8% and 4.0% and 5.9%, respectively from Figure 8. Centrifugal cotton and rock wool show a large change rate at high humidity. Their change rates under the lowest humidity condition, the highest humidity condition and the absolute dry condition are −1.8% and 38.6% and 0.3% and 24.0%, respectively. Aerogel-enhanced HGM has a large change rate of 18.1% under the lowest humidity condition. This is because it is still in the stage of development completion, and its price is high. Therefore, its value is small when calculating the optimal thickness, which leads to the need for more fossil energy heating and cooling during the building operation stage. The change rate of foamed cement will be larger as a whole. The change rates of foamed cement under the lowest and highest humidity conditions and absolute dry conditions are 12.1% and 27.1%, respectively. Cement-based materials can be greatly reduced in greenhouse gas emissions by measuring their thermal conductivity precisely. We explore the influence of the thickness change of the insulation layer on the SO_2 emission of typical buildings for the three materials with different carbon emissions caused by the change of humidity, as shown in Figure 9. We can observe that SO_2 emissions gradually decrease with the increase of the thickness of thermal insulation materials because the increase of thickness will reduce the energy demand, thus reducing the combustion of fossil fuels. By comparing the influence of three materials on SO_2 emission with the change of thickness under three working conditions, the basic change law is basically consistent with the change law of thermal conductivity with humidity, realizing three different change laws. Through the horizontal comparison of six materials, it can be concluded that EPS is the least affected by humidity and the lowest carbon emission.

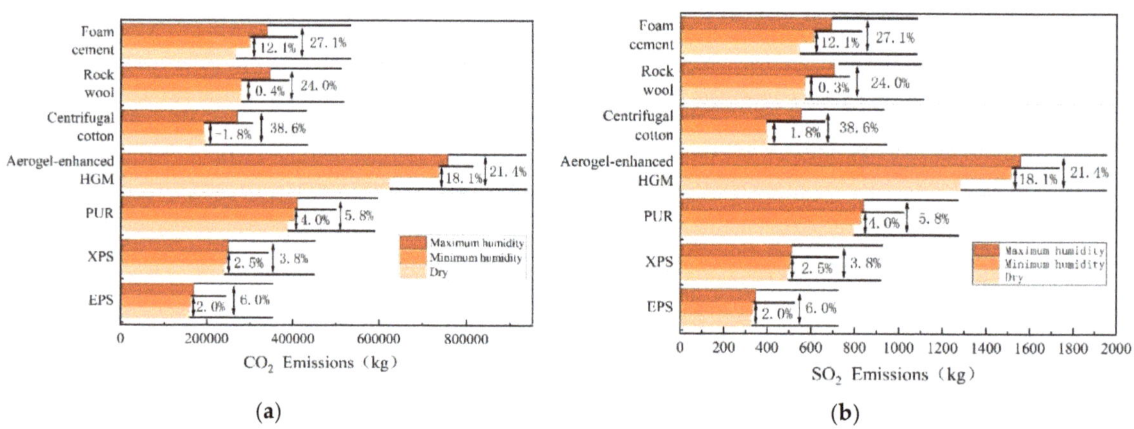

Figure 8. Carbon emissions of different materials under three working conditions: (**a**) CO_2 emissions; (**b**) SO_2 emission.

In the process of solving the optimal thickness of building thermal insulation materials, different results will not be caused by different building types, but greenhouse gas emissions will be caused by different building floors and window wall area ratios. We compared the SO_2 emissions of seven materials to three buildings, and the results are shown in Table 12. In the process of this algorithm, different layers will not lead to greenhouse gas emissions of each layer, so it is only the result of different layers. However, different window wall area ratios will lead to different exterior wall areas of each floor, leading to different greenhouse gas emissions of each floor, which will lead to different results of the whole building.

3.4. Results of the Energy Saving and Payback Period

3.4.1. Numerical Calculation Process

After determining the optimal insulation thickness under different relative humidity conditions, the energy saving under each condition can be calculated. Ozel [35] and Sisman [36] defined energy related savings as the difference between the total cost of a wall without insulation and the total cost of an insulated wall with the best insulation thickness.

$$ES = \frac{86400DD}{(R_{iaf} + R_{oaf} + R_w)\mu_f \eta_{hs}} C_f PWF - C_{tot} \quad (24)$$

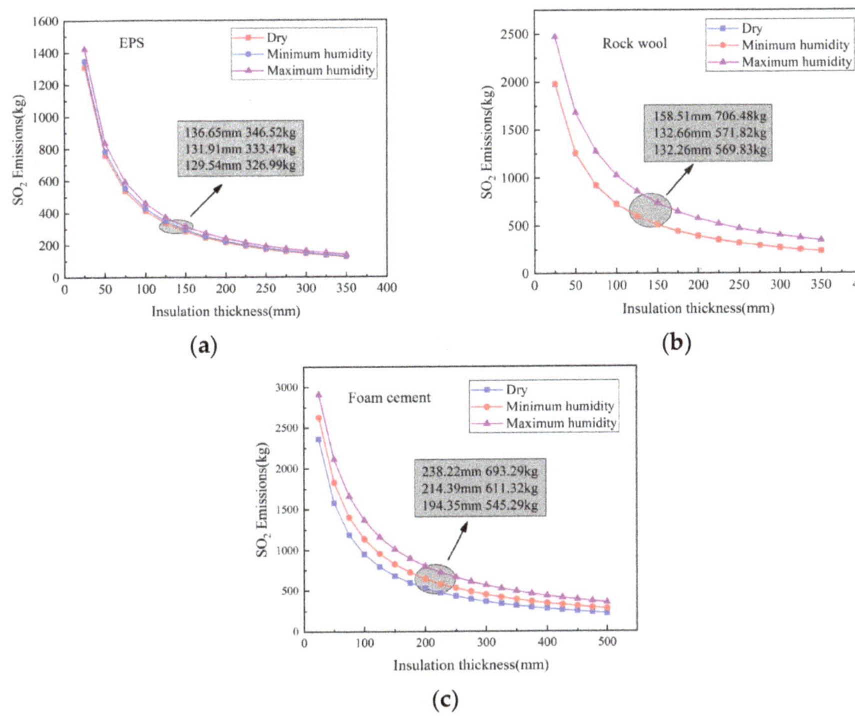

Figure 9. The influence of the thickness of thermal insulation materials of typical buildings in Chengshantou on SO_2 emission under different working conditions: (**a**) EPS; (**b**) rock wool; (**c**) foamed cement.

Table 12. Optimum insulation thickness of different insulation materials in ten typical cities under three working conditions (mm).

Material	SO_2 Emission of the Whole Building			SO_2 Emission Per Layer		
	Typical Group	Control Group A	Control group B	Typical Group	Control Group A	Control Group B
EPS	327.2	218.1	393.6	13.6	13.6	16.4
XPS	492.3	328.2	592.3	20.5	20.5	24.7
Centrifugal cotton	399.4	266.2	480.5	16.6	16.6	20.0
PUR	794.1	529.4	955.4	33.1	33.1	39.8
Rock wool	569.9	379.9	685.6	23.7	23.7	28.6
Aerogel-enhanced HGM	1280.2	853.5	1540.2	53.3	53.3	64.2
Foam cement	545.3	363.5	656.0	22.7	22.7	27.3

Payback period is another important term in the economic analysis of the optimal insulation thickness. Payback period is defined as the length of time required to recover the investment cost. The payback period is one of the most important things to consider when deciding whether the project can be realized. The investment payback period covers a wide range because a longer payback period is usually undesirable, especially in the construction industry. Sisman [36] calculated the payback period (PP) of the investment in thermal insulation materials as follows:

$$PP = \frac{C_{in}}{ES} PWF \tag{25}$$

3.4.2. Data and Analysis

According to the results observed by Ozel [35] in the work on traditional thermal insulation materials, the saving effect increases with the increase of the thickness of thermal insulation materials and reaches the maximum value at the optimal thermal insulation thickness. Cuce pointed out that electricity is the most suitable energy in terms of energy conservation [32], so we use electricity as our calculated energy. Taking Chengshantou city as an example, Figure 10 shows that the energy-saving change rate of EPS, XPS and PUR building insulation materials is not large, which is basically in the range of 0.3–2.4%, while centrifugal cotton, rock wool, aerogel-enhanced HGM and foamed cement change greatly. The change rates of minimum humidity and maximum humidity and absolute dry working conditions are −0.3% and 7.1%, 0.1% and 6.6%, 13.3% and 15.6% and 3.2% and 7.1%, respectively. We explore the impact of the thickness change of the insulation layer on the energy efficiency of typical buildings for the three materials with different carbon emissions caused by the change of humidity. We can observe that with the increase of the thickness of the insulation material, the energy saving first increases and then decreases, and the maximum value is obtained at the optimal thickness. By comparing the influence of three materials on energy saving with the change of thickness under three working conditions, it can be seen that the basic change law is basically consistent with the change law of thermal conductivity with humidity, realizing three different change laws. Through horizontal comparison, EPS is the most energy-saving material among the seven materials.

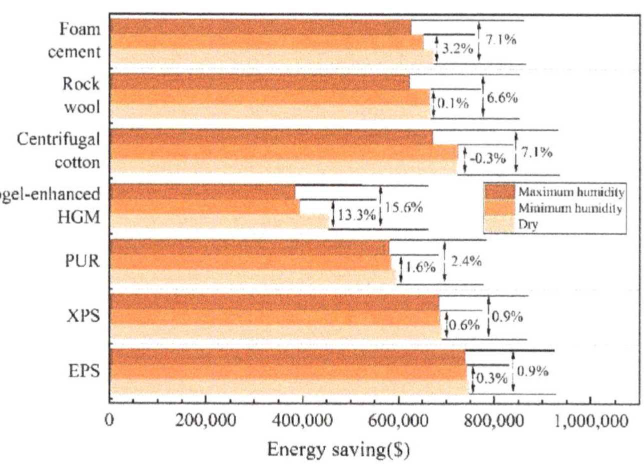

Figure 10. Energy saving of different materials of typical buildings in Chengshantou under three working conditions.

PP is not only related to the variables given in Formula (25), but also to the energy used. Cuce also points out that electricity is the most appropriate energy when considering the payback period because its energy saving is the highest [32]. Therefore, electricity was selected as the main energy source to judge the impact of relative humidity on the PP of building insulation materials. We still take Chengshantou city as an example. In Figure 11, with the increase of relative humidity, the PP value of the same material will also increase, but the change range is basically within half a year. According to Formula (25), PP is inversely proportional to energy saving, so the more energy-saving cities have, the shorter the payback period will be. Through horizontal comparison, EPS is the material with the shortest recovery period.

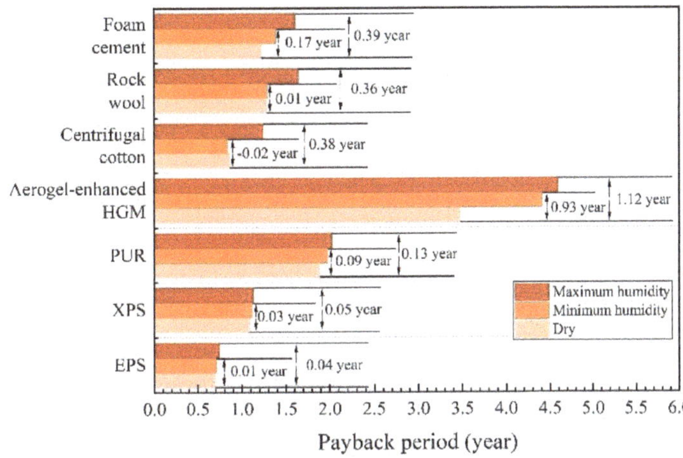

Figure 11. Recovery period of different materials under three working conditions in Chengshantou.

4. Conclusions

Residential buildings can improve their economic performance by using the best insulation thickness. In this study, considering the influence of relative humidity on the thermal conductivity of building insulation materials, an accurate method to determine the optimal thickness of insulation layer in external wall insulation considering the influence of relative humidity is proposed. This paper selects typical buildings in ten typical cities with high humidity in five thermal zones in China as the research object and adds two groups of contrast to explore the effect of shape coefficient and window wall area ratio. Taking the external wall insulation of typical residential buildings as an example, the optimal thickness of EPS, XPS, PUR, rock wool, centrifugal cotton, aerogel-enhanced HGM and foamed cement building insulation materials under absolute dry condition, minimum humidity condition and maximum humidity condition of monthly average of daily average over the years are calculated by degree day method and an LCCA economic model. Then, the carbon emissions under various working conditions are determined by the input–output ratio method. Finally, the recovery period and energy saving under various working conditions are obtained through calculation. The conclusion is that EPS is the material that is least affected by humidity, and when compared with other materials at the best thickness, the carbon emission is the smallest, the energy saving is the highest, and the recovery period is the shortest. The specific results are as follows:

1. TPS technology was used to measure the relationship between seven materials and relative humidity. EPS, XPS, PUR and aerogel-enhanced HGM have a linear relationship between their thermal conductivity and relative humidity, while their growth is relatively flat. Centrifugal cotton and rock wool have linear thermal conductivities before relative humidity reaches 70%, the growth is relatively gentle but increases sharply after 70% relative humidity. The thermal conductivity of foamed cement increases twice when the relative humidity is 30% and 70%, and it is at a stable growth stage when the relative humidity is 30–70%. The change rates of thermal conductivity of EPS, XPS, PUR, centrifugal cotton, rock wool, aerogel-enhanced HGM and foamed cement under 98% relative humidity and absolute dry state are 16.8%, 9.4%, 13.3%, 167.9%, 95.3%, 52.1% and 85.4%, respectively.
2. The optimum thickness of seven kinds of building insulation materials in ten typical cities was obtained by the degree day method and LCCA economic model analysis. The variation rates of the optimum thickness of EPS, XPS, PUR, centrifugal cotton, rock wool, aerogel-enhanced HGM and foamed cement in Chengshantou, a typical city with the highest humidity, under three working conditions are 0–5.5%, 0–3.4%,

0–4.6%, −1.6−33.4%, 0–19.8%, 0–11.6% and 0–22.6%, respectively. Compared with Haikou and Hanzhong, where the maximum relative humidity has little difference but the DD value has a large difference, the difference values of EPS and centrifugal cotton under absolute dry condition and maximum humidity condition are 165.8% and 168.2% and 175% and 186.1%, respectively. For Tonghe and Haikou with the largest difference in DD value, the optimal thickness difference of EPS and centrifugal cotton under absolute dry state is 369.6% and 344.0%, respectively.

3. The carbon emissions of seven building insulation materials in ten typical cities under different working conditions were obtained by using the input–output method. The carbon emission change rates of EPS, XPS, PUR, centrifugal cotton, rock wool, aerogel-enhanced HGM and foamed cement of Chengshantou, a typical city with the highest humidity, under three working conditions were 0–6.0%, 0–3.8%, 0–5.9%, −1.8–38.6%, 0–24.0%, 0–21.4% and 0–27.1%. By comparing the emissions between different cities with the continuous growth of DD value in cities, the carbon emissions of materials are also gradually increasing.
4. An energy saving and payback period can be determined based on the optimal thickness under various operating conditions. Among them, the typical city Chengshantou with the highest humidity has a small energy saving change rate of 0–2.4% under the three working conditions of EPS, XPS and PUR. However, the energy-saving change rates of centrifugal cotton, rock wool, aerogel-enhanced HGM and foamed cement under the three working conditions are large, which are −0.3–7.1%, 0–6.6%, 0–15.6% and 0–7.1%, respectively. However, except for aerogel reinforced materials, the recovery period of other materials increased within half a year.
5. To sum up, by comparing the thermal conductivity, optimal thickness, carbon emission, energy saving and recovery period of the six materials under three working conditions, we can conclude: Humidity has little effect on the thermal conductivity, optimal thickness, carbon emission, energy saving and recovery period of EPS, XPS and PUR materials. However, humidity shows great differences between centrifugal cotton and rock wool. Under low humidity conditions, the thermal conductivity, optimal thickness, carbon emission, energy saving and recovery period of centrifugal cotton and rock wool reflect the smallest change rate, even almost unchanged. However, it shows a large rate of change under high humidity conditions. The change rate of humidity on thermal conductivity, optimal thickness, carbon emissions, energy saving and recovery period of aerogel reinforced materials varies greatly under low humidity conditions but does not change much under high humidity conditions. Foamed cement is constantly affected by humidity. With the gradual increase of humidity, the change rates of thermal conductivity, optimal thickness, carbon emission, energy saving and recovery period are larger in these seven materials. In addition, among the seven building insulation materials, EPS is the most ideal building material. Because it has the minimum change of carbon emission, energy saving and recovery period before and after humidity change, and compared with other materials, it has the minimum carbon emission and recovery period and the maximum energy saving.

Author Contributions: W.Y.: data curation, conceptualization, methodology, funding acquisition; G.Z.: writing—original draft preparation, writing—reviewing and editing; W.H.: conceptualization, supervision; J.L.: resources. All authors have read and agreed to the published version of the manuscript.

Funding: This research was funded by China Postdoctoral Science Foundation Project (grant number 2021M702551), the independent research and development project of the State Key Laboratory of green building in Western China (grant number LSZZ202201), the key projects of Shaanxi Provincial Department of Education (grant number 20JS080), and the major project of the National Natural Science Foundation of China (grant number 51590910).

Data Availability Statement: Not applicable.

Acknowledgments: This study is provided by Xi'an University of architecture and technology with reference to urban basic meteorological data.

Conflicts of Interest: The authors declare that they have no known competing financial interests or personal relationships that could have appeared to influence the work reported in this paper.

References

1. Wang, Y.; Huang, J.; Wang, D.; Liu, Y.; Zhao, Z.; Liu, J. Experimental investigation on thermal conductivity of aerogel-incorporated concrete under various hygrothermal environment. *Energy* **2019**, *188*, 115999. [CrossRef]
2. Zach, J.; Korjenic, A.; Petránek, V. Performance evaluation and research of alternative thermal insulations based on sheep wool. *Energy Build.* **2012**, *49*, 246–253. [CrossRef]
3. Papadopoulos, A.M. State of the art in thermal insulation materials and aims for future developments. *Energy Build.* **2005**, *37*, 77–86. [CrossRef]
4. *ASTM C168-97 (1997)*; Standard Terminology Relating to Thermal Insulating Materials. ASTM International (Pennsylvania, America). Available online: www.astm.org (accessed on 10 October 2022).
5. Berardi, U.; Naldi, M. The impact of the temperature dependent thermal conductivity of insulating materials on the effective building envelope performance. *Energy Build.* **2017**, *144*, 262–275. [CrossRef]
6. Budaiwi, I.; Abdou, A. The impact of thermal conductivity change of moist fibrous insulation on energy performance of buildings under hot–humid conditions. *Energy Build.* **2013**, *60*, 388–399. [CrossRef]
7. Kontoleon, K.J.; Giarma, C. Dynamic thermal response of building material layers in aspect of their moisture content. *Appl. Energy* **2016**, *170*, 76–91. [CrossRef]
8. Budaiwi, I.; Abdou, A.; Al-Homoud, M. Variations of thermal conductivity of insulation materials under different operating temperatures: Impact on envelope induced cooling load. *J. Architect. Eng.* **2002**, *8*, 125–132. [CrossRef]
9. Khoukhi, M.; Tahat, M. Effect of operating temperatures on thermal conductivity of polystyrene insulation material: Impact on envelope-induced cooling load. *Appl. Mech. Mater.* **2014**, *564*, 315–320. [CrossRef]
10. Hoseini, A.; Bahrami, M. Effects of Humidity on Thermal Performance of Aerogel Insulation Blankets. *J. Build. Eng.* **2017**, *13*, 107–115. [CrossRef]
11. Alvey, J.B.; Patel, J.; Stephenson, L.D. Experimental Study on the Effects of Humidity and Temperature on Aerogel Composite and Foam Insulations. *Energy Build.* **2017**, *144*, 358–371. [CrossRef]
12. Nosrati, R.H.; Berardi, U. Hygrothermal Characteristics of Aerogel-Enhanced Insulating Materials under Different Humidity and Temperature Conditions. *Energy Build.* **2018**, *158*, 698–711. [CrossRef]
13. Mahlia, T.M.I.; Taufiq, B.N.; Ismail; Masjuki, H.H. Correlation between thermal conductivity and the thickness of selected insulation materials for building wall. *Energy Build.* **2007**, *39*, 182–187. [CrossRef]
14. Comakli, K.; Yuksel, B. Optimum insolation thickness of external walls for energy saving. *Appl. Therm. Eng.* **2003**, *23*, 473–479. [CrossRef]
15. Ucar, A.; Balo, F. Effect of fuel type on the optimum thickness of selected insulation materials for the four different climatic regions of Turkey. *Appl. Energy* **2009**, *86*, 730–736. [CrossRef]
16. Huang, H.; Zhou, Y.; Huang, R.; Wu, H.; Sun, Y.; Huang, G.; Xu, T. Optimum insulation thicknesses and energy conservation of building thermal insulation materials in Chinese zone of humid subtropical climate. *Sustain. Cities Soc.* **2020**, *52*, 101840. [CrossRef]
17. *GB/50176-2016*; Ministry of Housing and Urban Rural Development of the People's Republic of China, Code for Thermal Design of Civil Buildings. China Planning Press: Beijing, China, 2016.
18. Jing, Y.; Chai, B.; Song, Y. Influence of external wall insulation on building load in cold areas and determination of economic insulation layer thickness. *Build. Sci.* **2007**, *12*, 39–41.
19. Shi, Y.; Xie, Y. Analysis of the impact of building envelope on building energy consumption in Harbin. *Shanxi Archit.* **2019**, *45*, 145–146.
20. *10J121*; Exterior Wall External Insulation Building Structure. China Planning Press: Beijing, China, 2010.
21. Cui, Y.; Jiang, S.; Liu, Y. Research on green transformation strategy of urban old industrial areas—Taking Jinan Iron and Steel coking plant as an example. *Build. Energy Effic.* **2021**, *49*, 121–126.
22. Wang, S. Research on Energy-Saving Construction Technology of Top-Level Buildings of Urban Collective Housing in Xining. Master's Thesis, Xi'an University of Architecture and Technology, Xi'an, China, 2021.
23. Villasmil, W.; Fischer, L.J.; Worlitschek, J. A review and evaluation of thermal insulation materials and methods for thermal energy storage systems. *Renew. Sustain. Energy Rev.* **2019**, *103*, 71–84. [CrossRef]
24. Yang, W.; Liu, J.; Wang, Y.; Gao, S. Experimental study on the thermal conductivity of aerogel-enhanced insulating materials under various hygrothermal environments. *Energy Build.* **2020**, *206*, 109583. [CrossRef]
25. Anh, H.; Duong, L.; Pásztory, Z. An Overview of Factors Influencing Thermal Conductivity of Building Insulation Materials. *J. Build. Eng.* **2021**, *44*, 102604.
26. Cai, S.S.; Zhang, B.X.; Cremaschi, L. Review of moisture behavior and thermal performance of polystyrene insulation in building applications. *Build. Environ.* **2017**, *123*, 50–65. [CrossRef]

27. Yang, W.; Wang, Y.; Liu, J. Optimization of the thermal conductivity test for building insulation materials under multifactor impact. *Constr. Build. Mater.* **2022**, *332*, 127380. [CrossRef]
28. Hasan, A. Optimizing insulation thickness for buildings using life cycle cost. *Appl. Energy* **1999**, *63*, 115–124. [CrossRef]
29. Zhang, H.; Li, M.-J.; Fang, W.-Z.; Dan, D.; Li, Z.-Y.; Tao, W.-Q. A numerical study on the theoretical accuracy of film thermal conductivity using transient plane source method. *Appl. Therm. Eng.* **2014**, *72*, 62–69. [CrossRef]
30. Doty, S.; Turner, W.C. *Energy Management Handbook*, 7th ed.; Fairmont Press Inc.: Lilburn, GA, USA, 2009; pp. 377–410.
31. Alsayed, M.F.; Tayeh, R.A. Tayeh. Life Cycle Cost Analysis for Determining Optimal Insulation Thickness in Palestinian Buildings. *J. Build. Eng.* **2019**, *22*, 101–112. [CrossRef]
32. Cuce, E.; Cuce, P.M.; Wood, C.; Riffat, S.B. Optimizing insulation thickness and analysing environmental impacts of aerogel-based thermal superinsulation in buildings. *Energy Build.* **2014**, *77*, 28–39. [CrossRef]
33. Comakli, K.; Yuksel, B. Environmental impact of thermal insulation thickness inbuildings. *Appl. Therm. Eng.* **2004**, *24*, 933–940. [CrossRef]
34. Ucar, A.; Balo, F. Determination of environmental impact and optimum thickness of insulation for building walls. *Environ. Prog. Sustain. Energy* **2011**, *30*, 113–122. [CrossRef]
35. Ozel, M. Cost analysis for optimum thicknesses and environmental impacts of different insulation materials. *Energy Build.* **2012**, *49*, 552–559. [CrossRef]
36. Sisman, N.; Kahya, E.; Aras, N.; Aras, H. Determination of optimum insulation thicknesses of the external walls and roof (ceiling) for Turkey's different degree day regions. *Energy Policy* **2007**, *35*, 5151–5155. [CrossRef]

buildings

MDPI

Article

Shaking Table Tests of a Novel Flat Slab-Flanged Wall (FSFW) Coupled System with Embedded Concrete-Filled-Steel-Tubes in Wall Piers

Xin-Yu Zhao [1], Xiao-Dan Fang [2], Fan Wang [1,2,*] and Jing Zhou [1]

1 State Key Laboratory of Subtropical Building Science, South China University of Technology, Guangzhou 501641, China
2 Architectural Design & Research Institute of SCUT Co. Ltd., South China University of Technology, Guangzhou 501641, China
* Correspondence: wangfan@scut.edu.cn; Tel.: +86-137-1033-8386

Abstract: The flat slab-flanged wall (FSFW) coupled system has gained popularity in recent years; however, its seismic performance remains an issue, as beams and columns in it are commonly eliminated. To tackle this problem, embedding concrete-filled steel tubes (CFSTs) in wall piers has been proposed to strengthen the system; the viability of this approach has been verified at the member level. Along this line, this study embarks on a shaking table testing of a 1/8-scale five-story FSFW structure equipped with CFSTs in walls, with an aim to understand the overall seismic behavior of such an enhanced system. As with the practice in many countries, the plan layout of the test structure consisted of four rows of wall piers, thus presenting a 'fish-bone' floor configuration that relied only upon the walls to resist gravity and lateral loads. The structure was subjected to a suite of input ground motions along with white-noise excitations. By so doing, its damage progression, pattern and dynamic characteristics were clearly identified. Furthermore, a non-linear time history analysis was conducted using PERFORM-3D, and the goodness-of-fit of the computed responses to the experimental records was examined. Findings indicated that the application of CFSTs was instrumental in resisting the simulated earthquake loads acting on the FSFW system, hence the global response limits required by codes of practice were met, even in the case of extremely strong earthquakes. Nevertheless, the junction between the shear walls and floor slabs was found to be the weakest links in the whole system. Designers are thus cautioned to implement proper detailing in those regions to prevent local distress, though it did not appear to acutely impair the system's collapse-resisting capacity.

Keywords: slab-wall structure; concrete-filled steel tubes; shear walls; shaking table test; seismic performance

Citation: Zhao, X.-Y.; Fang, X.-D.; Wang, F.; Zhou, J. Shaking Table Tests of a Novel Flat Slab-Flanged Wall (FSFW) Coupled System with Embedded Concrete-Filled-Steel-Tubes in Wall Piers. *Buildings* **2022**, *12*, 1441. https://doi.org/10.3390/buildings12091441

Academic Editors: Xin Ren, Teng Shao and Yaolin Lin

Received: 23 July 2022
Accepted: 8 September 2022
Published: 13 September 2022

1. Introduction

The reinforced concrete (RC) flat slab-flanged wall (FSFW) coupled system has emerged as an appealing structural form to address ever-increasing design challenges. Historically, it evolves from the dual wall-frame system. The difference between the two is that in the former—as its name suggests—structural columns and beams are often removed, leading to the unique feature of shear walls only, connected by floor slabs, bearing the lateral forces caused by wind or seismic disturbances.

Yet a lot of advantages can be achieved by adopting such an alternative: (1) clear ceiling height can be maximized; (2) utility ducts and pipes do not need to pass through girders anymore; (3) both architectural flexibility and unobstructed space can be obtained, accompanied by an enhanced indoor aesthetic; and (4) the FSFW buildings have a great opportunity to be panelized via prefabricated modular elements, thereby resulting in a bolstered construction efficiency. Due to these benefits, FSFW structures have been popular for years, especially for office and residential buildings [1–5].

However, there is still a great concern regarding the system's seismic safety, due simply to the complete elimination of beams and columns in it. Under seismic conditions, the load transfer between the floor diaphragms and wall groups will get increasingly complex, which could even provoke slab punching failure. Moreover, if slender wall piers are used (a common practice now), the system would be predisposed to a high risk of suffering severe damages under strong earthquake excitations [6,7].

It is worth noting that shear wall failures observed in recent seismic events (the 2010 Chile and 2011 New Zealand earthquakes) indicate a dire need for enhancing the robustness of shear wall buildings. These failures, with limited ductility, were found to be related to a number of inadequacies [8–11], such as high wall-to-floor-area ratio, poor boundary element confinement, and high axial stresses in walls. As a result, the stability and lateral deflection capacity of these shear walls declined drastically during the earthquakes, resulting in rebar buckling and concrete crushing in the wall boundaries, and, worst, the total collapse of several shear wall buildings [12,13].

Analytical studies have been conducted extensively to apprehend the structural behavior of slab-wall sub-assemblages or entire buildings. The first systematic study was reported by Qadeer and Smith [14], who employed the finite-difference method to determine the coupling stiffness, effective width, and stresses in uniformly spaced plane walls. Then, Coull and his colleagues [15–18] contributed greatly to the understanding of the interactions between laterally loaded walls and connecting floor slabs. They indicated that the dimensions and shape of walls (plane walls, flanged walls, and box cores), wall spacing, and slab dimensions have significant influence on the effective width and stiffness of the slabs. Their appraisal also led to a design procedure for checking against punching shear failure of the slabs. Balkaya and Kalkan [3] numerically analyzed the three-dimensional effect of a slab-wall system with openings. They showed that the stress flow and cracking pattern around the openings of the so-called pierced walls are vastly different from those observed with 2D models. This was attributed to the nonexistence of contra-flexure points when modeling the 3D behavior. In addition, within the walls, the tension-compression coupling mechanism was highly affected by the size and locations of openings. More recently, two existing shear wall buildings representative of Chilean practice were analyzed by Ugalde et al. [19] using finite-element models. They observed that the total inter-story drift deformation of the buildings takes considerable values at some shear walls, and the axial-flexural demands and strains at wall boundaries are indeed significant. Similarly, Ramos and Hube [7] assessed via numerical simulation the seismic response of an RC shear wall building with coupling slabs. It was found that the shear forces transferred through slabs result in a significant variation in the axial load of walls. Furthermore, a larger amount of slab reinforcement tends to transfer a higher axial load to the compressed walls, generating remarkable changes in their flexural capacity. Pavel et al. [20] considered a numerical model that represents a class of slab-core wall structures in order to evaluate their seismic performance. The structure modeled was shown to be able to sustain significant seismic loads without suffering much damage; a large lateral deformation could be borne by the slab-wall connections as well.

In terms of experimental research, there are a few efforts devoted specifically to slab-wall structures. An effective slab-wall connection was proposed and experimentally validated by Pantazopoulou and Imran [21]. It was revealed that the vertical loads affect the in-plane stiffness and shear resistance of the floor slabs, particularly in the vicinity of the slab-wall connections. Moreover, the continuous reinforcement through the connections is capable of preventing sliding shear failure in these critical regions. The results of a shake table testing of a slice from a full-scale seven-story shear wall building were reported by Panagiotou et al. [22]. The building, which was tested in a single direction parallel to the web wall, responded satisfactorily to the ground motions reproduced by the shake table and met all performance objectives. Nevertheless, the interaction effects between the walls and slabs caused a notable increase in the system overturning moment as well as in the shear force demand in the web wall. Greeshma and Jaya [23] addressed the reinforcement detailing within and around slab-wall connections. They observed that the slab bars bent at 90 degrees in the joint core is a superior method of connecting slabs to walls than the conventional U hooks. Brunesi et al. [24] conducted a cyclic test on a full-scale one-story wall-slab-wall specimen. During the testing, severe damage of concrete occurred at both the wall base and slab-wall connections, showing the vulnerability of a large stock of slab-wall structures which have been designed with little seismic consideration. A shaking table test of a five-story coupled wall structure, representing part of a typical apartment building with a high wall-to-floor area ratio, was performed by Fischinger et al. [25]. It was observed that the lightly confined free edges of the wall flanges (using U hooks) behaved satisfactorily in the case of moderate seismic demand. However, the interactions between the slabs and walls had a negative impact, causing a brittle shear failure of the wall piers under bidirectional excitations.

It should be stressed that concrete confinement in wall boundaries is highly desirable in shear wall buildings [9–11], especially at the extremities of walls having T, L, or other similarly shaped cross sections [2]. This is understandable because it has been demonstrated in [7] that the compressive strain in wall boundaries reaches the failure strain earlier when considering the coupling effects of slabs. In addition, a larger flexural strength of slabs increases the axial load compression in wall piers, which increases their deformation demand considerably.

To get to grips with these problems, the authors of [26–28] have proposed a novel type of shear wall that incorporates a number of stiff concrete-filled steel tubes (CFSTs) at the corners and/or at the web–flange intersections of the walls. These CFSTs are purposely arranged at these key regions so that the entire wall unit can exhibit excellent seismic performance, even under tension-shear loadings [26]. Such a steel-concrete composite wall provides a new and compelling alternative to conventional concrete walls. Note, however, that this concept has only been discussed at the member level. Therefore, in order to examine the overall lateral load-resisting capacity of FSFW structures equipped with such composite walls, a large-scale system-level test is genuinely needed.

Motivated by the aforementioned considerations, a shake table testing of an FSFW structure with CFSTs embedded in its wall piers was conducted, and the main findings are presented in this paper. The remainder of this paper is organized as follows: first, the design of the test structure is introduced, followed by the description of the experimental procedure and loading sequence. Test results are then discussed with emphasis on describing the local and global behaviors of the structure observed in the tests. A numerical study was finally performed to reproduce the dynamic responses of the test structure, and to compare the seismic behaviors of the structures with and without embedded CFSTs in wall piers. It is believed that the outcomes of this work will contribute to the understanding of the overall seismic performance of the FSFW systems concerned, and, ultimately, promote the application of the highly ductile wall components embedded with CFSTs.

2. Test Program

2.1. Description of the Prototype and Test Structure

A 1/8 scale five-story FSFW structure was designed in this study, based on an RC prototype office building located in Guangzhou with seismic intensity degree 7 as per Chinese seismic code provisions [29] (China's earthquake intensity is roughly equivalent to the Modified Mecalli Intensity [30]). The site soil belonged to Category III, defined in [29] as that the thickness of the site soil is greater than 50 m, and the average shear wave velocity in the soil is between 150 m/s and 250 m/s.

The prototype structure had five stories with a total height of 30 m, each story being 6 m high. All of the floor slabs were of the same thickness, 400 mm. The maximum span of the slabs was 7.2 m. There were no beams in the structure. According to the length scaling factor (1/8), the test structure was hence 3.75 m high (0.75 m high per story), adopting

50 mm thick flat slabs that had the role of coupling all the wall piers with specially-shaped cross sections (L or T).

To accommodate architectural requirements, the prototype structure presented a 'fish-bone' floor configuration consisting of four rows of wall piers, with a longitudinal corridor running through the middle of each floor, as is the practice in many countries [10,11]. Figure 1 shows the plan layout of a typical floor of the prototype structure. Each wall pier possessed the same cross-sectional size throughout the building height. The wall-to-floor-area ratio was 3.75%. A 3D view of this structure is displayed in Figure 2.

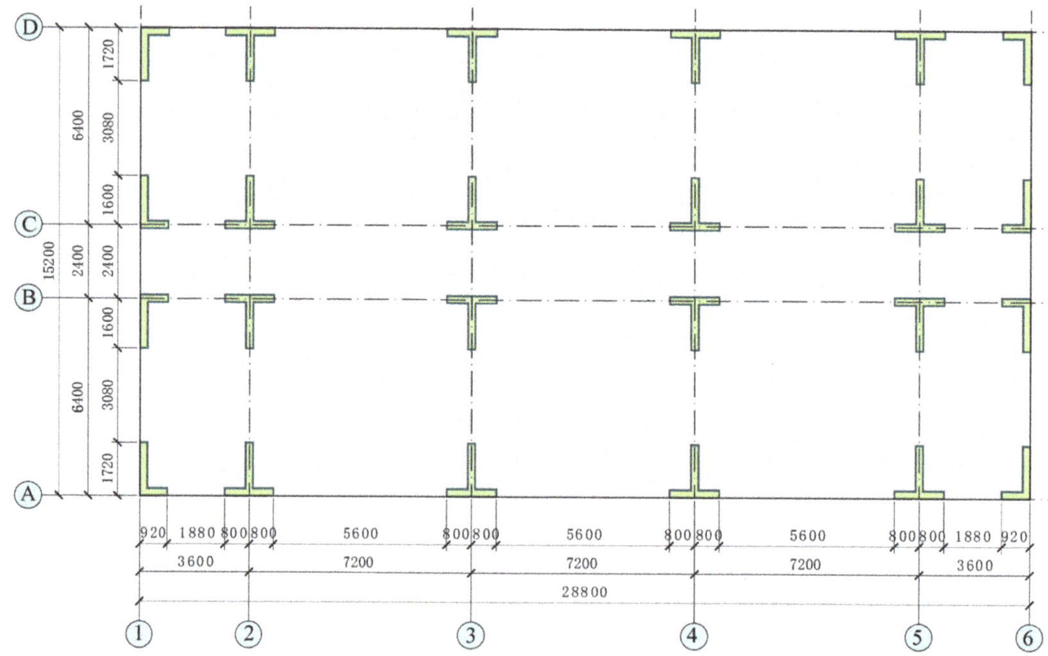

Figure 1. Plan layout of the prototype structure (Unit: mm).

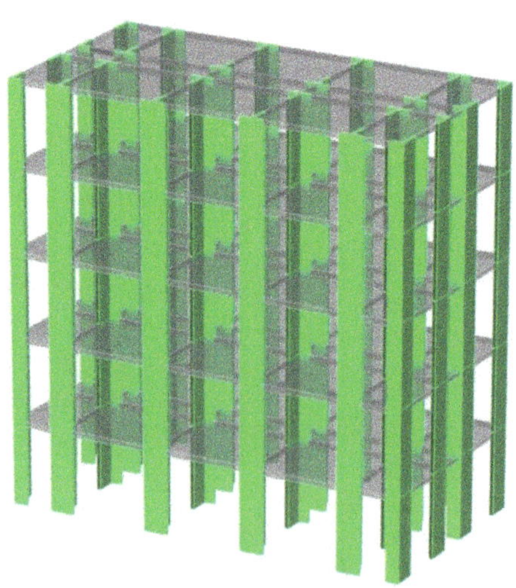

Figure 2. Three-dimensional rendering of the prototype structure.

Each wall pier was characterized by a T or L-sectional shape. For the prototype structure, the wall thickness was uniformly 240 mm, irrespective of the wall shape. Two curtains of steel bars were used to reinforce these walls, while CFSTs were arranged in them to serve as strengthening elements. Grade C30 concrete was used to cast both the slabs and walls, while C40 concrete was applied to fill the CFSTs.

As detailed in Figure 3, the CFSTs were inserted at the wall boundaries, web–flange intersection, and center of the longer wall segment. The circular steel tube of the CFSTs was made with Q235 plate. The tube's outer diameter and thickness were 160 mm and 8 mm, respectively. Due to the absence of columns, the axial load ratio of the walls on the ground floor was about 0.5; thus the CFSTs were expected to carry a significant portion of the building's gravity and lateral loads.

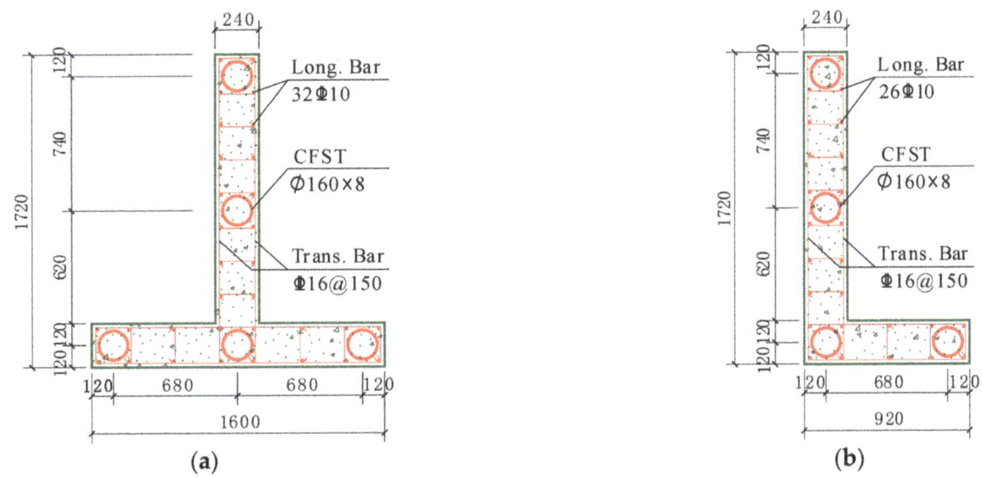

Figure 3. Dimensions and reinforcement details of the wall piers in the prototype structure (Unit: mm): (**a**) T-shaped wall pier; (**b**) L-shaped wall pier.

Steel bars made of HRB400 (i.e., Hot-rolled Ribbed Bars [31]) were used as reinforcement for the prototype structure. Figure 3 shows their details in the walls. As for each floor slab, two-way D16@150 and D14@150HRB400 steel bars were used as the top and bottom reinforcement, respectively.

2.2. Scaling Factors, Construction Materials and Manufacture of the Test Structure

According to the similitude theory [32,33], the primary scaling factors for the current tests, which are independent of each other, were determined as follows: (i) the length scaling factor was S_l = 1/8; (ii) the stress scaling factor was S_σ = 1/4.24; and (iii) the acceleration scaling factor was S_a = 2.5. Other scaling factors could be readily derived from the above three factors [32].

The length scaling factor S_l was set to 1/8 in this work for two reasons: (i) the plane size of the prototype structure was 28.8 m × 15.2 m, which meant that the length scaling factor should not be greater than 1/8, otherwise the shaking table (4 m × 4 m, as detailed later) could not accommodate the scaled-down structure; (ii) the total weight of the test structure was 12.809 t, which was exactly within the maximum payload of the shaking table (20 t).

Micro-aggregate concrete (i.e., cement mortar) and fine meshes were used to construct the RC elements. The round tube of the CFSTs was simulated by a #304 stainless steel pipe due to the latter's similar yielding properties to Q235 steel. Table 1 reports the measured mechanical properties of the construction materials used for the test structure. Figure 4 shows some construction scenarios and the appearance of the specimen before testing.

Table 1. Main mechanical properties of the construction materials.

Materials of Prototype Structure	Materials of Test Structure	Properties of Test Materials
C30 concrete	M7.5 cement mortar	$E_c = 7090$ MPa, $f_c = 7.28$ MPa
C40 concrete	M10 cement mortar	$E_c = 7670$ MPa, $f_c = 9.71$ MPa
HRB400 steel bar	Galvanized wire mesh	$E_s = 200$ GPa, $f_y = 270$ MPa
Q235 steel tube	#304 stainless steel tube	$E_s = 193$ GPa, $f_y = 218$ MPa

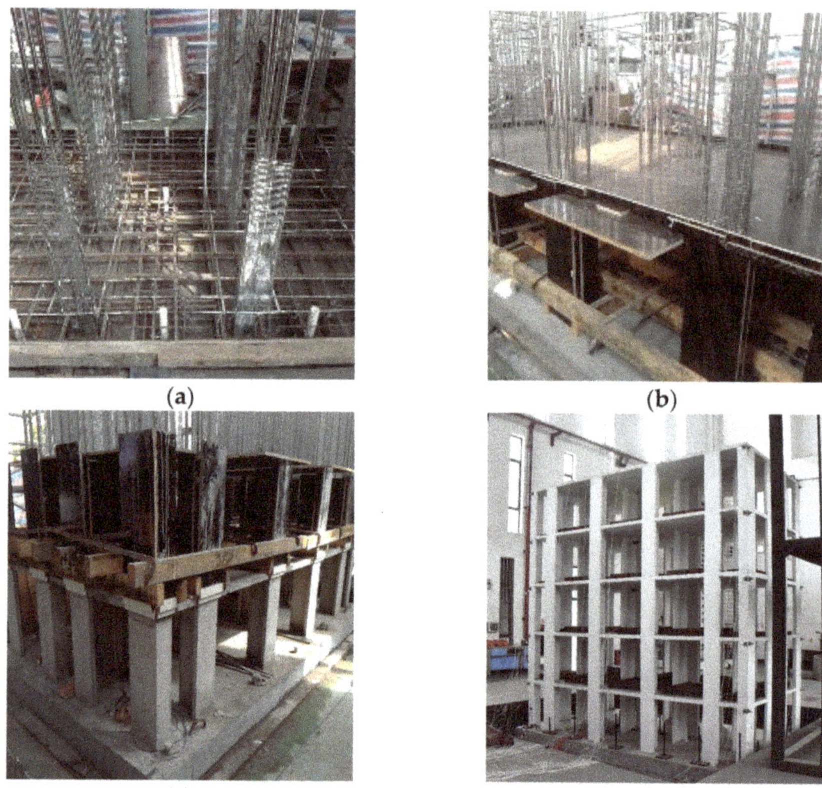

(a) (b) (c) (d)

Figure 4. Fabrication of the test structure: (**a**) tying of steel reinforcement; (**b**) casting of concrete; (**c**) installation of forms and shores for upper story; (**d**) appearance of the structure before testing.

2.3. Capacity of the Shake Table and Mass Applied to the Test Structure

The testing was carried out on a shake table at the South China University of Technology (SCUT). The table is capable of generating six-degree-of-freedom excitations. With a projection dimension of 4 × 4 m, the shake table is driven by eight electro-hydraulic servo actuators. The maximum payload is 200 kN. The maximum acceleration is 1.0 g with full payload, and the maximum velocity is 1000 mm/s. The effective frequency ranges from 0.1 to 50 Hz. Table 2 lists the main parameters of this facility.

Table 2. Parameters of the shake table at SCUT.

Parameters		Capacity
Maximum payload		20 t
DOF generation		SixDOF motions
Frequency		0.1~50 Hz
Maximum displacement		X&Y: ±125 mm; Z: ±80 mm
Maximum velocity		X&Y&Z: 1.0 m/s
Maximum acceleration	X	2.4 g (No load); 1.0 g (20 t)
	Y	2.4 g (No load); 1.0 g (20 t)
	Z	4.8 g (No load); 2.0 g (20 t)

To mimic the gravity load applied to the prototype structure, a number of lead weights were attached to each floor such that the weight of the test structure plus the subsidiary lead produced nominal axial stresses approximately equivalent to those produced by gravity loads (dead load plus half of live load) in the prototype building. The resulting axial load ratio was approximately 0.5 in the first-story walls. Figure 5 shows the distribution of the lead blocks on a typical floor. Table 3 gives the mass of the prototype and the test structures.

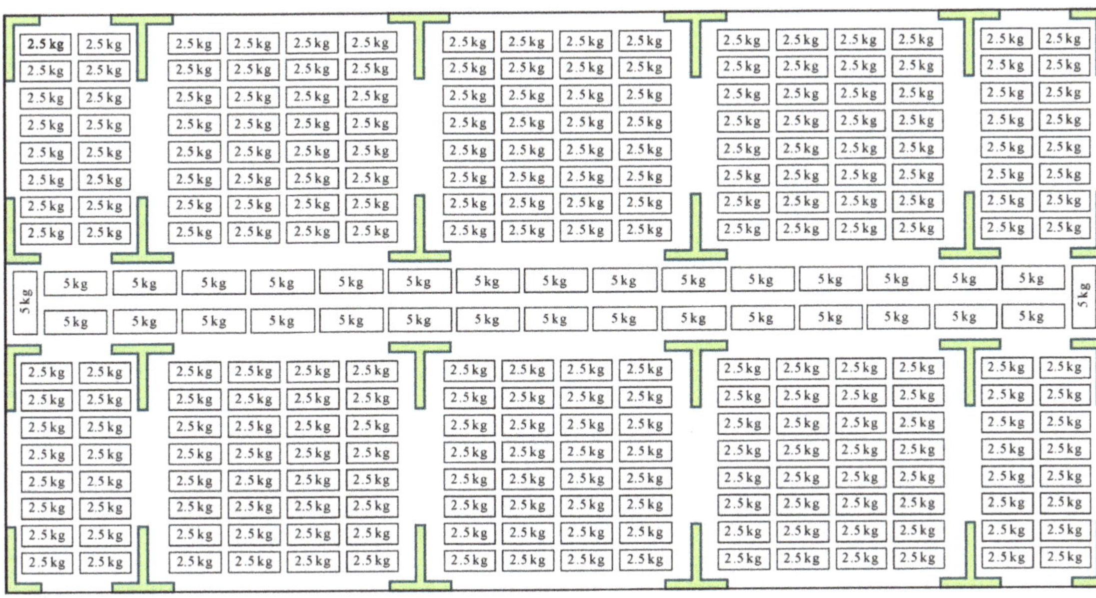

Figure 5. Distribution of the lead blocks on a typical floor.

Table 3. Mass of the prototype and test structures.

Story	m_p (t) [1]	m_m (t) [1]	m_{ma} (t) [1]
1	1741.8	2.569	1.317
2	1741.8	2.569	1.317
3	1741.8	2.569	1.317
4	1741.8	2.569	1.317
5	1741.8	2.532	1.280
Total	8683.9	12.809	6.548

[1] 'm_p' represents the total mass of the prototype structure; 'm_m' represents the total mass of the test (model) structure; and 'm_{ma}' denotes the additional mass (i.e., the mass of the lead blocks) applied to the test structure.

2.4. Input Ground Motions

According to China's seismic code [29], buildings in seismic areas should have the ability to withstand earthquakes of *minor*, *moderate*, and *major* levels. The corresponding 50-year probabilities of exceedance for those different hazard levels are approximately 63.2%, 10%, and 2%, with the return periods of 50, 475, and 2475 years, respectively. In specific terms, when a building is designed to resist a minor (*frequent*) earthquake, it should be intact or only slightly damaged; when subjected to an earthquake of design intensity (i.e., a *moderate* earthquake), it is expected to be damaged but can still be serviceable after limited repair; finally, in the event of a major (*rare*) earthquake, the building should neither collapse nor suffer life-threatening damages.

In compliance with these codified requirements, one artificial ground motion wave and three natural seismic records were used as the input excitations for the test structure, as shown in Table 4. The RGW wave, dominated by low-frequency long-period components, was generated by fitting the design response spectrum specified in [29]. The acceleration-time history curves of these selected ground motions are plotted in Figure 6.

Table 4. Input ground motions for the shake table testing.

Seismic Wave	Station and Direction	
RGW (i.e., the artificial wave)	–	
Northridge, CA, USA, 1994	West Covina	SOA225 SOA315
Imperial Valley, CA, USA,1940	El Centro Array #9	ELC180 ELC270
Chi-Chi, Taiwan, China, 1999	CHY056	E-W S-N

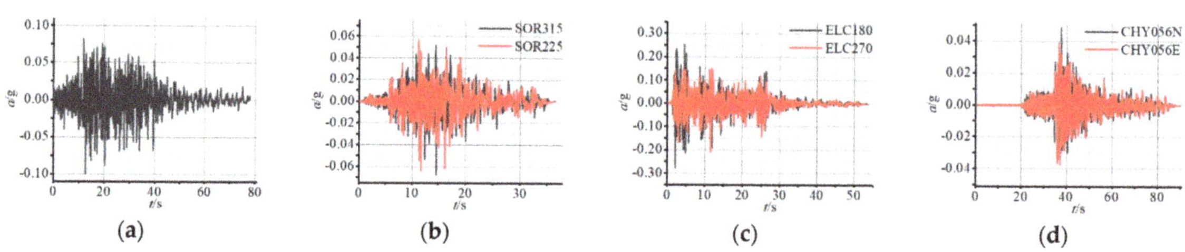

Figure 6. Acceleration-time history curves of the selected ground motions: (**a**) RWG; (**b**) Northridge; (**c**) El Centro; (**d**) Chi-Chi.

2.5. Instrumentation

An array of accelerometers (66 in total) along the X, Y and Z directions of the test structure were installed, at its centroid and at the four corners of each floor. In addition, linear variable displacement transducers (LVDTs) were deployed at the base beam of each floor to measure the vibration displacement of the specimen.

In order to monitor the dynamic strain response of the walls and embedded CFSTs near the ground level, strain gauges were attached to each steel tube of the interior T-shaped wall, and of the L-shaped wall in the southwest corner, as marked in Figure 7. In addition, strain gauges were mounted on the concrete surface of the aforementioned walls. Therefore, a total of 13 strain monitoring points were established.

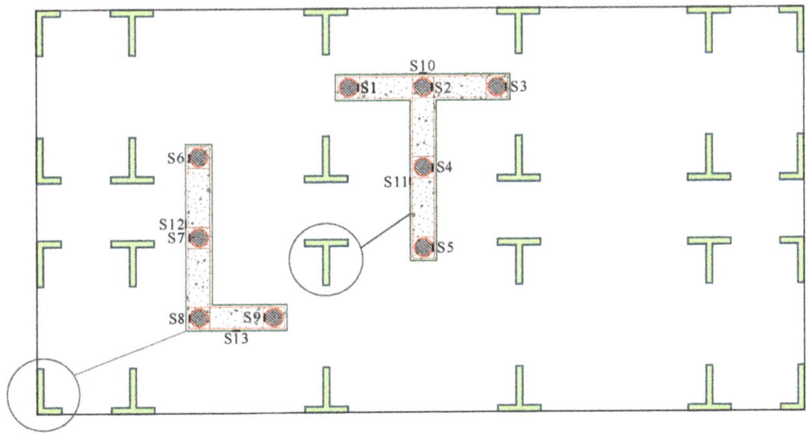

Figure 7. Positions of the strain gauges on the first-story walls (S# is the number of strain gauges).

2.6. Loading Protocol

The shaking table testing was conducted in four phases (I~IV), with increasing magnitudes of input ground motions corresponding, in turn, to the *frequent*, *moderate*, and *rare* levels of intensity 7, and all the way to the *rare* level of intensity 8. In each phase, the four seismic waves aforesaid were successively input as multi-directional excitations into the test structure, thus resulting in a total of 28 loading cases, as summarized in Table 5. Moreover, the low-amplitude white-noise loading cases were applied to the structure to obtain its changes in dynamic characteristics.

Table 5. Input excitations in the shake table testing.

No.	Loading Case [1]	Intensity	Excitation	Duration (s)	PGA (g)		
					X	Y	Z
1	W1	White-noise test		40	0.0500	0.0500	–
2	F7RXY	Frequent level of intensity 7 (Phase I)	RGW	16	0.0833	0.0708	–
3	F7RYX			16	0.0708	0.0833	–
4	F7NXY		Northridge	16	0.0833	0.0708	–
5	F7NYX			16	0.0708	0.0833	–
6	F7EXY		El Centro	16	0.0833	0.0708	–
7	F7EYX			16	0.0708	0.0833	–
8	F7CXY		Chi-Chi	20	0.0833	0.0708	–
9	F7CYX			20	0.0708	0.0833	–
10	W2	White-noise test		40	0.0500	0.0500	–
11	B7RXY	Moderate level of intensity 7 (Phase II)	RGW	16	0.2500	0.2125	–
12	B7RYX			16	0.2125	0.2500	–
13	B7NXY		Northridge	16	0.2500	0.2125	–
14	B7NYX			16	0.2125	0.2500	–
15	B7EXY		El Centro	16	0.2500	0.2125	–
16	B7EYX			16	0.2125	0.2500	–
17	B7CXY		Chi-Chi	20	0.2500	0.2125	–
18	B7CYX			20	0.2125	0.2500	–

Table 5. *Cont.*

No.	Loading Case [1]	Intensity	Excitation	Duration (s)	PGA (g)		
					X	Y	Z
19	W3	White-noise test		40	0.0500	0.0500	–
20	R7RXY	Rare level of intensity 7 (Phase III)	RGW	16	0.5000	0.4250	0.3250
21	R7RYX			16	0.4250	0.5000	0.3250
22	R7NXY		Northridge	16	0.5000	0.4250	–
23	R7NYX			16	0.4250	0.5000	–
24	R7EXY		El Centro	16	0.5000	0.4250	–
25	R7EYX			16	0.4250	0.5000	–
26	R7CXY		Chi-Chi	20	0.5000	0.4250	0.3250
27	R7CYX			20	0.4250	0.5000	0.3250
28	W4	White-noise test		40	0.0500	0.0500	–
29	R8RXY	Rare level of intensity 8 (Phase III)	RGW	16	1.0000	0.8500	0.6500
30	R8RYX			16	0.8500	1.0000	0.6500
31	R8NXY		Northridge	16	1.0000	0.8500	–
32	R8NYX			16	0.8500	1.0000	–
33	W5	White-noise test		40	0.0500	0.0500	–

[1] The naming rule for each loading case is: 'F' represents the frequent level of seismicity, 'B' the moderate level and 'R' the rare level. The number following, '7' or '8', means the seismic intensity. 'W', 'R', 'N', 'E' and 'C' denote the type of input ground motions, namely, the white-noise excitation, the artificial seismic wave, the Northridge, El Centro, and Chi-Chi seismic records, respectively. 'XY' means that the X direction of the test structure is the main direction of the input excitations, and vice versa, 'YX' means that the Y direction corresponds to the main loading direction.

In the tests, the effect of prior seismic damages on subsequent responses was not particularly considered, just as in previous shaking table tests (e.g., [22,30,34]).

3. Test Results

3.1. General Observations

Based on the inspections during the course of testing, the damage progression of the test structure can be summarized as follows:

(1) After the excitations equivalent to the frequent level of intensity 7 (Phase I, PGA below 0.1 g), no visible cracks were found on the concrete surface of the entire test structure. The results of the white-noise test (i.e., W2 in Table 5) also indicated that after this level of shaking, the first three natural frequencies of the specimen were reduced only slightly, by 4.88%, 5.39%, and 4.41% in turn, compared to the results of W1. Thus it can be said that the FSFW structure was well within the elastic state in this phase, satisfying the performance objective (remaining almost undamaged) specified by the seismic design code [29];

(2) Upon the loadings equivalent to the moderate level of intensity 7 (Phase II, the design intensity, with PGA $\leq$ 0.25 g), the white-noise test (W3) indicated that the natural frequencies were reduced by 17.38%, 14.26%, and 8.65%, relative to those obtained after Phase I. As shown in Figure 8, a few minor cracks were observed in this stage, mainly concentrated on the first and second stories, in the vicinity of slab-wall connections. Nevertheless, those visible damages were generally not severe, and still within the expected damage level stipulated by the seismic code;

(3) After a series of shaking representing the rare level of intensity 7 (Phase III, PGA $\leq$ 0.50 g), the natural frequencies were dropped by 25.17%, 26.37%, and 13.89%, with respect to the prior white-noise test (W3). As shown in Figure 9, cracks formed in the previous

phase now extended and widened; new cracks were also observed. However, the structure still performed well, though some local damages were appreciable;

(4) To look at how the structure would fare under extremely strong earthquakes, excitations equivalent to the rare level of intensity 8 were exerted on the test structure (Phase IV, PGA $\leq$ 1.0 g). At the end of this final phase, the first three natural frequencies were decreased by 10.14%, 10.81%, and 17.36% in turn, in comparison with the previous measurements (W4). This suggested a further reduction in the structural stiffness. As shown in Figure 10, previous damages developed more severely. At the proximity of some slab-wall connections, spalling of concrete was obvious (Figure 10a,b). Several cracks even spread from one wall pier to another along the slab (Figure 10c). However, the structure and the walls still remained stable. Neither obvious tilt nor distortion of the specimen was noticed. It can therefore be concluded that the tested FSFW structure was able to resist the rare earthquakes far stronger than the design intensity.

(a) (b) (c)

Figure 8. Damages after excitations of the moderate level of intensity 7 (Phase II): (**a**) a crack at a slab-wall connection; (**b**) another similar crack; (**c**) a crack on an L-shaped wall.

(a) (b) (c)

Figure 9. Damages after excitations of the rare level of intensity 7 (Phase III): (**a**) an extended interface crack; (**b**) slight crushing of a T-shaped wall; (**c**) interior view of the structure.

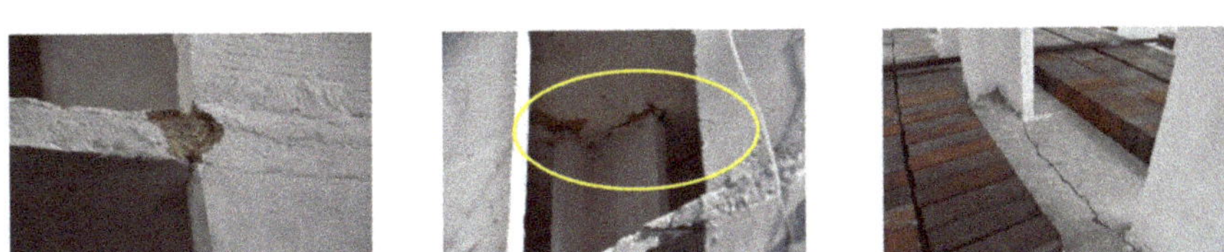

(a) (b) (c)

Figure 10. Damages after excitations of the rare level of intensity 8 (Phase IV): (**a**) crushing at the interface; (**b**) slab damages around a T-shaped wall; (**c**) a crack extending from one wall to another.

3.2. Variation of Dynamic Characteristics

Spectral analysis of the white-noise test results was performed to identify the dynamic characteristics of the test structure in different phases. The variations of the frequency and period of the specimen are shown in Table 6.

From this table, it is clear that as the excitation intensity increased, the main frequencies of the specimen decreased (as expected), signifying a reduction in the overall structural stiffness. In comparison with the intact structure, in Phase III (the rare level of intensity 7), the value of the first frequency decreased by as much as 40%. This significant loss was in correlation with the observation of severe local damages in the specimen. However, after the devastating seismic inputs (i.e., the rare level of intensity 8), the further loss in the frequency was only marginal (by about 47% compared with the initial value before the test). This could be attributed to most damages being localized only in the vicinity of certain slab-wall connections, whereas the composite walls themselves still performed well. As a result, this prevented the progression of failure into the walls. Note that according to Ramos and Hube's numerical simulations [7], a local failure of floor slabs has only a limited impact on the collapse behavior of FSFW structures. This is partly corroborated by the current real testing.

Table 6. Dynamic characteristics of the test structure in different phases.

Testing Phase	Frequency (Hz)/Period (s)		
	1st Order	2nd Order	3rd Order
Before test	3.690/0.271	5.190/0.193	5.440/0.184
Phase I	3.510/0.285	4.910/0.204	5.200/0.192
Phase II	2.900/0.345	4.210/0.238	4.750/0.211
Phase III	2.170/0.461	3.100/0.323	4.090/0.244
Phase IV	1.950/0.513	2.765/0.362	3.380/0.296

3.3. Drift Ratio Envelope

The inter-story drift ratio is one of the most useful engineering demand parameters (EDPs) that is routinely used to evaluate the overall seismic performance of structures. The envelope diagrams of the inter-story drift ratio along the test structure's height in different testing phases are shown in Figure 11.

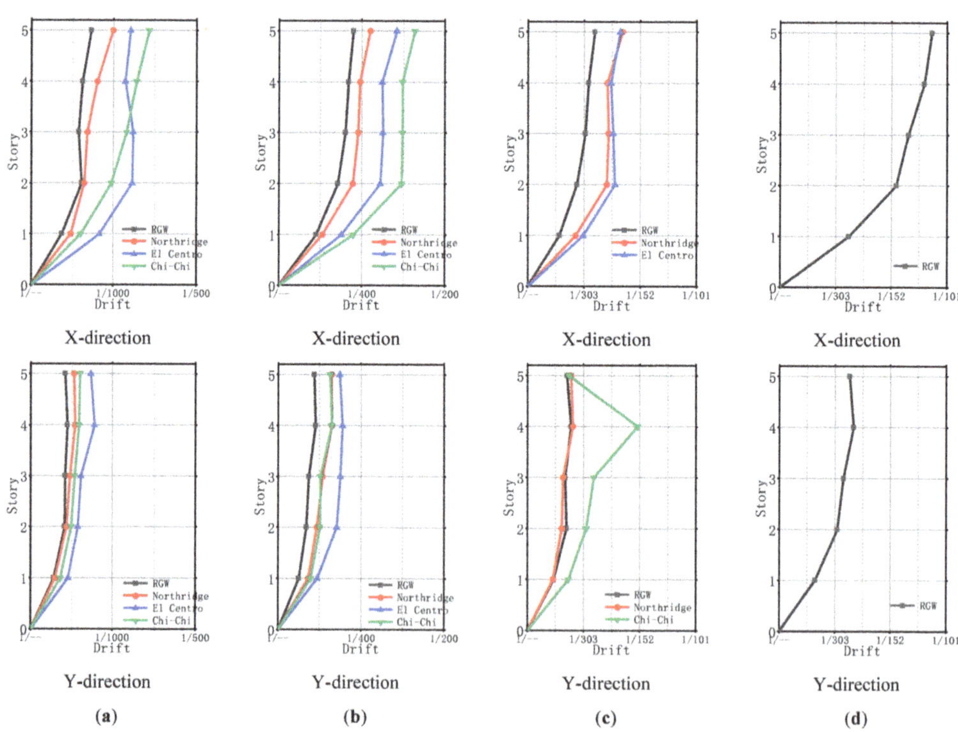

Figure 11. Development of inter-story drift ratio envelope: (**a**) Phase I; (**b**) Phase II; (**c**) Phase III; (**d**) Phase IV.

As per China's seismic code [29], the drift ratio limits for the slab-wall structure are 1/800 and 1/100 under frequent and rare earthquakes, respectively. Clearly, in Phase I the maximum drift ratio was significantly lower than the limit of 1/800, with the exception of the loading case of Chi-Chi seismic wave along the X-direction. This indicates that, overall, an elastic response was exhibited for the specimen. More favorably, in Phase III, and even in Phase IV, the maximum drift ratios were still well below 1/100. This apparently met the performance objective of preventing structural collapse under major (or even larger) earthquakes. In fact, the excessive lateral drift of flat-slab systems has long been a concern [35]. In this experiment, due partly to the strengthening effect of CFSTs, the lateral deformations of the FSFW structure were satisfactorily controlled.

3.4. Maximum Floor Shear Force

The maximum floor shear forces (MFSFs) registered in different testing phases are shown in Figure 12. It is evident that different earthquakes produced different magnitudes of MFSFs. The MFSFs along the Y-direction were generally larger than those along the X-direction. That was because the lateral stiffness along the Y-direction was stronger. Moreover, from Phase III to Phase IV, the MFSFs below floor 4 along the X-direction increased significantly, whereas the corresponding shears along the Y-direction increased more gradually and proportionally, showing a better seismic-resisting capacity.

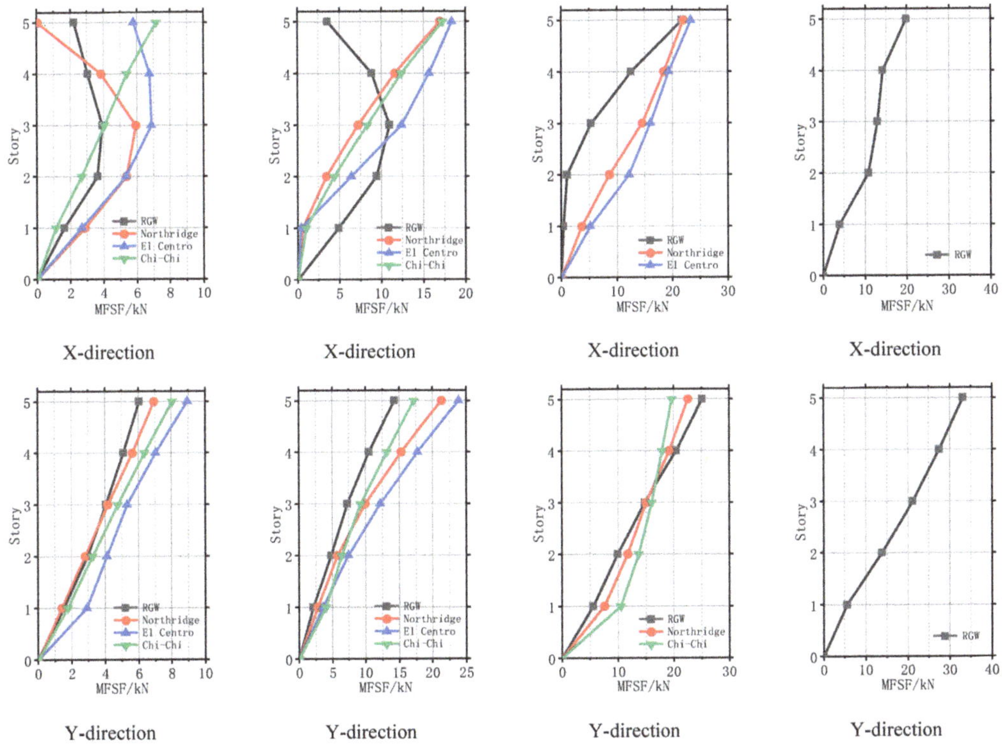

(a) (b) (c) (d)

Figure 12. Development of MFSFs: (**a**) Phase I; (**b**) Phase II; (**c**) Phase III; (**d**) Phase IV.

The MFSFs increased along the height of the structure, particularly in the later testing phases. Due to the so-called 'whiplash' effect [36], the dynamic response at floor 5 was slightly amplified when the specimen entered into the inelastic state. The MFSFs, nonetheless, were proportionally distributed along the height in most excitation cases; no weak story was formed in the tests.

3.5. Strain Response of CFSTs

As mentioned previously, a total of nine measuring points (S1~S9, see Figure 7) were set up to monitor the strains of the CFSTs. Figures 13 and 14 compare the strain developments at S8 and S9, which corresponded to the web–flange intersection and boundary positions of the L-shaped wall indicated in Figure 7.

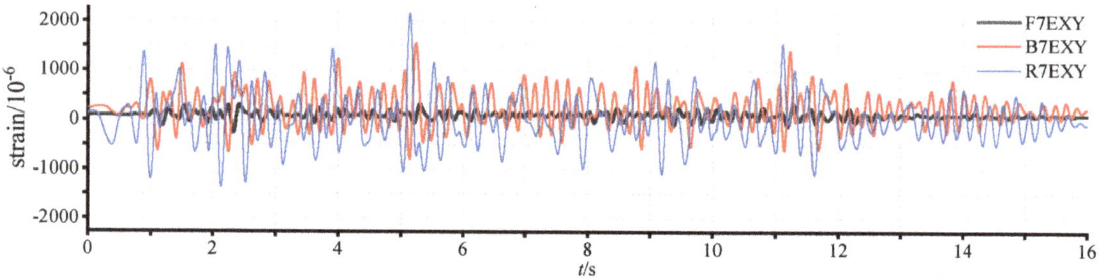

Figure 13. Comparison of time-history responses of strain at S8 under different loading cases.

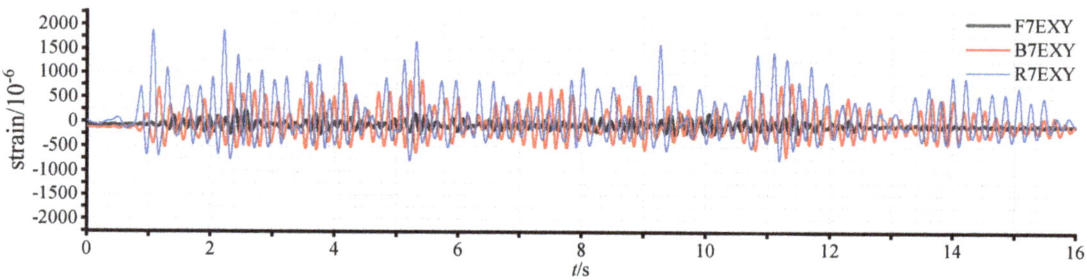

Figure 14. Comparison of time-history responses of strain at S9 under different loading cases.

The yield strain of the steel tubes used in the CFSTs was roughly 1500×10^{-6}. Clearly, in the case of the frequent level of intensity 7 (F7EXY in the figures), the strains at S8 and S9 were well within the elastic range. With the increase in excitation intensity, the strains increased markedly. Under the excitation equivalent to the moderate level of intensity 7 (B7EXY), the steel at S8 attained the yield strain, but that at S9 did not. This highlighted the larger strain demand at the web–flange intersection of the corner of the L-shaped wall. For the loading case of R7EXY (the rare level of intensity 7), both strains at S8 and S9 exceeded the yield strain, indicating the full mobilization of the CFSTs to resist lateral loads.

4. Numerical Modeling

4.1. General Descriptions of Modeling

A 3D non-linear finite-element (FE) model for the test structure was developed using the software PERFORM-3D [37]. The specially-shaped walls were modeled using the 4-node 'General Wall' element available in PERFORM-3D, which simplifies non-linear area elements intended for the specific purpose of modeling RC walls [38]. The cross-section of each wall pier was represented by fibers that account for the axial-bending non-linear interactions, but is uncoupled from the shear response.

No element is provided in PERFORM-3D for modeling the composite walls with embedded CFSTs. Thus an ad hoc treatment was adopted in this study. The two-noded fiber-based column (line) elements available in PERFORM-3D were used to model the individual CFSTs in a wall pier, which were connected by the 'embedded beam' model [39] to ensure the deformation compatibility of the CFSTs in the same wall pier.

The flat floor slabs were modeled using the elastic beam elements with non-linear hinges. This modeling approach has been proven to have adequate accuracy in capturing the overall out-of-plane non-linear response of floor slabs [13]. These equivalent beam elements have a rectangular section with its width and depth dimensions equal to the effective slab width and total slab thickness. In this study, the effective slab width was determined by the method recommended by Grossman [40]. This width was further multiplied by one-third to consider the impact of concrete cracking [13,40].

Figure 15 shows a schematic diagram of the FE elements used in this study. Figure 16 provides a 3D view of the test structure established in PERFORM-3D.

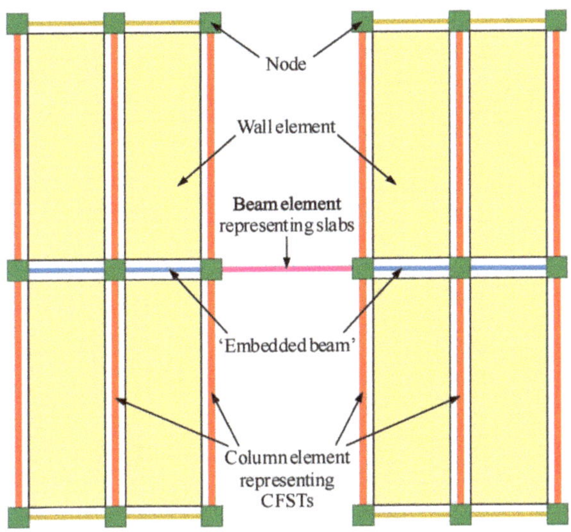

Figure 15. Schematic diagram of the FE elements used in this study.

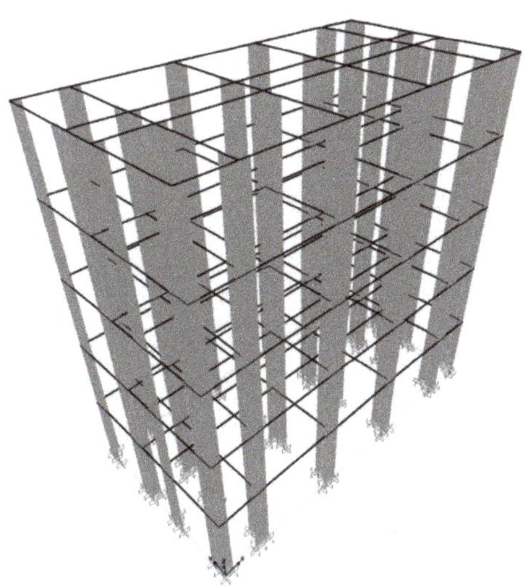

Figure 16. Three-dimensional view of the structural model in PERFORM-3D.

4.2. Material Constitutive Models

The concrete in CFSTs was represented by the model proposed by Tao et al. [41], which takes into account the confinement effect coming from the outer steel tube. The stress–strain relationship recommended in China's concrete design code [31] was adopted for the concrete in other parts of the test structure; in other words, no confinement effect was considered in these cases. The steel materials used in the test structure (including steel tube and rebar) were represented by the tri-linear stress–strain model [42], ignoring local buckling effect.

Figure 17 illustrates the material constitutive models used for concrete and steel. Note that in PERFORM-3D, all generalized force–displacement (F–D) curves can only be composed of a specific number of line segments. In this study, the principle of equal area criterion (that is, the area under the line segments and the area covered by the constitutive curve are the same) was followed to determine the F–D curves for PERFORM-3D. The fitting diagram of the concrete's constitutive relationship is shown in Figure 17a, and the tri-linear model for steel is shown in Figure 17b.

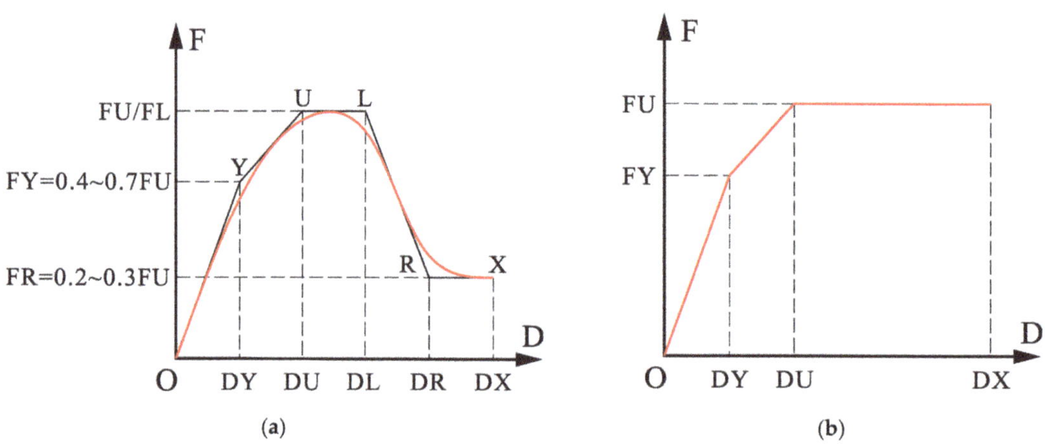

Figure 17. Generalized F–D curves used in PERFORM-3D for: (**a**) concrete; (**b**) steel.

4.3. Modeling Results

The first three periods obtained by the FE analysis were 1.132s, 0.912s, and 0.830s, while the corresponding periods of the test structure were measured as 1.212, 0.862, and 0.822s, respectively. The errors between them were less than 7%. This suggests that the FE model closely matched the dynamic characteristics of the test structure.

Figure 18 compares the experimental and numerical time-history displacements at the roof along the X or Y directions under different loading conditions. It is clear that, in general, the calculated results agreed well with the experimental records. However, in the later testing phase (i.e., R7EXY in the figure), the numerical model overestimated the actual roof displacements. This may be due to the limitations of using PERFORM-3D to simulate shear walls, as discussed in depth in [43], or it might likely be ascribed to the excessive reduction in the effective width of the slabs adopted in the modeling. It is noteworthy that although simulations of shear wall buildings have been extensively investigated (e.g., [44–47]), there are still challenges to accurately reproduce their non-linear behavior, especially when the walls are severely damaged.

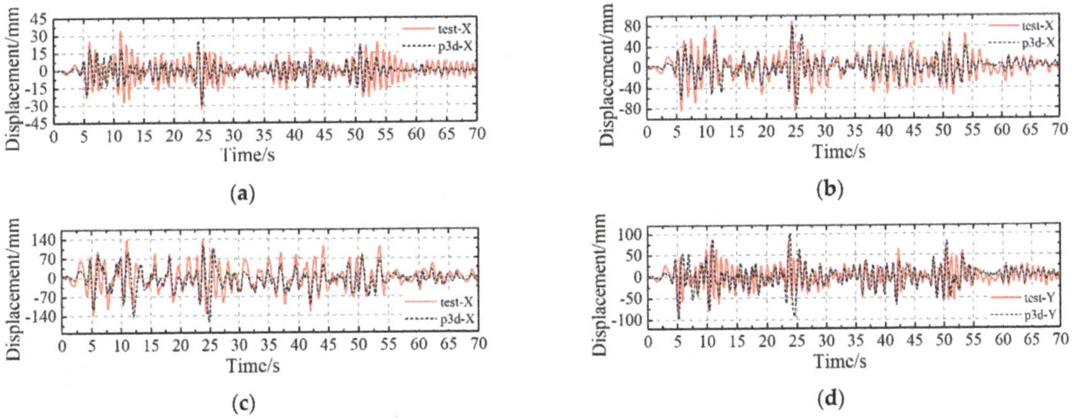

Figure 18. Comparisons of the measured and calculated roof displacements in different loading cases: (**a**) F7EXY-X; (**b**) B7EXY-X; (**c**) R7EXY-X; (**d**) R7EXY-Y.

4.4. Comparison of FSFW Systems with and without CFSTs

Finally, to highlight the efficacy of CFSTs embedded in wall piers, two FSFW systems equipped with and without CFSTs were compared in terms of their seismic responses.

Figure 19 shows the comparison results for two scenarios (namely, two FSFW systems with and without CFSTs embedment) in terms of their inter-story drift ratio envelope in different loading cases. Obviously, the presence of CFSTs significantly reduced the lateral deformation of the test structure. In the case of small earthquakes (F7CXY-X), the inter-story drift ratio of the system without CFSTs exceeded the limit value of 1/800. Further, in the case R7NXY-X, the inter-story drift ratio of that system exceeded the limit value of 1/100. By comparison, the FSFW system incorporating CFSTs did meet all those lateral drift requirements.

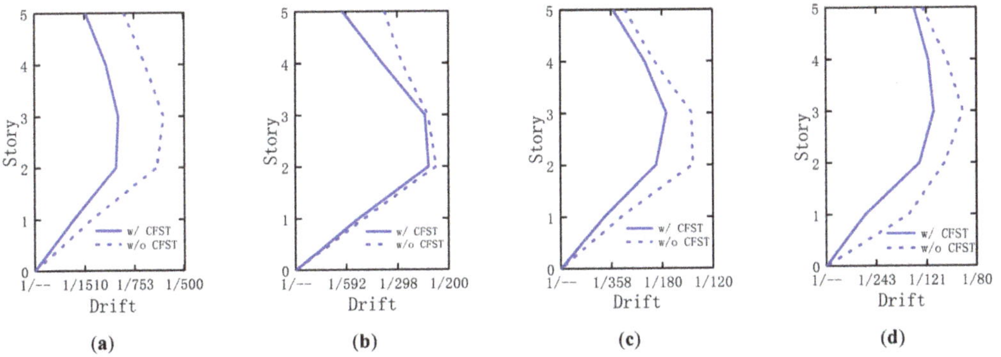

Figure 19. Comparison of inter-story drift ratio envelopes in different loading cases: (**a**) F7CXY-X; (**b**) B7EXY-X; (**c**) R7NXY-X; (**d**) R8RXY-X.

A push-over analysis was further conducted to see how the two systems could behave differently. A monotonically increasing horizontal force with an inverted triangular distribution was applied to the systems [48]. Figure 20 gives the push-over results along the X and Y directions. Despite their similar initial overall stiffness, the CFST-equipped system had considerably higher lateral load-carrying capacity than the system without CFSTs. This significant discrepancy clearly shows the utility of applying CFSTs in wall piers to resist seismic lateral loads.

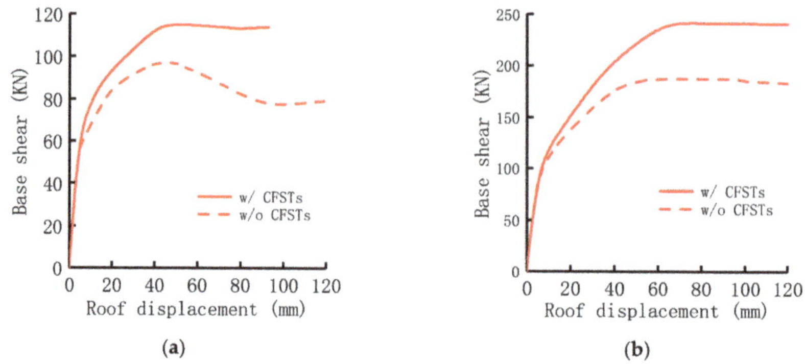

Figure 20. Push-over results along two main directions: (**a**) X direction; (**b**) Y direction.

5. Conclusions

In this paper, the results of a shake-table testing of a 1/8-scale five-story FSFW structure incorporating CFSTs as strengthening elements are briefly presented. Numerical modeling is also attempted to capture the structure's non-linear time-history responses. Conclusions can be summarized as the following:

(1) The application of CFSTs in flanged concrete walls helps to resist earthquake loads acting on the FSFW system. Consequently, the performance objectives specified in the seismic code are generally met, even under extremely strong earthquakes;

(2) The seismic damage is mainly localized in the vicinity of slab-wall connections of the tested structure, which leads to a significant loss of natural frequencies (up to 47%). Clearly, these regions are the most vulnerable part of the FSFW structure. As such, detailing of the connections between the walls and slabs should be adequately considered and further addressed;

(3) In spite of the above disadvantages, it is observed that the composite walls themselves performed fairly well, even when subjected to ground motions much stronger than the design intensity. This may partially explain why the FSFW model structure was still standing after the shake table tests;

(4) A 3D non-linear FE model is developed to simulate the dynamic responses of the tested structure. Adopting the method suggested herein for modeling the embedded CFSTs, a good match is found between the FE results and the experimental observations. Nevertheless, a detailed description of the non-linear behavior of slab-wall connections and a rational representation of the floor contribution deserve further investigation.

Author Contributions: Conceptualization, X.-Y.Z., X.-D.F. and F.W.; methodology, X.-Y.Z. and F.W.; software, X.-Y.Z.; validation, X.-Y.Z.; formal analysis, X.-Y.Z. and F.W.; investigation, X.-Y.Z. and F.W.; resources, X.-Y.Z. and F.W.; data curation, X.-Y.Z., F.W. and J.Z.; writing—original draft preparation, X.-Y.Z.; writing—review and editing, F.W.; visualization, X.-Y.Z. and F.W.; supervision, F.W.; project administration, F.W.; funding acquisition, X.-Y.Z. and F.W. All authors have read and agreed to the published version of the manuscript.

Funding: This research was funded by Guangdong Provincial Key Laboratory of Modern Civil Engineering Technology, grant number 2021B1212040003; and by Architectural Design & Research Institute of SCUT Co. Ltd., grant number 2017-001.

Data Availability Statement: Data sharing is not applicable to this article.

Acknowledgments: The authors thank Ze-Rong Lai for his assistance in this study while pursuing his master's degree.

Conflicts of Interest: The authors declare no conflict of interest.

References

1. Paulay, T.; Taylor, R.G. Slab coupling of earthquake-resisting shearwalls. *ACI J. Proc.* **1981**, *78*, 130–140.
2. Wallace, J.W.; Moehle, J.P. Ductility and detailing requirements of bearing wall buildings. *J. Struct. Eng.-ASCE* **1992**, *118*, 1625–1644. [CrossRef]
3. Balkaya, C.; Kalkan, E. Three-dimensional effects on openings of laterally loaded pierced shear walls. *J. Struct. Eng.-ASCE* **2004**, *130*, 1506–1514. [CrossRef]
4. Mwafy, A. Analytically derived fragility relationships for the modern high-rise buildings in the UAE. *Struct. Des. Tall. Spec.* **2010**, *21*, 824–843. [CrossRef]
5. Mwafy, A.; Khalifa, S. Effect of vertical structural irregularity on seismic design of tall buildings. *Struct. Des. Tall. Spec.* **2017**, *26*, e1399. [CrossRef]
6. Massone, L.M.; Sayre, B.L.; Wallace, J.W. Load–deformation responses of slender structural steel reinforced concrete walls. *Eng. Struct.* **2017**, *140*, 77–88. [CrossRef]
7. Ramos, L.; Hube, M.A. Seismic response of reinforced concrete wall buildings with nonlinear coupling slabs. *Eng. Struct.* **2021**, *234*, 111888. [CrossRef]
8. Massone, L.M.; Bonelli, P.; Lagos, R.; Lüders, C.; Moehle, J.P.; Wallace, J.W. Seismic design and construction practices for RC structural wall buildings. *Earthq. Spectra.* **2012**, *28* (Suppl. 1), S245–S256. [CrossRef]
9. Sritharan, S.; Beyer, K.; Henry, R.S.; Chai, Y.H.; Kowalsky, M.; Bull, D. Understanding poor seismic performance of concrete walls and design implications. *Earthq. Spectra.* **2014**, *30*, 307–334. [CrossRef]
10. Jünemann, R.; de La Llera, J.C.; Hube, M.A.; Cifuentes, L.A.; Kausel, E. A statistical analysis of reinforced concrete wall buildings damaged during the 2010, Chile earthquake. *Eng. Struct.* **2015**, *82*, 168–185. [CrossRef]
11. Jünemann, R.; de la Llera, J.C.; Hube, M.A.; Vásquez, J.A.; Chacón, M.F. Study of the damage of reinforced concrete shear walls during the 2010 Chile earthquake. *Earthq. Eng. Struct. Dyn.* **2016**, *45*, 1621–1641. [CrossRef]
12. Song, C.; Pujol, S.; Lepage, A. The collapse of the Alto Río building during the 27 February 2010 Maule, Chile, earthquake. *Earthq. Spectra.* **2012**, *28* (Suppl. 1), S301–S334. [CrossRef]
13. Deger, Z.T.; Wallace, J.W. Collapse assessment of the Alto Río building in the 2010 Chile earthquake. *Earthq. Spectra.* **2015**, *31*, 1397–1425. [CrossRef]
14. Qadeer, A.; Smith, B.S. The bending stiffness of slabs connecting shear walls. *ACI J. Proc.* **1969**, *66*, 464–473.
15. Coull, A.; El Hag, A.A. Effective coupling of shearwalls by floor slabs. *ACI J. Proc.* **1975**, *72*, 429–431.
16. Wong, Y.C.; Coull, A. Interaction between floor slabs and shearwalls in tall buildings. *ACI Spec. Publ.* **1980**, *63*, 543–574.
17. Coull, A.; Chee, W.Y. Stresses in slabs coupling flanged shear walls. *J. Struct. Eng.-ASCE* **1984**, *110*, 105–119. [CrossRef]
18. Coull, A.; Chee, W.Y. Cracked coupling slabs in shear wall buildings. *J. Struct. Eng.-ASCE* **1990**, *116*, 1744–1748. [CrossRef]
19. Ugalde, D.; Parra, P.F.; Lopez-Garcia, D. Assessment of the seismic capacity of tall wall buildings using nonlinear finite element modeling. *Bull. Earthq. Eng.* **2019**, *17*, 6565–6589. [CrossRef]
20. Pavel, F.; Vacareanu, R.; Marcu, D. Seismic performance assessment and rating for a flat-slab RC core wall structure in Bucharest, Romania. *Structures* **2021**, *31*, 1006–1016. [CrossRef]
21. Pantazopoulou, S.; Imran, I. Slab-wall connections under lateral forces. *ACI Struct. J.* **1992**, *89*, 515–527.
22. Panagiotou, M.; Restrepo, J.I.; Conte, J.P. Shake-table test of a full-scale 7-story building slice. Phase I: Rectangular wall. *J. Struct. Eng.-ASCE* **2011**, *137*, 691–704. [CrossRef]
23. Greeshma, S.; Jaya, K.P. Effect of slab shear reinforcement on the performance of the shear wall–floor slab connection. *J. Perform. Constr. Fac.* **2013**, *27*, 391–401. [CrossRef]
24. Brunesi, E.; Peloso, S.; Pinho, R.; Nascimbene, R. Cyclic testing and analysis of a full-scale cast-in-place reinforced concrete wall-slab-wall structure. *Bull. Earthq. Eng.* **2018**, *16*, 4761–4796. [CrossRef]
25. Fischinger, M.; Kante, P.; Isakovic, T. Shake-table response of a coupled RC wall with thin T-shaped piers. *J. Struct. Eng.-ASCE* **2017**, *143*, 04017004. [CrossRef]
26. Zhou, J.; Fang, X.D.; Yao, Z.Q. Mechanical behavior of a steel tube-confined high-strength concrete shear wall under combined tensile and shear loading. *Eng. Struct.* **2018**, *171*, 673–685. [CrossRef]
27. Zhou, J.; Li, P.; Guo, N.F. Seismic performance assessment of a precast concrete-encased CFST composite wall with twin steel tube connections. *Eng. Struct.* **2020**, *207*, 110240. [CrossRef]
28. Zhou, J.; Fang, X.D.; Jiang, Y. Cyclic behavior of concrete-encased high-strength concrete-filled steel tube composite walls: An experiment. *Struct. Concr.* **2021**, *22*, 691–708. [CrossRef]
29. *GB50011-2016*; Code for Seismic Design of Buildings. China Architecture and Building Press: Beijing, China, 2016.
30. Lu, X.L.; Zou, Y.; Lu, W.S.; Zhao, B. Shaking table model test on shanghai world financial center tower. *Earthq. Eng. Struct. Dyn.* **2007**, *36*, 439–457. [CrossRef]
31. *GB50010-2010*; Code for Design of Concrete Structures. China Architecture and Building Press: Beijing, China, 2010.
32. Harris, H.G. *Structural Modeling and Experimental Techniques*, 2nd ed.; CRC Press: Boca Raton, FL, USA, 1999.
33. Maekawa, K.; Ishida, T.; Kishi, T. *Multi-Scale Modeling of Structural Concrete*; Taylor & Francis: London, UK, 2009.
34. Lu, X.L.; Chen, C.; Jiang, H.J.; Wang, S.S. Shaking table tests and numerical analyses of an RC coupled wall structure with replaceable coupling beams. *Earthq. Eng. Struct. Dyn.* **2018**, *47*, 1882–1904. [CrossRef]
35. Erberik, M.A.; Elnashai, A.S. Fragility analysis of flat-slab structures. *Eng. Struct.* **2004**, *26*, 937–948. [CrossRef]

36. Hassan, M.; El-Tawil, S. Inelastic dynamic behavior of hybrid coupled walls. *J. Struct. Eng.-ASCE* **2004**, *130*, 285–296. [CrossRef]
37. Computers & Structures Inc. *Perform3D V5 Nonlinear Analysis and Performance Assessment for 3D Structures*; Computers & Structures Inc.: Berkeley, CA, USA, 2015.
38. Correal, J.F.; Hidalgo, V.; Reyes, J.C.; Ángel, C.C. A comparative study of seismic diaphragm design forces for RC dual system buildings. *Bull. Earthq. Eng.* **2020**, *18*, 4515–4540. [CrossRef]
39. Kolozvari, K.; Kalbasi, K.; Orakcal, K.; Wallace, J. Three-dimensional model for nonlinear analysis of slender flanged reinforced concrete walls. *Eng. Struct.* **2021**, *236*, 112105. [CrossRef]
40. Grossman, J.S. Verification of proposed design methodologies for effective width of slabs in slab-column frames. *ACI Struct. J.* **1997**, *94*, 181–196.
41. Tao, Z.; Wang, Z.B.; Yu, Q. Finite element modelling of concrete-filled steel stub columns under axial compression. *J. Constr. Steel Res.* **2013**, *89*, 121–131. [CrossRef]
42. Hu, H.S.; Xu, L.; Guo, Z.X.; Shahrooz, B.M. Behavior of eccentrically loaded square spiral-confined high-strength concrete-filled steel tube columns. *Eng. Struct.* **2020**, *216*, 110743. [CrossRef]
43. Kolozvari, K.; Arteta, C.; Fischinger, M.; Gavridou, S.; Hube, M.; Isakovic, T.; Lowes, L.; Orakcal, K.; Vásquez, J.; Wallace, J. Comparative study of state-of-the-art macroscopic models for planar reinforced concrete walls. *ACI Struct. J.* **2018**, *115*, 1637–1657. [CrossRef]
44. Wallace, J.W. Modelling issues for tall reinforced concrete core wall buildings. *Struct. Des. Tall. Spec.* **2007**, *16*, 615–632. [CrossRef]
45. Shin, M.; Kang, T.H.K.; Grossman, J.S. Practical modelling of high-rise dual systems with reinforced concrete slab-column frames. *Struct. Des. Tall. Spec.* **2010**, *19*, 728–749. [CrossRef]
46. Alwaeli, W.; Mwafy, A.; Pilakoutas, K.; Guadagnini, M. A methodology for defining seismic scenario-structure-based limit state criteria for RC high-rise wall buildings using net drift. *Earthq. Eng. Struct. Dyn.* **2017**, *46*, 1325–1344. [CrossRef]
47. Terzic, V.; Kolozvari, K.; Saldana, D. Implications of modeling approaches on seismic performance of low-and mid-rise office and hospital shear wall buildings. *Eng. Struct.* **2019**, *189*, 129–146. [CrossRef]
48. Kilar, V.; Fajfar, P. Simple push-over analysis of asymmetric buildings. *Earthq. Eng. Struct. Dyn.* **1997**, *26*, 233–249. [CrossRef]

buildings MDPI

Article

Freezing Effect of Enhancing Tubes in a Freeze-Sealing Pipe Roof Method Based on the Unsteady-State Conjugate Heat Transfer Model

Shengjun Deng [1,2], Dong Hu [1], Siyuan She [3], Zequn Hong [4], Xiangdong Hu [5] and Feng Zhou [1,*]

1 Institute of Geotechnical Engineering, Nanjing Tech University, Nanjing 210009, China
2 Research Center of Coastal and Urban Geotechnical Engineering, Zhejiang University, Hangzhou 310058, China
3 CapitaLand Investment Management, Shanghai 201114, China
4 State Key Laboratory for Geomechanics and Deep Underground Engineering, School of Mechanics and Civil Engineering, China University of Mining and Technology, Xuzhou 221116, China
5 Key Laboratory of Geotechnical and Engineering of Ministry of Education, Department of Geotechnical Engineering, Tongji University, Shanghai 200092, China
* Correspondence: zhoufeng@njtech.edu.cn; Tel.: +86-139-1384-4622

Abstract: The freeze-sealing pipe roof (FSPR) method was applied as an innovative construction technology to the Gongbei Tunnel of the Hong Kong–Zhuhai–Macau Bridge. A freezing scheme involving master freezing tubes, enhancing freezing tubes, and limiting freezing tubes is the key component of the freezing effect of the FSPR method during the construction process under various working conditions. This is related to whether the thickness and temperature of the frozen soil meet the design requirements under various complex working conditions, and it is also related to frost heave control and energy saving. Based on the unsteady-state conjugate heat transfer model, different freezing schemes of enhancing freezing tubes—that is, the shape, layout, operating duration, and heat preservation—were simulated to analyze the freezing effect, which can be measured by the thickness of frozen soil around the steel pipes and the average temperature of the frozen soil curtain. The results show that the greater the contact area between the enhancing tube and the inner wall of the steel pipe, the better the freezing effect, and that the semicircle enhancing freezing tube scheme is superior to the other three shapes of freezing tubes. The arrangement of enhancing freezing tubes far away from the excavation surface, without heat preservation measures, has a better freezing effect due to the function of the hollow pipe as a freezing pipe. Moreover, the enhancing freezing tube can be operated intermittently to control frost heave. Our research simulated the temperature fields of different media—such as steel pipes, frozen soil, and air—providing a design basis for similar projects, such as the combination of the pipe-roofing method and artificial freezing method.

Keywords: Gongbei Tunnel; freeze-sealing pipe roof method; heat transfer model; freezing scheme; enhancing freezing tube

Citation: Deng, S.; Hu, D.; She, S.; Hong, Z.; Hu, X.; Zhou, F. Freezing Effect of Enhancing Tubes in a Freeze-Sealing Pipe Roof Method Based on the Unsteady-State Conjugate Heat Transfer Model. *Buildings* **2022**, *12*, 1373. https://doi.org/10.3390/buildings12091373

Academic Editor: Suraparb Keawsawasvong

Received: 4 August 2022
Accepted: 30 August 2022
Published: 2 September 2022

1. Introduction

The freeze-sealing pipe roof (FSPR) method, as an innovative pre-supporting method, was first applied in the Gongbei Tunnel of the Hong Kong–Zhuhai–Macao Bridge. The roofing pipes play the role of bearings, while the ground freezing mainly plays the role of sealing water. The freezing design of the FSPR method includes three types of freezing tubes—master freezing tubes, enhancing freezing tubes, and limiting freezing tubes—which is different from conventional freezing designs such as the one applied in the Berlin Metro Line 5 project [1]. Among these types of freezing tubes, enhancing freezing tubes are the most important to ensure the freezing effect. This is related to whether the thickness and temperature of the frozen soil meet the design requirements under various complex working conditions, and it is also related to frost heave control and energy saving. However, there are only a few studies that specifically deal with enhancing freezing tubes.

To explore the effects of various design schemes of enhancing freezing tubes on the freezing effect, various research methods can be employed, such as analytical solutions, numerical simulations, model tests, and field tests. Since the freeze-sealing pipe roof method is a new type of tunnel pre-support method, it was first applied to the Gongbei Tunnel in China, and there have been few cases of its use in other countries. Therefore, the relevant technical standards and specifications were mostly formulated in China, and the research results have mainly been reported by Chinese scholars. Temperature is one of the important indicators to evaluate the freezing effect. In terms of analytical solutions, a new typical freezing temperature model of the FSPR method was proposed, with the freezing tubes arranged inside jacking pipes, and then an analytical solution to the steady-state temperature field was obtained by transforming the circular boundary condition to a linear boundary condition via conformal mapping and superposition methods [2]. Then, the FSPR model based on the Gongbei Tunnel was proposed, considering the operation of limiting freezing tubes, and the analytical solution of the steady-state temperature field was first investigated using the superposition method and boundary separation method of the Laplace equation [3]. However, the analytical solution can only solve the simplified model of the FSPR method, which concerns regularly symmetric objects far from the actual project. Model tests can achieve studies of similar scale and size. A physical model test on temperature field of a large-scale FSPR structure was conducted to validate the freezing effect [4]. Similar mechanical model tests of the FSPR method based on two and three steel pipes were also carried out to optimize the design of suitable freezing temperatures from the perspective of water-sealing performance [5]. The influence of thermal disturbance of FSPR structures during construction on the freezing effect was explored by model tests [6]. Compared with model tests, field tests can effectively reduce the size effect. A field test on the active freezing scheme of the FSPR method was conducted to analyze the freezing effect under different freezing modes [7]. The freezing effect and optimal freezing scheme of the FSPR method in different phases was also explored [8]. However, field tests are time-consuming and overly expensive, so it is difficult to conduct all tests that consider various parameters. Therefore, numerical simulation is a relatively good method, and the freezing effect can be visually observed through the simulation results. Temperature simulation of the FSPR method can be carried out during the active and maintained freezing phase, which is similar to the work described in [9,10]. The influence of operating and stop duration of enhancing freezing tubes on the freezing effect of the FSPR method was analyzed by numerical simulation [11]. Combined with the monitoring method of the excavation surface, the construction safety factor can be further improved [12,13].

In the above numerical simulation model, frozen soil is regarded as the only heat transfer medium, and the temperature boundary is directly assigned to the outer surface of the steel pipe. However, the heat conduction process in the real project is designed with three media: steel pipes, frozen soil, and air. Therefore, this research concerns the numerical simulation of a specific shape of enhancing freezing tubes and the heat conduction between different media, i.e., steel pipes, frozen soil, and air. An optimized design proposal requires the comparison of different shapes of enhancing freezing tubes through numerical simulation, and it is necessary to consider unsteady-state conjugate heat transfer between various media. The research on enhancing freezing tubes is crucial to the reliability of water sealing and frost heave control of the FSPR method. This study takes the Gongbei Tunnel of the Hong Kong–Zhuhai–Macao Bridge as the engineering background, and explores the freezing effect of enhancing freezing tubes with different design shapes, layout, operating duration, and heat preservation measures, based on the unsteady-state conjugate heat transfer theory.

2. Engineering Background

The Gongbei Tunnel is a key link of the Hong Kong–Zhuhai–Macao Bridge. The layout of the FSPR method is shown in Figure 1. During construction, 36 steel pipes with a diameter of Φ1620 mm and thickness of 20 mm were jacked; among them, the odd-numbered steel jacking pipes were full of concrete, while the even-numbered jacking pipes were hollow. The two types of pipes were arranged alternately, with a spacing of 357 mm. The two types of jacking pipes were arranged in a staggered manner inside and outside; the height difference of the circle center was 300 mm, and the buried depth of the tunnel was 4~5 m.

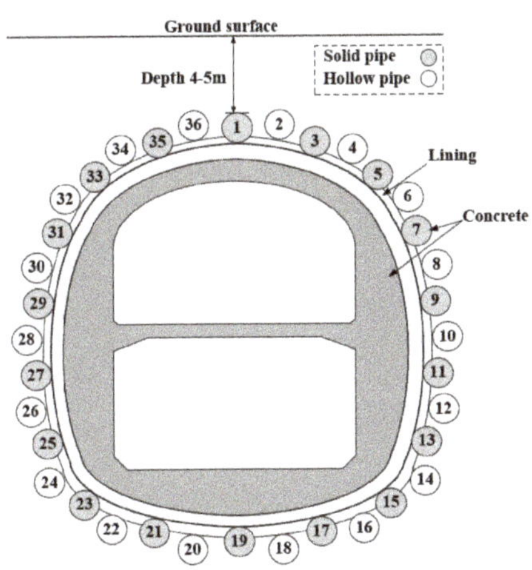

Figure 1. Layout scheme of the freeze-sealing pipe roof method.

The layout of the freeze-sealing pipe roof method in the Gongbei Tunnel includes three types of freezing tubes: master freezing tubes with a diameter of 133 mm and thickness of 4 mm, limiting freezing tubes with a diameter of 133 mm and thickness of 4 mm, and enhancing freezing tubes with a diameter of 159 mm and thickness of 4.5 mm; the layout of the enhancing freezing tubes is shown in Figure 2.

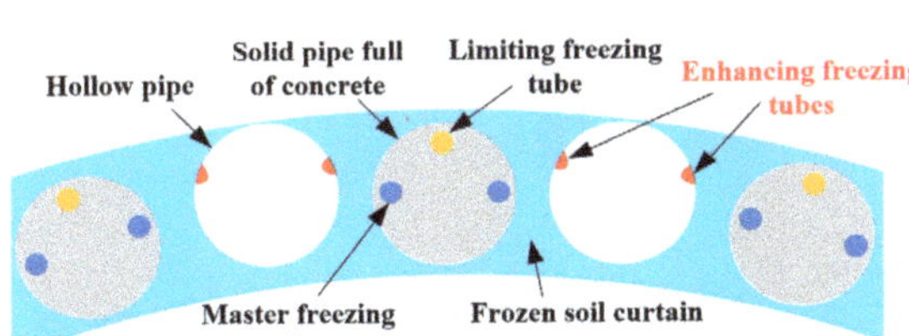

Figure 2. Layout of the freezing tubes in the pipe-roofing.

Among the freezing tubes, the function of the master freezing tubes is to freeze the soil between the jacking pipes. The limiting freezing tubes mainly limit the thickness of the frozen soil to control frost heave by increasing the temperature. There are two main functions of enhancing freezing tubes: the first is to strengthen the freezing effect; the second is to resist the hydration heat during the pouring of concrete to maintain the thickness of the FSPR structure. Therefore, further research on the appearance, layout, and freezing scheme of enhancing freezing tubes is necessary to ensure the freezing effect and frost heave control.

3. Numerical Model

3.1. Assumptions and Computational Model

The Gongbei Tunnel is a typical engineering case of the FSPR method, and is used as a case study in this paper. To establish an efficient computational model, the typical parts of the FSPR structure are taken as the computational model, as shown in Figure 3. The size of the model is 11 m $\times$ 1.977 m. For the unsteady-state conjugate heat conduction model, the left and right sides and the lower part of the model are the soil, which can be regarded as the second temperature boundary condition. The upper part of the model is the ground surface, which can be regarded as the third boundary condition [14,15]. Inside the model, the freezing tube as a cold source can be regarded as the first temperature boundary condition, and the temperature of the brine flowing into the main tube is used as the boundary temperature [16]. The steel pipe is a conjugate heat transfer surface [17]. In this computational model, the strong coupling integral computational method is selected, and the general control equation is used to find a global solution. Therefore, the initial values of the temperature field, velocity field, and pressure field need to be given for the whole region. For the initial value of the velocity field, both the solid domain and fluid domain are recorded as 0 m/s; for the initial pressure distribution, the fluid and solid domains are denoted as 1 atm. The upper boundary that represents the ground surface is regarded as the convective heat transfer boundary. The heat flux q in the lower boundary is 0.4 W/m^2, as expressed by Equation (1). The left and right boundaries are the adiabatic boundaries.

$$q = \lambda \frac{\partial T}{\partial y} \tag{1}$$

where λ is the thermal conductivity of the soil, and $\frac{\partial T}{\partial y}$ is the geothermal gradient, with a value of 0.03 °C/m.

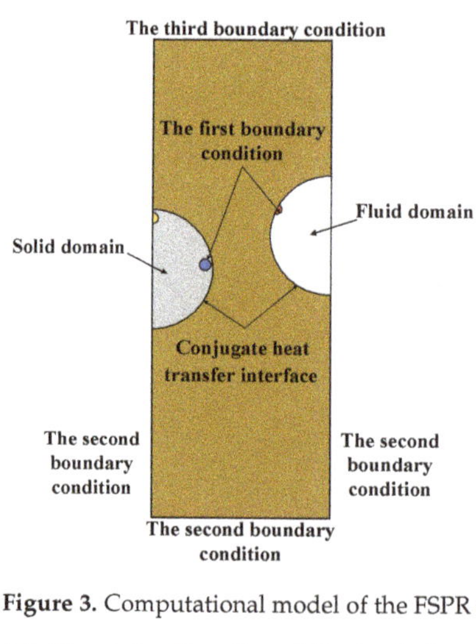

Figure 3. Computational model of the FSPR method based on unsteady-state conjugate heat transfer.

3.2. Governing Equation of Unsteady-State Conjugate Heat Transfer Model

Ground freezing is an unsteady-state heat transfer process with complex phase transition. In the solidification process of pure substances such as water, solidification occurs at a single temperature, and the solid phase and liquid phase are separated by a clear moving interface. However, the soil freezing occurs in a larger temperature range, and there is a separation of the solid and liquid phases by moving regions of two phases in the process [18]. The phase transition problem is mathematically strongly nonlinear, meaning that the governing equation is linear, but the position of the two-phase interface must always be determined, and the energy conservation condition of the interface is nonlinear. It is not possible to use the superposition principle of solutions. Therefore, most of these problems are treated by numerical simulation methods [19]. When using numerical methods to solve phase transition problems, there are generally two methods to deal with the moving boundary in the process of phase transition: The first focuses on the solution of the phase transition interface. After determining the interface position, the temperature distributions in the solid and liquid regions are solved. The second method is to assume the problem as a single-phase nonlinear heat conduction problem, determine the temperature or enthalpy distribution in the whole solution region, and then determine the position to reach the phase transition temperature as the phase transition interface [20–22]. The second method is convenient and practical, and is more suitable for the soil phase transformation process, which has no clear interface. For the second method, the sensible heat capacity method is used. Assuming that the physical properties of the solid and liquid phases are spatially invariant, ignoring the possible natural convection in the liquid phase, the conjugate heat transfer interface between soil and air is the steel pipe. The sensible heat capacity method takes temperature as the function to be solved, without introducing the concept of enthalpy, and establishes a unified energy equation for the whole region. For the treatment of phase transition, the specific heat is expressed in the form of equivalent specific heat [23,24]. For convenience of explanation and comparison with the enthalpy method, the equivalent specific heat of phase transition that occurs at a given temperature Tm is as expressed in Equation (2):

$$\begin{gathered}\tilde{c}(T) = c(T) + L\delta(T - T_m); \\ c(T) = \begin{cases} c_s(T) & T < T_m \\ c_l(T) & T > T_m \end{cases}; \ \delta(T - T_m) = \begin{cases} 1 & T = T_m \\ 0 & T \neq T_m \end{cases}\end{gathered} \tag{2}$$

where $\delta(T - T_m)$ is a Dirac function and, thus, has a heat capacity model as shown in Equation (3):

$$\rho\tilde{c}\frac{\partial T}{\partial t} = \mathrm{div}(\lambda \ \mathrm{grad}\ T\Big) \tag{3}$$

References [25,26] proved the equivalence between Equation (3) and the commonly used equations describing the phase transition problem.

For the phase transition that occurs in the temperature range near $T_{\mathrm{m}}(T_m \pm \Delta T)$, the influence of T should be taken into account when constructing the equivalent specific heat. The expression $\tilde{c}$ should be expressed as shown in Equation (4):

$$\begin{gathered}\tilde{c}(T) = \begin{cases} c_s(T) & T < (T_m - \Delta T) \\ c_l(T) & T > (T_m + \Delta T) \end{cases}; \\ \int_{T_m-\Delta T}^{T_m+\Delta T} \tilde{c}(T)\mathrm{d}T = L + \int_{T_m-\Delta T}^{T_m} c_s(T)\mathrm{d}T + \int_{T_m}^{T_m+\Delta T} c_l(T)\mathrm{d}T\end{gathered} \tag{4}$$

When the specific heat and coefficient of thermal conductivity of the solid phase and liquid phase are constant, Equation (5) can be obtained:

$$\begin{gathered}\lambda = \begin{cases} \lambda_s & T < (T_m - \Delta T) \\ \lambda_s + \frac{\lambda_l - \lambda_s}{2\Delta T}[T - (T_m - \Delta T)] & (T_m - \Delta T) \le T \le (T_m + \Delta T); \\ \lambda_l & T > (T_m + \Delta T) \end{cases} \\ \tilde{c}(T) = \begin{cases} c_s & T < (T_m - \Delta T) \\ \frac{L}{2\Delta T} + \frac{c_s + c_l}{2} & (T_m - \Delta T) \le T \le (T_m + \Delta T) \\ c_l & T > (T_m + \Delta T) \end{cases}\end{gathered} \tag{5}$$

References [18,27,28] note that the phase change of water in frozen soil can be divided into three regions:

(1) Severe phase transition zone: when the temperature in this zone changes by 1 °C, the variation in unfrozen water content is greater than or equal to 1%;

(2) Transition zone: when the temperature in this zone changes by 1 °C, the variation in unfrozen water content is between 0.1% and 1%;

(2) Frozen zone: when the temperature in this zone decreases by 1 °C, the amount of the water phase becoming ice is less than 0.1%.

Accordingly, when the sensible heat capacity method is used to deal with the phase change problem of the soil freezing process, the change in the unfrozen water in the soil should be divided into at least three sections according to the experimental data—a violent phase change zone, transition zone, and frozen solid zone—and then the phase change should be treated with the equivalent specific heat in each section.

Similar to references [29,30], the specific heat of the soil region is as shown in Equation (6):

$$c(T) = \begin{cases} c_f & T < T_b \\ c_f + \frac{c_u - c_f}{T_a - T_b}(T - T_b) + \frac{L}{(1+w)}\frac{\partial w_i}{\partial T} & T_a \le T \le T_b \\ c_u & T > T_a \end{cases} \tag{6}$$

where c_u, c_f represent the specific heat of unfrozen soil and frozen soil, respectively (unit: J/(kg·K)); L is the latent heat of the phase change of water; w and w_i are the total water content and ice content of the frozen soil, respectively; and T_a and T_b are the upper and lower boundary temperatures of the frozen soil's phase transition zone, respectively.

In the hollow pipe, when the air pressure is low and the temperature is high, the air can be treated as an ideal gas [31]. The air in the fluid domain can also be regarded as a compressed fluid and a viscous fluid. At this time, the continuity equation and motion equation (Navier–Stokes equation) are changed. Considering the causes of fluid movement, the flow state of air can be assumed to be laminar flow. The continuous condition of heat flux can also be treated according to laminar flow in the conjugate heat transfer interface [32].

According to the above assumptions, the general strong coupling control equations of the computational model can be obtained, containing the energy equation, ideal gas state equation, continuity equation, and motion equation.

The energy equation can be expressed with Equation (7):

$$\rho c \frac{\partial T}{\partial t} = \mathrm{div}(\lambda\, \mathrm{grad}\, T) + \Phi \tag{7}$$

The ideal gas state equation can be expressed with Equation (8):

$$p = \rho R T \tag{8}$$

The continuity equation can be expressed with Equation (9):

$$\frac{\mathrm{D}\rho}{\mathrm{D}t} + \rho \mathrm{div}\, \mathbf{v} = 0 \tag{9}$$

The motion equation can be expressed with Equation (10):

$$\rho \frac{\mathrm{D}v}{\mathrm{D}t} = \mathbf{F} - \mathrm{grad}\, p + \mathrm{div}(2\mu \mathbf{S}) - \frac{2}{3}\mathrm{grad}(\mu\, \mathrm{div}\, v) \tag{10}$$

where ρ represents the fluid density, c represents the specific heat shown in Equation (6), v represents the velocity vector, λ represents the thermal conductivity, T represents the temperature, t represents time,μ represents the coefficient of viscosity, Φ represents the intensity of the internal heat source, and R represents the gas constant, $R = R_0/M$.

3.3. Model Parameter

The properties of the typical soil layer in Gongbei Tunnel are shown in Table 1. The thermal conductivity of the soil and air is shown in Table 2. The air viscosity is shown in

Table 3. The specific heat at constant pressure of air is shown in Table 4. The properties of steel and concrete are shown in Table 5.

Table 1. Properties of typical soil layers in Gongbei Tunnel.

Soil Layer Number	Lithology	Moisture Content w (%)	Density ρ (kg/m³)	Dry Density ρ_d (kg/m³)
①	Artificial fill	16.05	1660	1470
③-3	Pebble sand	13.54	2000	1760
④-3	Muddy silty clay	47.6	1820	1230
⑤-1	Silty clay	26.37	2010	1590
⑤-2	Fine sand	18.25	1950	1650
⑤-3	Muddy silty clay	38.27	1880	1360
⑥-2	Medium sand	17.92	2020	1720
⑦-1	Gravel clay	31.98	1890	1430
⑧-1	Completely decomposed granite	17.31	2040	1740
⑧-2	Highly weathered granite	19.65	1980	1650

Table 2. Thermal conductivity of soil and air (W/(m·K)).

Material Names	Thermal Conductivity at Different Temperatures						
	−30 °C	−20 °C	−10 °C	0 °C	10 °C	20 °C	30 °C
①-Artificial fill	/	1.962	1.690	1.511	1.398	1.109	/
③-3 Pebble sand	/	1.925	1.758	1.538	1.217	1.066	/
④-3 Muddy silty clay	/	2.047	1.772	1.614	1.485	1.206	/
⑤-2 Fine sand	/	2.019	1.775	1.719	1.497	1.266	/
⑤-3 Muddy silty clay	/	1.994	1.893	1.643	1.402	1.344	/
⑦-1 Gravel clay	/	2.030	1.790	1.623	1.442	1.319	/
Air	0.022	0.0228	0.0236	0.0244	0.0251	0.0259	0.0267

Table 3. Air viscosity (μPa·s).

−40 °C	−20 °C	0 °C	10 °C	20 °C	30 °C
15.60	16.83	17.09	17.59	18.08	18.56

Table 4. Specific heat at constant pressure of air (J/(kg·K).

−30 °C	−20 °C	−10 °C	0 °C	10 °C	20 °C	30 °C
1011	1009	1009	1010	1012	1013	1014

Table 5. Properties of steel and concrete.

Material	Density (kg/m³)	Thermal Conductivity (W/(kg·K))	Specific Heat (J/(kg·K))
Steel	7850	44.7	459.8
Concrete	2344	1.835	419.8

3.4. The Shape of the Enhancing Freezing Tubes

Research on the shape of enhancing freezing tubes, which has a significant effect on the freezing effect, is lacking. This is the basis of the follow-up work in this paper. Since the air flow in the hollow pipes has a great impact on the enhancing freezing tubes [33], enhancing freezing tubes with different cross-sectional shapes were compared to improve the heat transfer efficiency. The enhancing freezing tubes of different shapes—such as circular, crescent, groove, and semicircular—are shown in Figure 4. To amplify and compare the computational results of the freezing effect, these four types of enhancing freezing tubes were given the same cross-sectional area and tripled to 0.047 m^2.

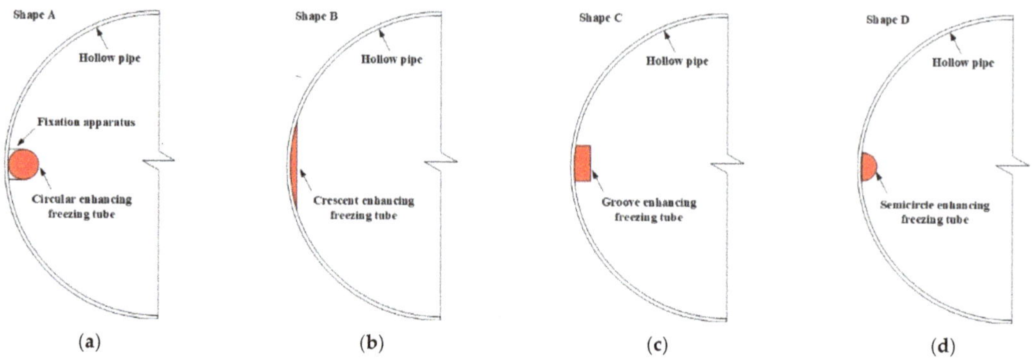

Figure 4. Cross-sectional shapes of enhancing freezing tubes: circular tube (**a**); crescent tube (**b**); groove tube (**c**); semicircular tube (**d**).

The temperature cloud graphs of the enhancing tubes with different shapes over the course of 30 days during the active freezing phase are shown in Figure 5. After 30 days of active freezing, the frozen soil curtain formed between the two pipes based on the different shapes of the enhancing tubes. The frozen soil curtain of Shape A and Shape D was more uniform, while the non-uniformity of the frozen soil curtain of Shape B and Shape C was greater than that of Shape A and Shape D, which may lead to frost heaving.

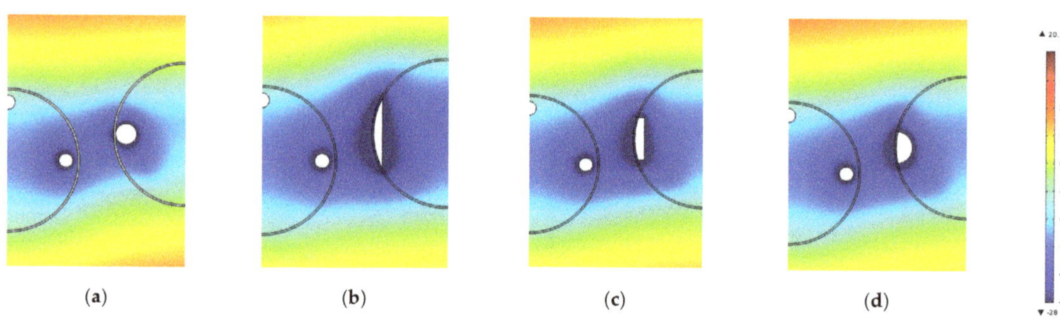

Figure 5. Temperature cloud graphs of enhancing freezing tubes with different shapes (freezing for 30 days): Shape A (**a**), Shape B (**b**), Shape C (**c**), Shape D (**d**).

As shown in Figure 6, the average temperature of the hollow pipe with a circular tube was the highest, while that of the hollow pipe with a crescent tube was the lowest. If the contact area between the enhancing freezing tube and the inner surface of the hollow pipe is increased as much as possible, the cooling capacity of the enhancing freezing tubes can be effectively and quickly transferred to the soil. The average temperature of the circular enhancing freezing tubes of Shape A was the highest; the freezing effect of these tubes was weaker than that of Shape D due to the small contact surface between the enhancing freezing tubes and the steel pipe. Therefore, the scheme of Shape A should be abandoned.

Although the average temperature of the enhancing freezing tubes of Shape B and Shape C was lower, the non-uniformity of their frozen soil curtain thickness was relatively greater. Taking the area below −10 °C as the strength of the frozen soil curtain [34,35], it can be concluded from Table 6 that the thickness of the frozen soil curtain with Shape B was the thickest, and had the greatest degree of unevenness. Uneven frost heaving is detrimental to the pipeline and the surrounding environment [36,37]. To avoid uneven frost heaving, Shape B should be abandoned. The thickness of the frozen soil curtain with Shape C and Shape D was similar, but the degree of unevenness in the thickness of Shape D was far less than that of Shape C. Therefore, the semicircular enhancing freezing tube is the best scheme in terms of the freezing effect and the degree of unevenness in the thickness of the frozen soil curtain.

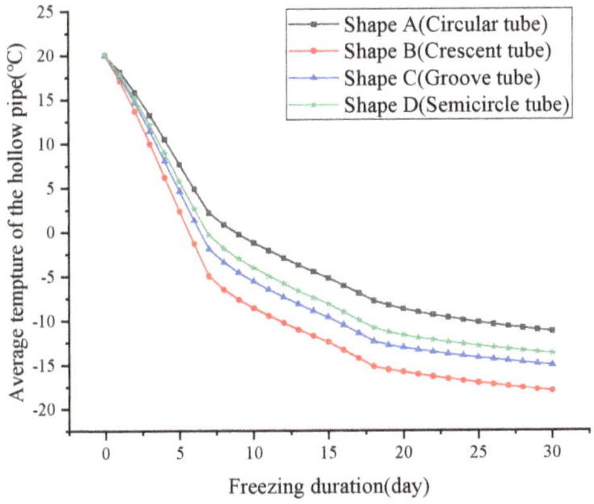

Figure 6. Average temperature of the hollow pipes.

Table 6. Comparison of the freezing effects of enhancing tubes with different shapes.

	Shape A		Shape B		Shape C		Shape D	
Active Freezing for 30 Days	**Left Side**	**Right Side**	**Left Side**	**Right Side**	**Left Side**	**Right Side**	**Left Side**	**Right Side**
Thickness of frozen soil curtain (m)	1.076	0.938	1.274	1.477	1.225	1.288	1.193	1.195
Differentials (m)	0.138		0.203		0.063		0.002	
Thickness unevenness degree	12.825%		15.934%		5.143%		1.676%	

Remark: The thickness unevenness degree is the ratio of the differentials and the thickness of the frozen soil curtain on the left side.

3.5. Calculation Scheme

Research on the layout of enhancing freezing tubes is the most important factor in the FSPR freezing scheme. The layout of the enhancing freezing tubes has a great influence on the freezing effect and time, which is the most concerned part in construction sites. To understand the influence of the layout, operating duration, and heat preservation measures of the semicircular enhancing freezing tubes on the freezing effect, three different simulation schemes were set up, as shown in Figure 7. The initial temperature of the model was set at 20 °C, and the surface boundary was set as the third boundary condition, with a surface heat transfer coefficient of 15 W/(kg·K).

The layout of the enhancing freezing tubes has great influence on the freezing effect [38]. A freezing scheme of enhancing freezing tubes in two different positions was considered, as shown in Figure 5a. The enhancing freezing tubes were arranged far from or near to the excavation side with an angle of 15°, recorded as Scheme A and Scheme B, respectively. The master freezing tubes and enhancing freezing tubes were operated from the beginning to the end of the freezing process.

The operating duration of the enhancing freezing tubes was also set as a simulation scheme involving weather, to make the enhancing freezing tubes work and maintain an active freezing duration of 60 days, recorded as Scheme C and Scheme D. The master freezing tubes continued working during this period.

Most of the engineering literature on the use of the artificial ground freezing method posits that air convection has a great influence on the freezing temperature field [39]. The air convection interface should be insulated to limit and reduce the loss of cooling capacity caused by air convection, so as to concentrate the cooling capacity on the formation and development of frozen soil. Fortunately, each section of the hollow jacking pipes has thermal insulation treatment in the longitudinal direction, and the air in the jacking pipe did not undergo directly convect with the atmosphere during the freezing construction of the Gongbei Tunnel. The convective heat transfer in the limited space has limited heat dissipation; even this kind of air convection can homogenize the cold capacity of the enhancing freezing tubes, and make the hollow pipes play the role of "freezing pipe", which is beneficial to the freezing effect. Therefore, whether the strengthening of heat preservation measures is beneficial to the freezing effect must be discussed through simulation comparison. The simulation scheme is shown in Figure 7b, where the models with and without heat preservation measures are recorded as Scheme E and Scheme F, respectively.

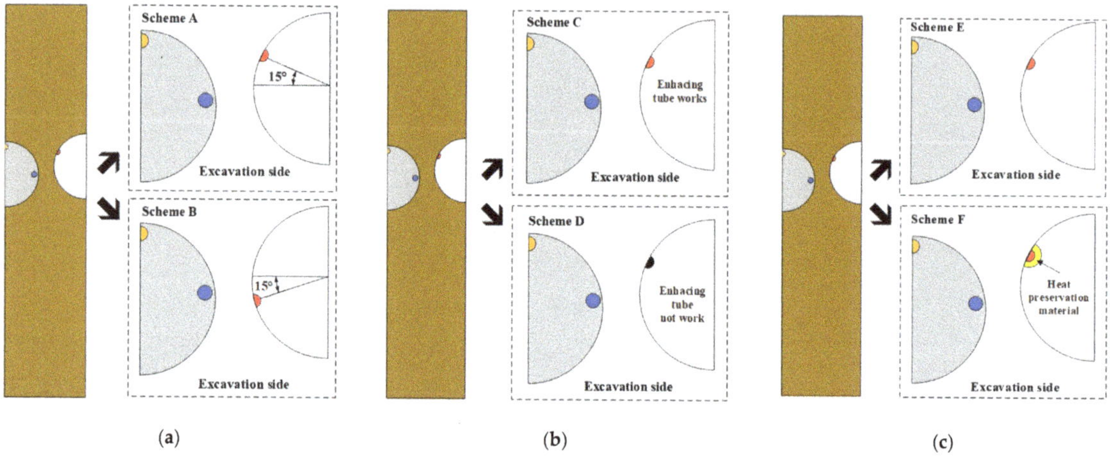

Figure 7. Calculation scheme of enhancing freezing tubes: layout angle (**a**), operating duration (**b**), heat preservation measures (**c**).

According to the engineering scheme [40], polyurethane foam was selected as the heat preservation material. The thickness of the heat preservation material was set to 0.03 m based on construction experience [41,42]. The thermal conductivity of the heat preservation material was set to 0.04 W/(kg·K), the density was 34 kg/m^3, and the specific heat was 2016 J/(kg·K).

4. Results and Analysis

4.1. The Layout of the Enhancing Freezing Tubes

Figure 8a,b show the temperature distribution in different layouts of enhancing freezing tubes over different durations. As the change in temperature causes the air density to change, the cooler air sinks to the bottom of the pipe, and the hotter air floats to the crown of the pipe. This air flow further aggravates the change in the temperature field, and cyclically reciprocates until it reaches an equilibrium state. The results show that the air convection velocity field under the two freezing schemes reached a state of distribution equilibrium after 30 days of active freezing. The arrangement of the enhancing freezing tubes in Scheme A is more conducive to the air flow. Therefore, Scheme A is better than Scheme B in terms of distribution range and flow rate, and can effectively ensure the full flow of air through the entire hollow pipe.

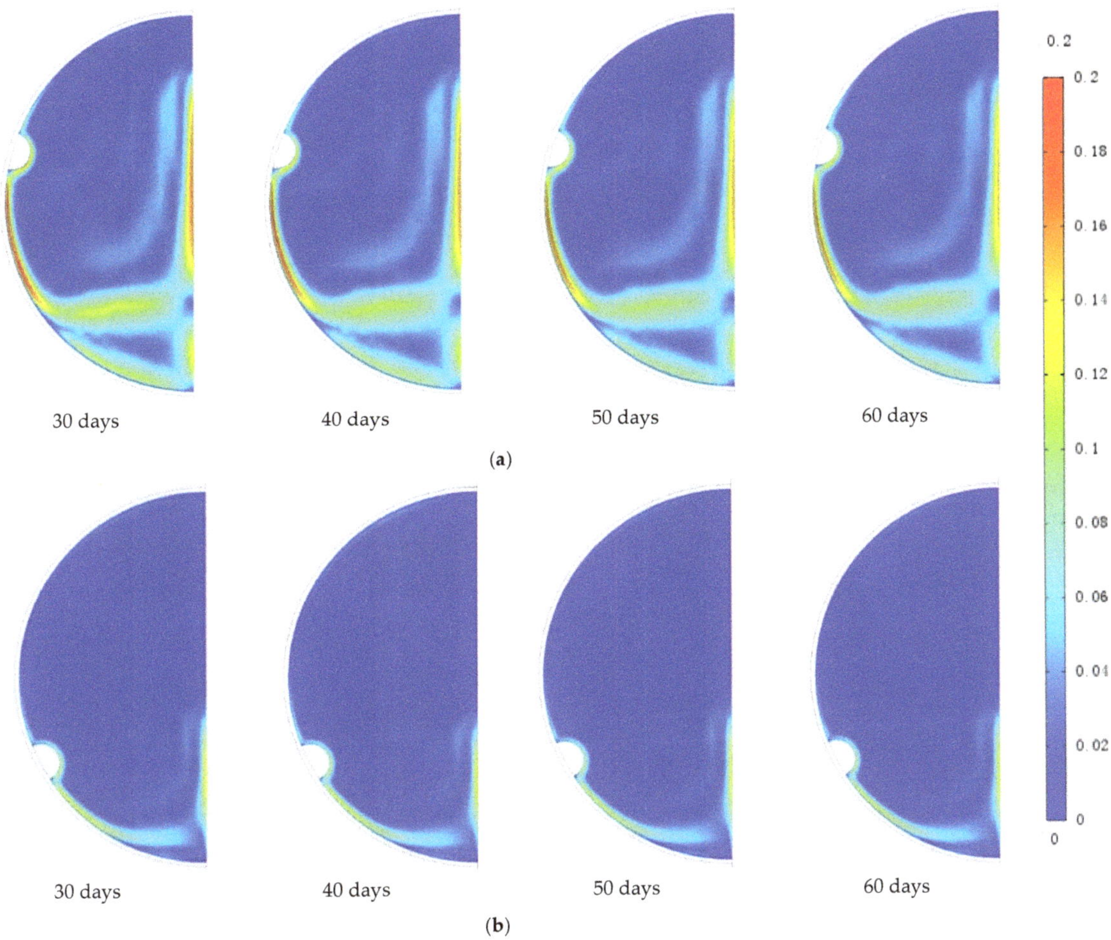

Figure 8. Temperature distribution cloud graphs of the different layouts of the enhancing freezing tubes: Scheme A (**a**); Scheme B (**b**).

When comparing the enhancing freezing tube schemes, the thickness of the frozen soil curtain is the most important criterion, including the thickness of frozen soil between the pipes and the thickness of frozen soil at the central axis of the steel pipe. Figure 9 shows the changes in the thickness of the frozen soil curtain of Scheme A and Scheme B over time.

Figure 10 shows that the frozen soil curtain formed after 7 days of the freezing process. For Scheme A, the frozen soil curtain reached the thickness of 2.0 m required by the design requirements after 33 days of active freezing, while it needed to freeze for 38 days to reach 2 m in Scheme B. Additionally, the thickness of the frozen soil curtain at the central axis of the steel pipe in Scheme A was also greater than that of Scheme B. Therefore, Scheme A is better than Scheme B. Figure 11 shows the thickness of the frozen soil at the central axis of the steel pipe, including the upper section and the bottom section.

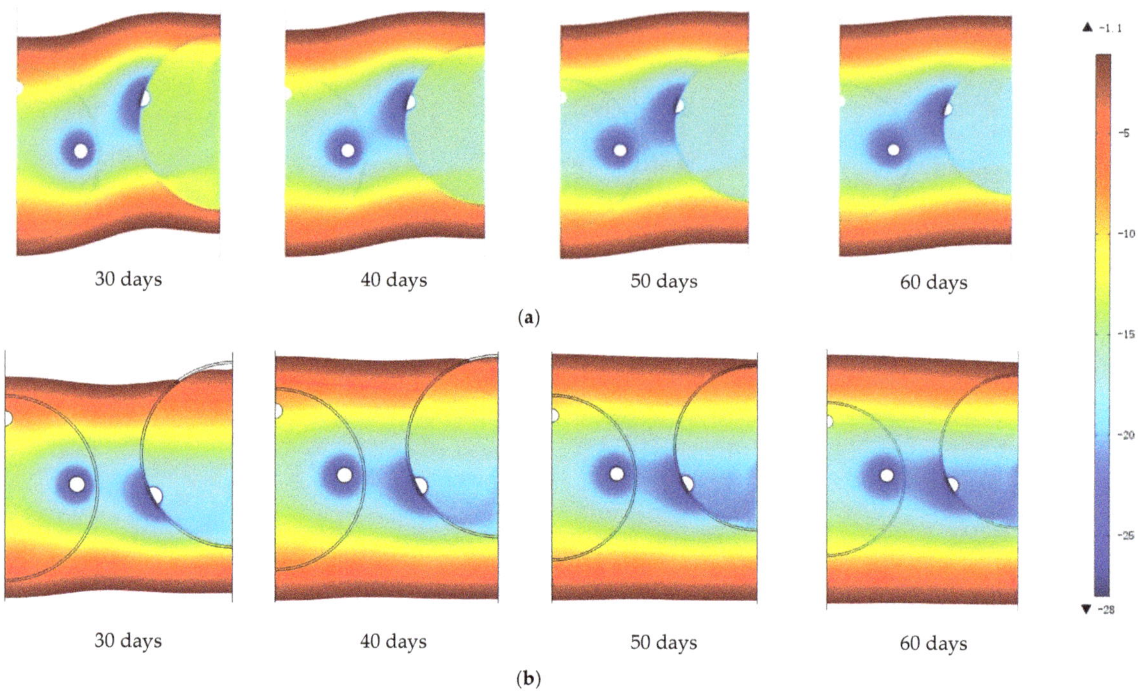

Figure 9. Temperature cloud graphs of the different layouts of the enhancing freezing tubes: Scheme A (**a**); Scheme B (**b**).

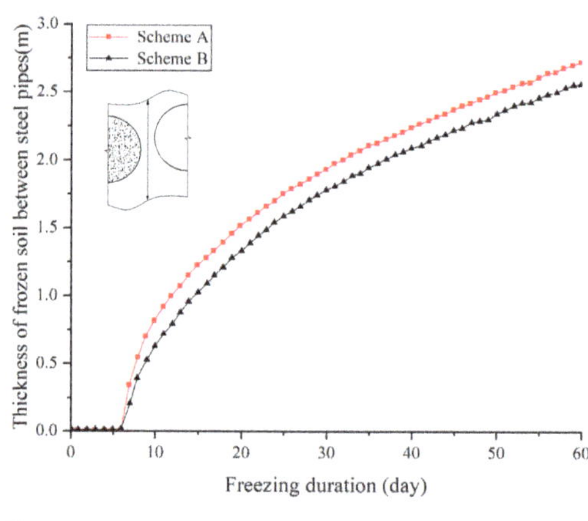

Figure 10. The thickness of the frozen soil between the pipes.

In summary, the layout of the enhancing freezing tubes should be fully considered in promoting the full flow of air in the hollow pipe. The basic law of air flow is that the colder air sinks to the bottom while the warmer air floats to the top. For the pipe jacking on the upper part of the FSPR structure, the layout of the enhancing freezing tubes should be slightly outside the excavation surface. The specific position can be fine-tuned according to the steel jacking process. Generally, it can be located at about 15 degrees counterclockwise to the horizontal radius of the hollow jacking pipe, as shown in Figure 7a.

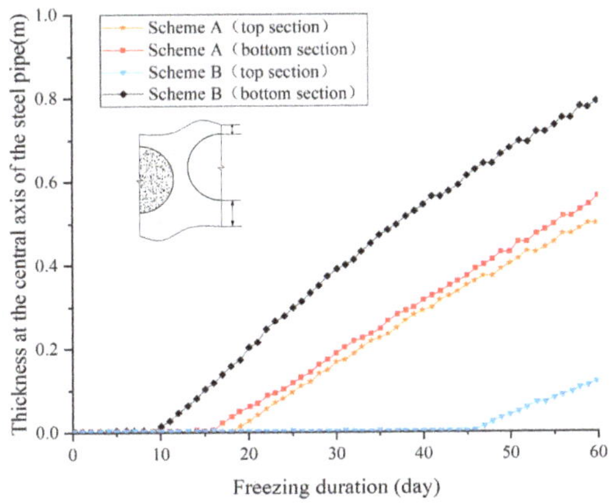

Figure 11. The thickness of the frozen soil at the central axis of the steel pipe.

4.2. The Operating Duration of the Enhancing Freezing Tubes

According to the above results, Scheme A is the best plan. The operating duration of the enhancing freezing tubes was also set as a simulation scheme, involving weather, to make the enhancing freezing tubes maintain an active freezing duration of 60 days, recorded as Scheme C and Scheme D. The master freezing tubes continued working during this period.

Figure 12 shows the temperature cloud graphs for different operating durations. After 60 days of freezing, the thickness of the frozen soil curtain in Scheme D still struggled to reach the thickness required by the design.

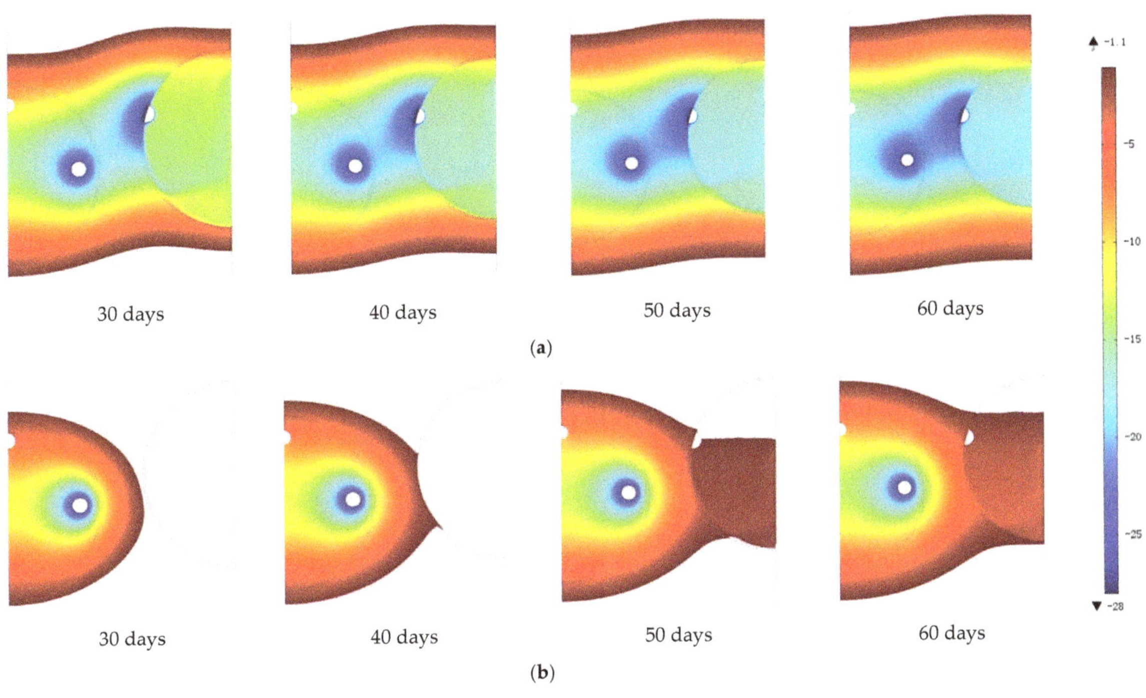

Figure 12. Temperature cloud graphs of the different operating durations of the enhancing freezing tubes: Scheme C (**a**); Scheme D (**b**).

Figure 13 compares the development conditions of the frozen soil curtains under the two schemes. In Scheme C, the frozen soil curtain intersected on the 6th day of freezing, and reached a thickness of 2 m in about 32 days. However, the frozen soil curtain intersected after 17 days of freezing in Scheme D, and the thickness of frozen soil reached only 1.75 m after 60 days of freezing, as shown in Figure 13. Moreover, the average temperature showed a similar trend, struggling to meet the design requirements in Scheme D, as shown in Figure 14. Therefore, Scheme C is better than Scheme D.

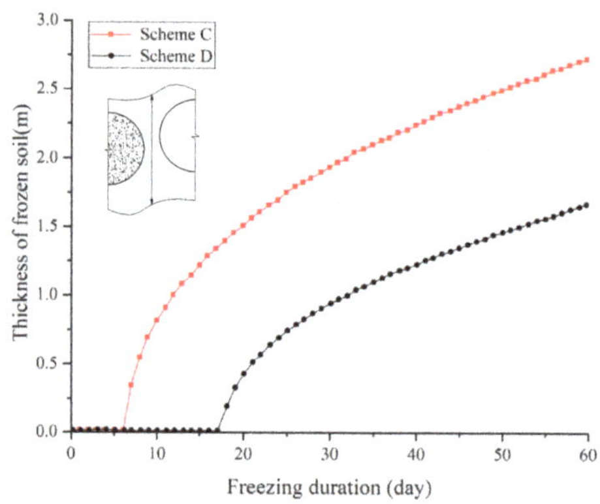

Figure 13. The thickness of the frozen soil between the pipes.

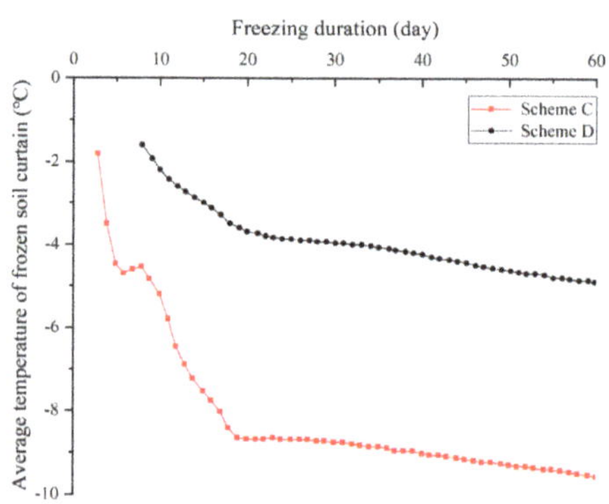

Figure 14. The average temperature of the frozen soil curtains.

Taking the construction period of Gongbei Tunnel into account, the master freezing tubes and the enhancing freezing tubes should be operated at the same time during the construction. However, this may cause the frozen soil area in the longitudinal partial area to be too large during construction. Therefore, from the perspective of controlling frost heave, it may be necessary to carry out intermittent operating and closing treatments for the enhancing freezing tubes. In summary, Scheme C, in which the two types of freezing tubes are working at the same time, should be the recommended scheme.

4.3. The Heat Preservation Measures of the Enhancing Freezing Tubes

In the construction of the freezing section of the pipe curtain in Gongbei Tunnel, the hollow pipe had heat preservation measures in the longitudinal direction. The air convection was not in an open state connected with the atmosphere, and the heat dissipation caused by convection heat transfer in limited space was objectively limited. More importantly, the air convection in the hollow pipe can homogenize the cooling capacity of the enhancing freezing tubes, so that the pipe can play the role of freezing. Therefore, whether the strengthening of heat preservation measures is beneficial to the freezing effect needs to be studied through simulation comparison.

Figures 15 and 16 show that Scheme E (without heat preservation measures) has a better freezing effect than Scheme F (with heat preservation measures). Because the heat preservation measures promote the suppression of the cold source in the hollow pipe, the thickness of the frozen soil in Scheme E is obviously greater than that in Scheme F, including the thickness of the frozen soil between the steel pipes and that at the central axis of the steel pipe, as shown in Figures 15 and 16, respectively. Therefore, Scheme E is more effective.

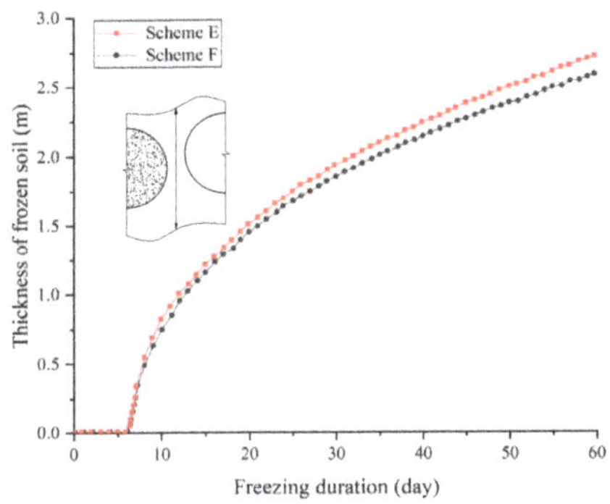

Figure 15. The thickness of the frozen soil between the steel pipes.

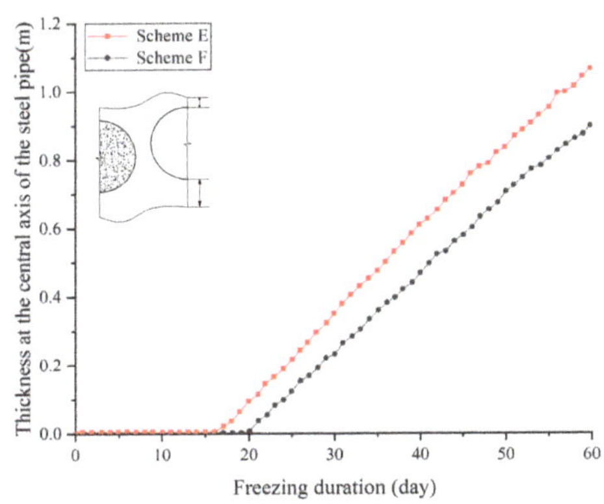

Figure 16. Thickness of the frozen soil at the central axis of the steel pipe.

The average temperature of the frozen soil curtain in Scheme E was 5 °C lower than that in Scheme F, as shown in Figure 17. The state of active freezing for 40 days is shown in Figure 18. In Scheme F, the value of the air velocity field was slightly smaller than that of Scheme E, due to the addition of heat preservation measures. According to the above findings, the heat preservation material makes the temperature in the hollow pipe decrease slowly, and the hollow pipe does not fully play the role of a large freezing pipe, reducing the freezing efficiency.

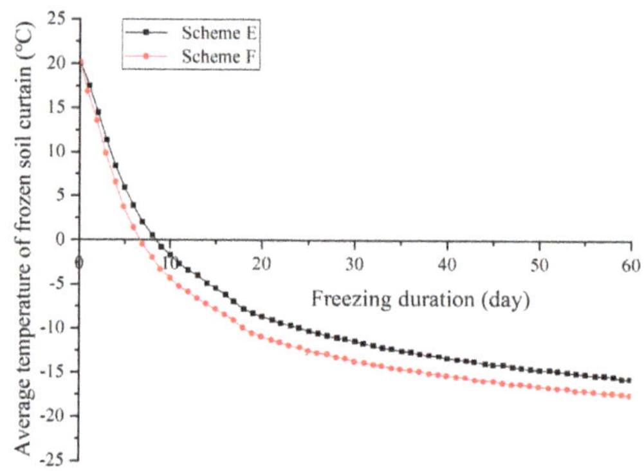

Figure 17. Average temperature of the frozen soil curtains.

(a) (b)

Figure 18. Velocity cloud diagrams of the enhancing freezing tubes with or without the heat preservation measures (active freezing for 40 days): Scheme E (**a**); Scheme F (**b**).

Based on the above analysis, it can be determined that the heat preservation measures of the enhancing freezing tubes cannot significantly reduce the loss of cooling capacity of the tubes due to air convection heat transfer, but the freezing effect is significantly weakened after the heat preservation of the enhancing freezing tubes. Therefore, there is no need to install heat preservation measures during the freezing process.

5. Conclusions

In this paper, based on the unsteady conjugate heat transfer model, the temperature field of the FSPR project including three heat transfer media—steel pipe, frozen soil, and air—was successfully simulated, and the simulation accuracy can be used for engineering guidance. Then, the simulation and analysis of the layout, operating duration, and heat preservation of the enhancing freezing tubes in Gongbei Tunnel were carried out; the following conclusions were obtained:

(1) The greater the contact area between the enhancing freezing tube and the inner wall of the steel pipe, the better the freezing effect. Considering both the freezing effect and frost heave control, the semicircular enhancing freezing tube scheme is superior to the other three shapes of freezing tubes.

(2) The enhancing freezing tubes arranged far away from the excavation surface (Scheme B) have a better freezing effect.

apply to the special pipe position layout of pipe-curtain freezing [7,8]. Because the frozen pipe in the FSPR method is set inside the steel pipe, its heat conduction mode is different from that of the traditional frozen pipe directly contacting the soil. The heat transfer process involves different media such as concrete, steel pipe, soil, air, etc. [9]; thus, the numerical calculation method is more complicated. Relevant scholars have used numerical calculation to analyze the freezing temperature field of the FSPR method [10]. An analytical solution for the temperature field of the FSPR method is based on the Laplace equation boundary conditions for the dislocation arrangement of the single circle pipes [11]. The above research greatly simplifies the model and does not fully consider the heat transfer process between steel pipe, concrete, soil, air, and other media. Therefore, unsteady-state conjugate heat transfer theory is applied to the numerical simulation of the FSPR method [12], which has been applied to the simulation of high-pressure turbines and certain fluid-related parts of aircraft [13]. The term conjugate heat transfer refers to a heat transfer process involving the interaction of heat conduction within a solid body with either the free [14], forced, or mixed convection from its surface to a fluid flowing over it [15]. It finds application in numerous fields, from thermal interaction between surrounding air and fins to thermal interaction between flowing fluid and turbine blades [16–18]. At present, there is no freezing simulation of engineering temperature field involving three media such as that of the FSPR structure. Based on the above, this paper takes the FSPR method of the Gongbei tunnel as the research object, proposes an unsteady-state conjugate heat transfer model, establishes a global solution algorithm of a strong coupling governing equation based on the virtual density method that can accurately describe this kind of heat transfer process, and analyzes the sensitivity of the factors such as initial formation temperature, different soil layers, and brine temperature affecting the FSPR effect.

2. Engineering Description of FSPR Method in Gongbei Tunnel

Gongbei Tunnel is the critical link connecting the Hong Kong–Zhuhai–Macao Bridge and Zhuhai [19]. It was constructed with the FSPR method, which is the combination of the pipe-roofing method and artificial ground-freezing method. The layout of the FSPR method is shown in Figure 1 [20]. It is composed of 36 steel jacking pipes with a diameter of 1620 mm and a thickness of 20 mm. Odd-numbered steel pipe jacking is filled with concrete. Even-numbered steel pipe jacking is empty pipes, which are in a staggered layout, and the spacing between pipes is 357 mm. Considering that it is convenient for the overlapping of temporary support during the construction, the two kinds of steel pipe jacking are in staggered layout. The height difference of the circle center is 300 mm, and the buried depth of the tunnel is 4–5 m.

The construction of Gongbei tunnel consisted of four stages, the pipe jacking stage, the freezing stage, the tunnel excavation stage, and the thawing stage. In the construction stage of the pipe jacking, 36 steel pipes were sequentially jacked into a ring-shaped support system by the pipe jacking machine. In the freezing stage, the low-temperature brine was circulated through the freezing tubes such as the circular freezing tube, the enhancing freezing tube, and the limiting freezing tube, so that the frozen soil reached the designed thickness, forming a waterproofing curtain. The tunnel excavation stage excavated the soil in different areas step by step, and the thawing stage restored the frozen soil to normal temperature soil by circulating high-temperature brine after the excavation was completed.

Considering the calculation efficiency and symmetry of the structure, the typical part of the above-mentioned FSPR was taken as the research object, including solid steel pipe, hollow pipe, and the frozen soil between pipes, as shown in Figure 2.

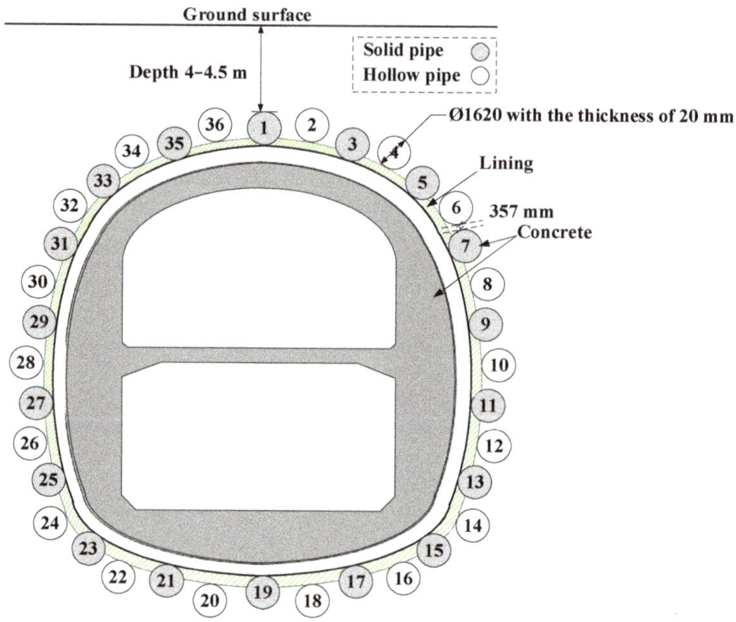

Figure 1. Layout of pipe roofing.

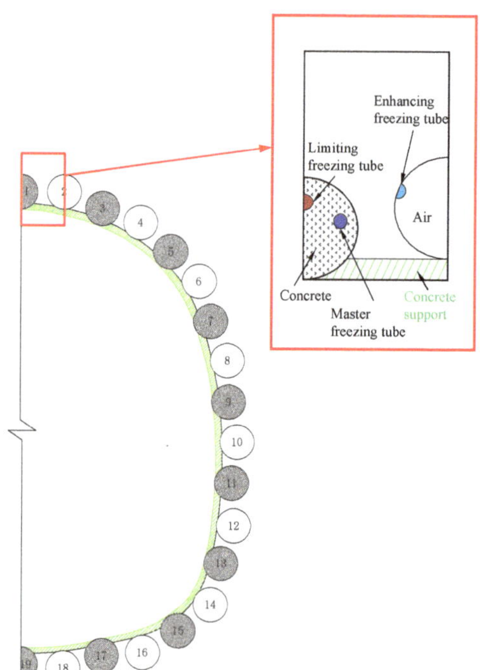

Figure 2. Local view of FSPR structure.

3. Calculation Model of Unsteady Conjugate Heat Transfer of FSPR Method

3.1. Mathematical Calculation Model

There are three types of freezing tubes in the FSPR method of the Gongbei Tunnel: master freezing tubes with a diameter of 133 mm and thickness of 4 mm, limiting freezing tubes with a diameter of 133 mm and thickness of 4 mm, and enhancing freezing tubes with a diameter of 159 mm and thickness of 4.5 mm. The layout of the freezing tubes is shown in Figure 3.

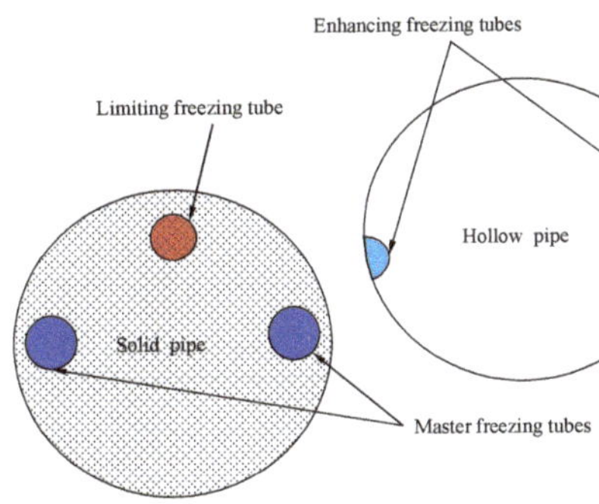

Figure 3. The layout of freezing tubes.

The freezing phase can be divided into the active freezing phase and the maintainable freezing phase. The construction conditions of the two phases are completely different. In the active freezing phase, the soil below the pipe-roofing has not been excavated, and the model can be simplified into two-dimensions for convenience of calculation, as shown in Figure 4a. In the maintainable freezing phase, the soil in the tunnel has been excavated, the boundary condition is shown in Figure 4b.

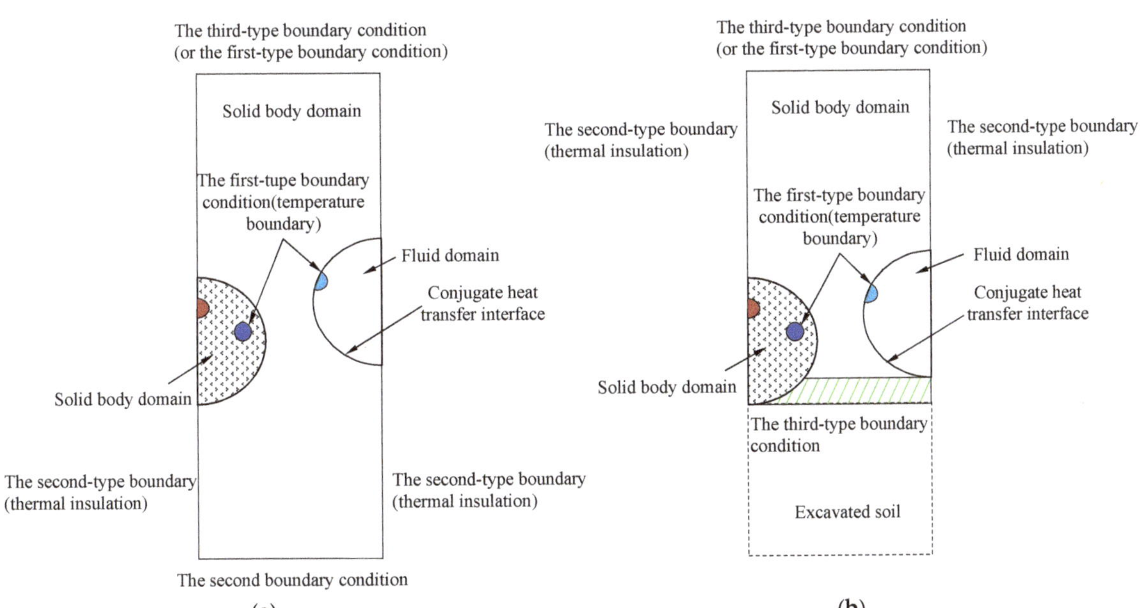

Figure 4. Calculation model of unsteady-state conjugate heat transfer of FSPR method. (**a**) Active freezing period. (**b**) Maintenance freezing period.

The left and right sides of the model are soil. According to the symmetry, it is set as the second-type boundary condition, corresponding to the Neumann boundary condition, which can be expressed as

$$q_w = -\lambda(\frac{\partial T}{\partial n})_w,\ t > 0 \tag{1}$$

where q_w is the heat flux value (W/m^2), λ is the thermal conductivity (W/(m·K)), n is the normal direction at the boundary, w is the interface, T is the temperature (K), and t is the time (s).

Since the left and right sides are symmetrical, it is assumed that all heat fluxes passing through the left and right sides are zero.

The upper edge of the model is the surface, which is set as the third-type boundary condition, corresponding to the Robin condition, which can be expressed as

$$-\lambda\left(\frac{\partial T}{\partial n}\right)_w = h\left(T_w - T_f\right) \tag{2}$$

where h is the surface heat transfer coefficient (W/(m^2·K)).

The freezing tube is a given temperature boundary, set as the first-type boundary condition, corresponding to the Dirichlet condition, which can be expressed as

$$T_w = const,\ T > 0 \tag{3}$$

The continuity condition of conjugate heat transfer interface is as follows: the conjugate heat transfer interface between soil and air is a steel pipe wall where the continuous conditions of temperature and heat flow shall be met; the expressions are

$$T_W|_{\mathrm{I}} = T_W|_{\mathrm{II}} \tag{4}$$

$$q_W|_{\mathrm{I}} = q_W|_{\mathrm{II}} \tag{5}$$

In Equations (4) and (5), I and II represent the two regions.

In this model, the air is assumed to be laminar. At this time, the heat transfer of the air next to the steel pipe is heat conduction, and the expression for the continuous condition of heat flow is

$$-\lambda_s(\frac{\partial T}{\partial n})_w\Big|_{\mathrm{I}} = h_f(T_w - T_f)\Big|_{\mathrm{II}} \tag{6}$$

where h_f is the coefficient of flow heat transfer, λ_s is the thermal conductivity of the solid, T_w is the interface temperature, and its expression is

$$h_f = \frac{q}{T_w - T_f} = -\frac{\lambda_f}{T_w - T_f}(\frac{\partial T}{\partial n})_w \tag{7}$$

where λ_f is the fluid thermal conductivity.

The heat transfer model in the solid adjacent to the interface is heat conduction, and the expression of the heat flow continuity condition is

$$-\lambda_s(\frac{\partial T}{\partial n})_w\Big|_{\mathrm{I}} = -\lambda_s\frac{T_w - T_s}{\Delta n}\Big|_{\mathrm{I}} \tag{8}$$

where T_s is the solid temperature and Δn is the distance between the interface and the center point of the solid unit.

3.2. Governing Equation

3.2.1. Fluid Domain Governing Equation

In this two-dimensional model, air heat transfer is regarded as fluid heat transfer, and the fluid domain control equation is represented by three types of control equations (continuity equation, motion equation, and energy equation) [21–23]. The description method adopts the Euler description method. Unless otherwise specified, rectangular coordinates are used to represent Euler coordinates.

The continuity equation is determined according to the Law of Conservation of Mass. In the rectangular coordinate system, the differential form of the continuity equation can be expressed as

$$\frac{\partial \rho}{\partial t}+\frac{\partial(\rho u)}{\partial x}+\frac{\partial(\rho v)}{\partial y}=0 \tag{9}$$

where ρ is the fluid density (kg/m^3), v is the velocity vector (m/s) and its components in the x, y directions are recorded as u, v, respectively, and t is time. The first term represents the mass increment per unit volume in unit time; the last three items represent the net outflow of mass per unit time and unit volume.

It can be expressed in vector form by using the divergence formula [24]:

$$\begin{gathered}\frac{\partial \rho}{\partial t}+\operatorname{div}(\rho v)=0 \\ v=ui+vj=\frac{\partial x}{\partial t}i+\frac{\partial y}{\partial t}j\end{gathered} \tag{10}$$

To describe the convenience and the simplicity of the motion and energy equations, the concept of the satellite derivative is introduced [25]. Under Euler coordinates, the satellite derivative is more complicated and can be expressed as

$$\frac{\mathrm{D}}{\mathrm{D}t}=\frac{\partial}{\partial x}\frac{\partial x}{\partial t}+\frac{\partial}{\partial y}\frac{\partial y}{\partial t}+\frac{\partial}{\partial t}=(v\cdot\nabla)+\frac{\partial}{\partial t} \tag{11}$$

where the del operator, also known as the nabla operator [26], and the divergence of the vector is $\operatorname{div}A=\nabla\cdot A$. Then, the continuity equation is expressed by the following derivative:

$$\frac{\mathrm{D}\rho}{\mathrm{D}t}+\rho\operatorname{div}v=0 \tag{12}$$

The motion equation is determined according to the law of momentum balance. The differential form of the motion equation can be expressed as

$$\rho\frac{\mathrm{D}v}{\mathrm{D}t}=\rho F_b+\frac{\partial p_x}{\partial x}+\frac{\partial p_y}{\partial y} \tag{13}$$

Similarly, the divergence formula can be used to change Equation (13) into

$$\rho\frac{\mathrm{D}v}{\mathrm{D}t}=F+\operatorname{div}P \tag{14}$$

where the left part is the inertial force of unit volume fluid; the first term on the right is the mass force F acting on the unit volume fluid, and the second term is the surface force acting on the unit volume fluid where P is the pressure (Pa).

If air, as a fluid with a simple molecular structure, can be regarded as a Newtonian fluid, the expression form of the motion equation will introduce the concept of Generalized Newton's law to obtain a special form of the motion equation. This equation is called the Navier–Stokes equation, and the expression form is

$$\rho\frac{\mathrm{D}v}{\mathrm{D}t}=F-\operatorname{grad}p+\operatorname{div}(2\mu S)-\frac{2}{3}\operatorname{grad}(\mu\operatorname{div}v) \tag{15}$$

In which, the left part represents the inertial force per unit volume of fluid, the first term at the right part represents the mass force per unit volume, the second term represents the pressure gradient force acting on the fluid per unit volume, the third term is the viscous deformation stress, and the fourth term is the viscous expansion force.

The expression of the energy conservation equation in thermodynamics is described as the first law of thermodynamics. Taking the rectangular coordinate system as the Euler coordinate system, the energy equation in differential form can be derived as [27]

$$\rho \frac{\mathrm{D}e_s}{\mathrm{D}t} = F \cdot v + \mathrm{div}(P \cdot v) + \mathrm{div}(\lambda \mathrm{grad} T) + \rho\sigma + \Phi \tag{16}$$

where σ represents the heat transferred from radiation unit mass fluid in unit time (W/m^3), λ is the thermal conductivity (W/m·K), Φ and is the internal heating rate (W/m^3).

If the air is regarded as ideal air, the energy equation can be expressed as

$$\rho c \frac{\mathrm{D}T}{\mathrm{D}t} = \mathrm{div}(\lambda \mathrm{grad} T) + \Phi \tag{17}$$

where c is specific heat, which is an ambient-pressure specific-heat meter for fluids here (J/kg·K). At this time, the three basic governing equations—continuity equation, motion equation, and energy equation—constitute the differential equations for solving the fluid domain. When solving the simultaneous equations, it is necessary to introduce the state equation connecting P and ρ, and the equations can be solved. The ideal air state equation is

$$p = \rho R T \tag{18}$$

where R is the air constant $R = R_0/M$, R_0 is the universal air constant, R_0 is 8.31 J/K·mol, M is the molar mass, and R is 287 J/kg·K for air.

3.2.2. Solid Domain Governing Equation

In this model, the heat conduction of soil and concrete is regarded as solid heat conduction. According to the Law of Energy Conservation and Fourier Law, the differential equation of heat conduction in the solid domain is established as follows:

$$\rho c \frac{\partial T}{\partial t} = \mathrm{div}(\lambda \mathrm{grad} T) + \Phi \tag{19}$$

Generally, when studying the heat transfer in the solid domain, only the heat conduction effect is approximately considered, and phenomena such as the convective heat transfer of a small amount of air and liquid that may be contained in the solid and the resulting water migration are ignored.

3.3. Numerical Implementation of Unsteady-State Conjugate Heat Transfer Model

Through the virtual density method, the global solution algorithm of a strong coupling governing equation is established. The energy equation expressed in physical density is as follows, and the implementation of the algorithm is based on multiphysics module of COMSOL software (Femlab3.2a, Svante Littmarck and Farhad Saeidi, originated from the MATLAB Toolbox, Stockholm, Sweden), which is a coupled tool for solving partial differential equations.

$$\rho \frac{\partial i}{\partial t} + \rho u \frac{\partial i}{\partial x} + \rho v \frac{\partial i}{\partial y} = \frac{\partial}{\partial x}\left(\frac{\lambda}{c}\frac{\partial i}{\partial x}\right) + \frac{\partial}{\partial y}\left(\frac{\lambda}{c}\frac{\partial i}{\partial y}\right) + \Phi \tag{20}$$

The density in Equation (20) was replaced by ρc to solve the coupling boundary of the steel pipe roof in unsteady-state conjugate heat transfer analysis to solve the problem of heat flux continuity at the coupling interface of steel pipe roof in the analysis of unsteady-state conjugate heat transfer [28,29]. Thus, the strong coupling governing equations' overall solution algorithm of the unsteady-state conjugate heat transfer model is as follows:

$$\rho c \frac{\partial i}{\partial t} + \rho c u \frac{\partial i}{\partial x} + \rho c v \frac{\partial i}{\partial y} = \frac{\partial}{\partial x}\left(\lambda \frac{\partial i}{\partial x}\right) + \frac{\partial}{\partial y}\left(\lambda \frac{\partial i}{\partial y}\right) + \Phi' \tag{21}$$

In the process of soil freezing with phase transition, the sensible heat capacity method was used to deal with the problem of soil freezing phase transition [30]. The sensible heat capacity method takes the temperature as the function to be solved and establishes a unified energy equation over the whole region, which is expressed in the form of "Equivalent Specific Heat" [31]. The "Equivalent Specific heat" when the phase transition occurs at a specific temperature T_m is

$$\tilde{c}(T) = c(T) + L\delta(T - T_m)$$
$$c(T) = \begin{cases} c_s(T) & T < T_m \\ c_l(T) & T > T_m \end{cases};\ \delta(T - T_m) = \begin{cases} 1 & T = T_m \\ 0 & T \neq T_m \end{cases} \quad (22)$$

where $\delta(T - T_m)$ is the Dirac function; a sensible heat capacity method model then emerges:

$$\rho\tilde{c}\frac{\partial T}{\partial t} = \mathrm{div}(\lambda \mathrm{grad} T) \quad (23)$$

The phase transition of water in the soil freezing process is relatively slow, and it is approximately completed within a certain temperature range. When using the sensible heat capacity method to deal with the phase transition problem of the freezing process, the unfrozen water in the soil must first be removed according to the experimental data. The changes are divided into three sections: the severe phase transition zone, transition zone, and frozen zone. Then, the phase transition is processed by the equivalent specific heat parameter. The equivalent specific heat of the soil area in the energy equation is taken as

$$c(T) = \begin{cases} c_f & T < T_b \\ c_f + \frac{c_u - c_f}{T_a - T_b}(T - T_b) + \frac{L}{(1+w)}\frac{\partial w_i}{\partial T} & T_a \leq T \leq T_b \\ c_u & T > T_a \end{cases} \quad (24)$$

where C_u and C_f are the specific heat of thawed soil and frozen soil, J/(kg·K); L is the latent heat of water phase transition, taken as 334,560 J/kg; w and w_i are the total water content and ice content of frozen soil; T_a and T_b are the upper and lower boundary temperatures (k) of the phase transition zone of frozen soil. For the convenience of calculation, the temperature is calculated in international Kelvin (k) and described in Celsius (°C).

4. Application of Unsteady Conjugate Heat Transfer Model of FSPR in Gongbei Tunnel

4.1. Assumptions and Calculation Model

The calculation models (a)(b) with a height of 11 m and a width of 6.15 m are shown in Figure 4, and the thickness of both is 1.977 m. In the calculation domain, the cold source freezing pipe wall was regarded as the first temperature boundary, namely the Dirichlet boundary condition, and the refrigerant inlet temperature was taken as the boundary temperature value [32]. When the air pressure is low and the temperature is high, it can be treated as an ideal gas, including the air in the fluid domain. The air in the fluid domain can also be regarded as a compressed fluid and considered as a viscous fluid. At this time, the continuity equation and motion equation (Navier–Stokes equation) are changed [33]. Considering the causes of fluid movement, the flow state of air can be assumed as laminar flow. The continuous condition of heat flux was also treated according to laminar flow in the conjugate heat transfer interface.

According to the above assumptions, the strong coupling general control equations of the calculation model were obtained [34]. The energy equation, ideal gas state equation, continuity equation, and motion equation can be expressed as follows:

$$\begin{array}{c} \rho c\frac{\partial T}{\partial t} = \mathrm{div}(\lambda \mathrm{grad}T) + \Phi \\ p = \rho RT \\ \frac{\mathrm{D}\rho}{\mathrm{D}t} + \rho \mathrm{div}v = 0 \\ \rho\frac{\mathrm{D}v}{\mathrm{D}t} = F - \mathrm{grad}p + \mathrm{div}(2\mu S) - \frac{2}{3}\mathrm{grad}(\mu \mathrm{div}v) \end{array} \tag{25}$$

In the numerical calculation and analysis, the multiphysics module of COMSOL was used for the secondary development of the software combined with the strong coupling governing equations.

4.2. Soil Mechanics and Thermal Physical Parameters

The soil layer of the Gongbei tunnel mainly includes miscellaneous fill, sludge, mucky silty clay, silty clay, silty sand, medium sand, and gravel sand. The soil layer crossed by tunnel concealed excavation construction is mainly ①, ③-3, ④-3, ⑤-2, ⑤-3, and ⑦-1. According to the test report on physical and mechanical parameters of artificially frozen soil in "Test Report on Physical and Mechanical Parameters of Artificial Frozen Soil at Gongbei Port of Zhuhai Link of Hong Kong-Zhuhai-Macao Bridge" [35], the properties of typical soil layers in Gongbei Tunnel are shown in Table 1, the freezing temperature of the soil is shown in Table 2, and the thermal conductivity of soil is shown in Table 3.

Table 1. Properties of typical soil layers in Gongbei Tunnel.

Soil Layer	Lithology	Moisture Content w (%)	Density ρ (kg/m^3)	Dry Density ρ_d (kg/m^3)
①	Artificial fill	16.05	1660	1470
③-3	Medium gravel sand	13.54	2000	1760
④-3	Muddy silty clay	47.6	1820	1230
⑤-2	Fine sand	18.25	1950	1650
⑤-3	Muddy silty clay	38.27	1880	1360
⑦-1	Sandy clayey soil	31.98	1890	1430

Table 2. The freezing temperature of the soil layer and unfrozen water content at −20 °C.

Soil Layer	Lithology	0 (°C)	Unfrozen Water Content at −20 °C (%)
①	Artificial fill	−0.5	0.02
③-3	Medium gravel sand	−0.4	0.03
④-3	Muddy silty clay	−1.1	0.4
⑤-2	Fine sand	−1.6	0.05
⑤-3	Muddy silty clay	−1.5	0.33
⑦-1	Sandy clayey soil	−1.5	0.25

Table 3. Thermal conductivity of soil layer (W/(m·K)).

Soil Layer	Thermal Conductivity at Different Temperatures				
	−20 °C	−10 °C	0 °C	10 °C	20 °C
① Artificial fill	1.962	1.690	1.511	1.398	1.109
③-3 Medium gravel sand	1.925	1.758	1.538	1.217	1.066
④-3 Muddy silty clay	2.047	1.772	1.614	1.485	1.206
⑤-2 Fine sand	2.019	1.775	1.719	1.497	1.266
⑤-3 Muddy silty clay	1.994	1.893	1.643	1.402	1.344
⑦-1 Sandy clayey soil	2.030	1.790	1.623	1.442	1.319

According to the calculation formula of "Equivalent Specific Heat", the unfrozen water content in frozen soil is the prerequisite for calculating "Equivalent Specific Heat" [36]. According to the relevant experimental ① conclusions of the Shanghai Yangtze River Tunnel in the same coastal area, the water content of each soil layer at −20 °C was assumed as shown in Table 2, and it was assumed that the phase transition of water in the soil had been completed at −20 °C.

According to the three regions of the phase change of the water in the frozen soil: the severe phase transition zone, transition zone, and freezing zone, each soil layer is divided into four sections from the freezing temperature to −20 °C, combined with the above-unfrozen water content. The specific heat calculation results are listed in Table 4. The values of dynamic viscosity of air at different temperatures are shown in Table 5.

Table 4. Equivalent Specific Heat of soil layer (J/(kg·K)).

Soil Layer	Temperature (°C)					
	>0 °C	0–−3 °C	−3–−5 °C	−5–−10 °C	−10–−20 °C	<−20 °C
① Artificial fill	1420	19,149.37	1902.62	1461.84	1374.49	1360
③-3 Medium gravel sand	1410	16,026.31	1810.90	1442.00	1357.27	1340
④-3 Muddy silty clay	1510	47,364.79	7326.67	2662.19	1643.43	1450
⑤-2 Fine sand	1420	29,747.69	5502.70	1938.33	1417.27	1360
⑤-3 Muddy silty clay	1630	45,990.22	9454.42	3023.12	1765.04	1560
⑦-1 Sandy clayey soil	1580	40,759.56	8285.30	2706.18	1638.75	1470

Table 5. Dynamic viscosity of air (μPa·s).

Temperature (°C)	−40 °C	−20 °C	0 °C	10 °C	20 °C	30 °C
Dynamic viscosity	15.60	16.83	17.09	17.59	18.08	18.56

Considering that the change of pressure can be approximately ignored in engineering, the specific heat of air is taken as the specific heat of constant pressure. Its value varies with temperature, thermal conductivity, and specific heat are shown in Table 6. Other material parameters such as steel pipe jacking and filled concrete related to thermal physical parameters are shown in Table 7.

Table 6. Thermal conductivity of air (W/(m·K)).

Catalogue	−30 °C	−20 °C	−10 °C	0 °C	10 °C	20 °C	30 °C
Thermal conductivity	0.022	0.0228	0.0236	0.0244	0.0251	0.0259	0.0267
Specific heat	1011	1009	1009	1010	1012	1013	1014

Table 7. Thermal physical parameters.

Catalogue	Density (kg/m^3)	Thermal Conductivity (W/(m·K))	Specific Heat (J/(kg·K))
Steel pipe	7850	44.7	459.8
Concrete	2344	1.835	419.8

4.3. Evaluation Criterion

The parameter value given above are brought into the strong coupling governing equation, and the COMSOL software is used to calculate the freezing state of frozen soil. The numerical model is shown in Figure 4a. There are two evaluation indexes to measure the freezing state, one is the thickness of the frozen soil curtain between steel pipes, and the other is the area of frozen soil, as shown in Figure 5. The cloud graph of the frozen temperature field is obtained as shown in Figure 6. It can be seen that the index of frozen soil develops with the freezing process.

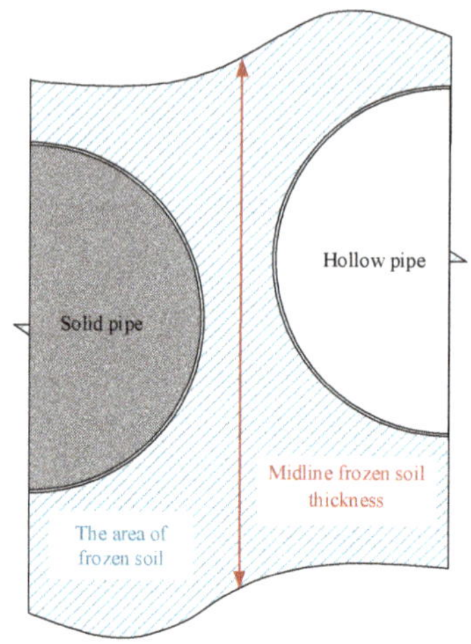

Figure 5. Evaluation index.

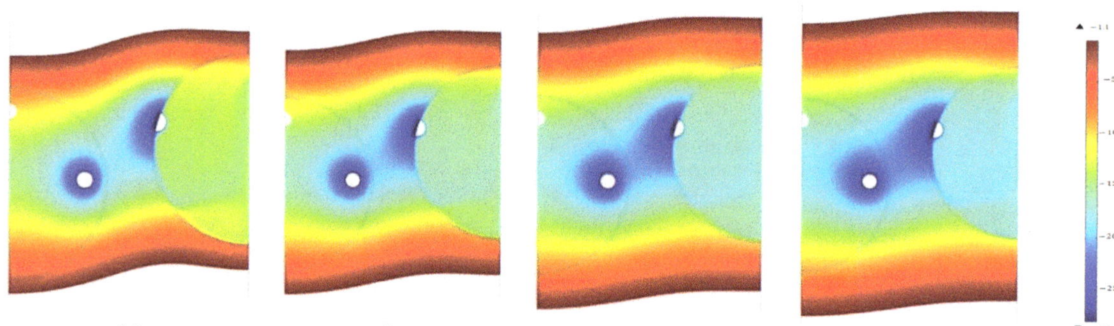

(a) (b) (c) (d)

Figure 6. The freezing temperature distribution: (**a**) 30 days, (**b**) 40 days, (**c**) 50 days, (**d**) 60 days.

5. Result

Figure 6 is the freezing temperature distribution for 30 days, 40 days, 50 days, and 60 days. When the initial temperature is 20 °C, the model needed to be frozen for 40 days to meet the design requirements of a 2 m thickness of frozen soil curtain between pipes.

Due to the uneven soil quality and uneven frozen soil boundary on the actual project, it is impossible to monitor the actual boundary position of frozen curtain. Therefore, the average thickness of the frozen soil is calculated through the temperature values of monitoring points. According to the field experience of freezing for 40 days to reach a thickness of 2 m, as shown in Figure 6, the model is basically consistent with the numerical simulation results. It is difficult for general methods to obtain such engineering-scale results. The simulation results are reasonable, and the model is validated. Therefore, due to the complexity of site construction conditions, it was necessary to analyze the parameter sensitivity of various factors affecting the freezing effect, such as different initial formation temperatures, different soil layers, and brine temperatures to improve freezing efficiency.

According to the thickness of frozen soil and the area of frozen soil curtain between jacking pipes, the sensitivity of initial formation temperature and different soil layers to the freezing effect was analyzed.

5.1. The Initial Formation Temperature

This analysis took the temperature change and the ground temperature gradient in the formation of Zhuhai City into account. Three different initial temperature conditions of 20 °C, 25 °C, and 30 °C were considered in the calculation model. When the third-type boundary condition was considered, the ambient temperature was the temperature change curve. Temperature change is generally expressed by trigonometric function [37]. The third-type boundary condition with a heat transfer coefficient of 15 W/(m^2·K) was actively frozen for 60 days, and the non-isothermal flow module in COMSOL was used to calculate and compare three different initial temperatures.

As shown in Figure 7, the development of frozen soil thickness between pipe jacking was quite different under different initial temperatures. Similarly, as shown in Figure 8, the development of frozen soil curtain areas is investigated. The difference in frozen soil thickness between pipes under similar initial temperatures for 40 days is about 0.1 m. The freezing duration required for the midline of frozen soil thickness to achieve a representative thickness at different initial temperatures is shown in Table 8. When the initial temperature was 20 °C, the water-sealing path of frozen soil had been completely wrapped with pipe jacking after freezing for 32 days. When the initial temperature was 25 °C, the fully wrapped pipe jacking needed 37 days of freezing, and when the initial temperature was 30 °C, it needed freezing for 40 days.

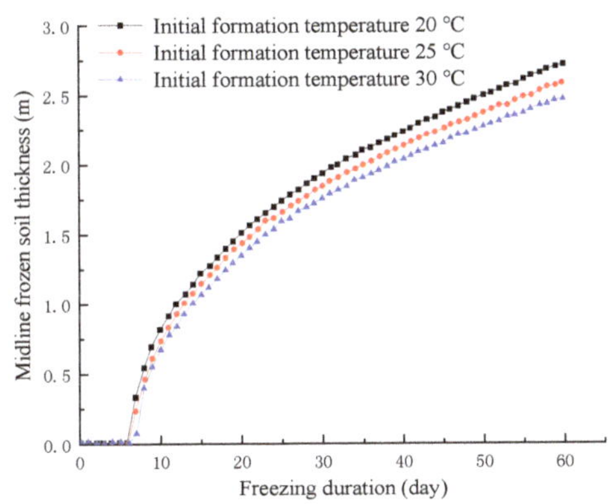

Figure 7. The thickness of frozen soil between steel pipes.

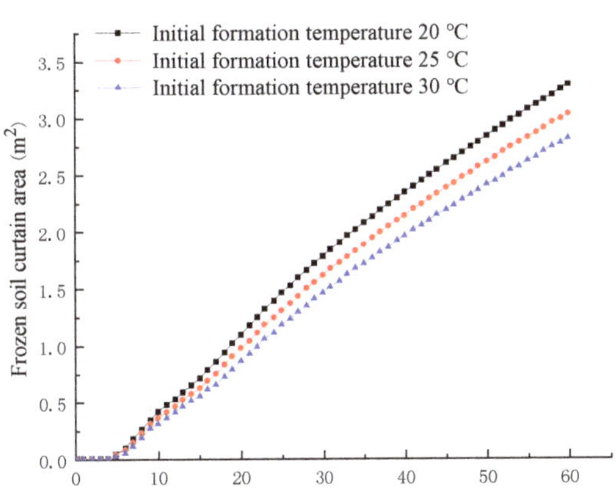

Figure 8. The area of frozen soil curtain for different initial formation temperature.

Table 8. Freezing duration of the frozen soil thickness midline at different initial temperatures (day).

Midline Thickness of Frozensoil (m)	Initial Formation Temperature		
	20 °C	25 °C	30 °C
0.5	8	9	10
1.0	12	13	14
1.5	20	22	23
2.0	32	37	40
2.5	50	56	60

As shown in Figure 9, for the development of the average temperature of the frozen soil curtain, after 35 days of freezing (at this time, the thickness of the central frozen soil under the three initial temperatures was about 2 m), the average temperature of the frozen soil was basically about −9 °C. The difference between adjacent cases was less than 0.1 °C, and tended towards consistency with increase in freezing duration. Therefore, in the construction of Gongbei Tunnel by the FSPR method, the initial temperature had a slight influence on the freezing effect. The higher the initial formation temperature, the longer the freezing process needs to last.

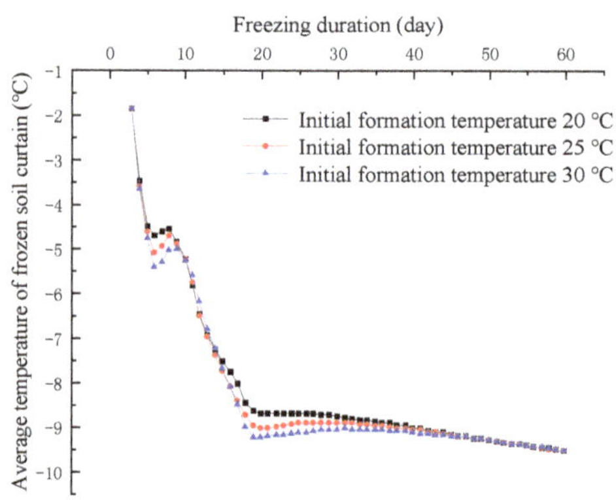

Figure 9. The average temperature of frozen soil curtain for different initial formation temperature.

5.2. Different Soil Layers

The main soil layers in the FSPR area of Gongbei Tunnel include six kinds of soil layers: ① artificial fill, ③-3 medium gravel sand, ④-3 muddy silty clay, ⑤-2 fine sand, ⑤-3 muddy silty clay, and ⑦-1 sandy cohesive soil. The initial temperature was set at 20 °C, and the surface boundary was set as the third-type boundary condition with a surface heat transfer coefficient of 15 $W/(m^2{\cdot}K)$. It was actively frozen for 60 days. The non-isothermal flow module in COMSOL was used to calculate and compare the above six soil layers.

As shown in Figure 10, there were considerable differences in the development of frozen soil thickness between pipe jacking of different soil layers, which can be focused on three typical soil layers: ③-3 medium gravel sand, ① artificial fill, ⑤-3 muddy silty clay. The freezing duration required to achieve the midline of pipe jacking of the three layers with representative different thicknesses is shown in Table 9.

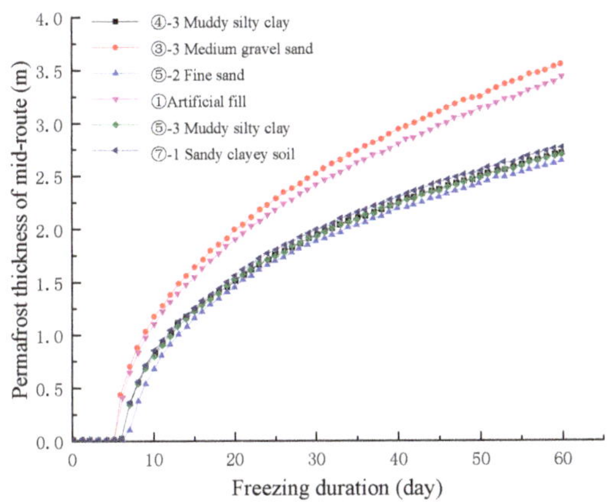

Figure 10. The thickness of frozen soil between jacking pipes.

Table 9. The freezing duration required for the frozen soil thickness midline in different soil layers (day).

Midline Thickness of Frozen Soil (m)	Soil Layers		
	③-3 Medium Gravel Sand	① Artificial Fill	⑤-3 Muddy Silty Clay
0.5	6	7	8
1.0	9	10	12
1.5	14	15	20
2.0	20	22	32
2.5	30	32	50

Correspondingly, the frozen soil curtain area was investigated, as shown in Figure 11. After the ③-3 medium gravel sand was frozen for 20 days, the water sealing path of frozen soil had been completely wrapped by pipe jacking, while the ① artificial fill completely wrapped by pipe jacking needed to be frozen for 22 days to replace the soil. The ⑤-3 muddy silty clay soil layer needed to be frozen for about 32 days.

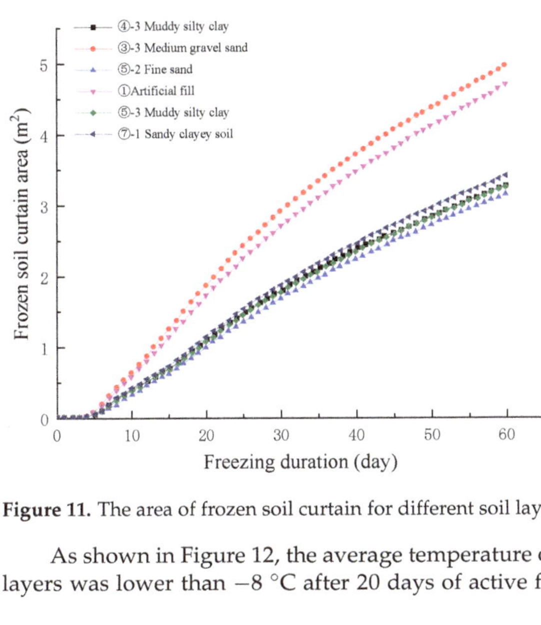

Figure 11. The area of frozen soil curtain for different soil layers.

As shown in Figure 12, the average temperature of the frozen soil curtain in six soil layers was lower than −8 °C after 20 days of active freezing and reached −10 °C after

60 days of active freezing. Although the average temperature difference of frozen soil in each soil layer was not large enough, there was also a certain difference. Essentially, the difference between soil layers lies in equivalent specific heat and thermal conductivity. When the temperature was −3–−5 °C, the equivalent specific heat of medium gravel sand was only 1810.9 J/(kg·K), much smaller than other soil layers, and some could reach up to 9454.42 J/(kg·K). The freezing effect of the medium gravel sand soil was relatively better.

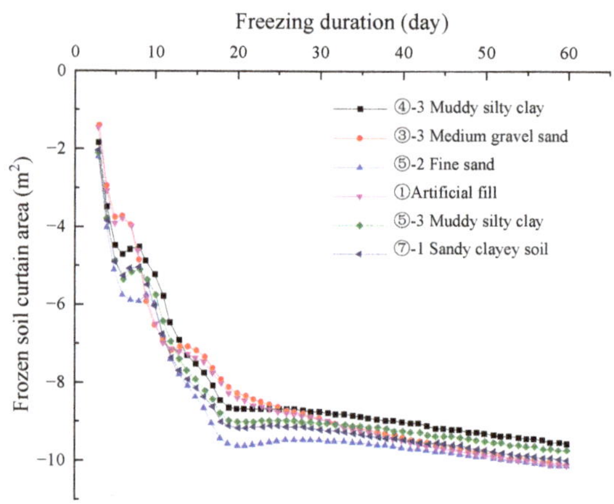

Figure 12. The average temperature of frozen soil curtain in different soil layers.

In conclusion, all soil layers could meet the freezing effect required by the design with the freezing duration; however, different soil layers have a great impact on the freezing effect. In practical freezing construction, it is necessary to pay more attention to the differences in soil layers and carry out a different freeze scheme design.

5.3. Brine Temperatures in the Freezing Tube

Here, ④-3 muddy silty clay was taken as the calculated soil layer, the initial temperature was set at 20 °C, the active freezing was set for 60 days, and the initial brine temperature pouring into the freezing tube was set at −20 °C, −30 °C, and −40 °C, respectively. The non-isothermal flow module in COMSOL was used to calculate and compare three different brine temperatures. There were significant differences in the midline of frozen soil thickness, as shown in Figure 13. The freezing effect was the best when the brine temperature is −40 °C. It can be seen that the freezing effect was better when the initial temperature of inflow brine in the freezing tube was lower. Therefore, in the case of the buried depth of Gongbei Tunnel, from the perspective of engineering accuracy requirements, it the brine temperature of the freezing tube can be set as low as possible.

Table 10 shows the freezing duration required for the midline of pipe jacking to achieve a representative thickness at three different brine temperatures in the freezing tube. If the brine temperature is −20 °C, it will take 44 days to meet the design requirements of 2 m. If the brine temperature is −30 °C, 27 days will be enough. If the brine temperature is −40 °C, then 20 days can meet the requirement.

Freeze duration depends on freezing intensity and medium. The freezing strength is the cold source, that is, the given brine temperature; the specific heat and thermal conductivity of the medium were directly related to the freezing efficiency, which is the essence of the soil layer difference in this paper.

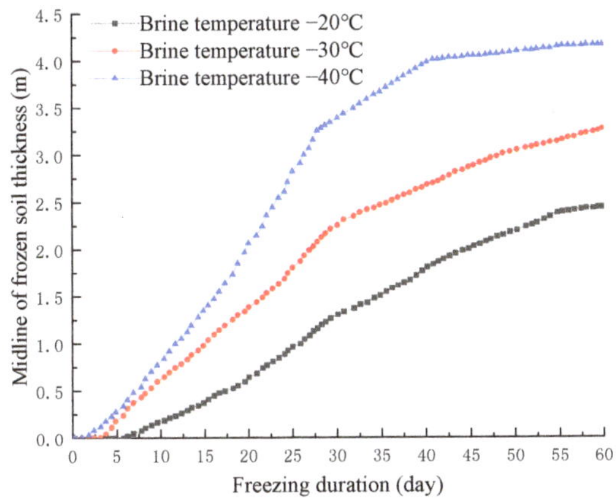

Figure 13. Thickness of frozen soil in the middle line between pipe jacking.

Table 10. The freezing duration required in different brine temperatures (day).

Midline of Frozen Soil Thickness (m)	Brine Temperature		
	−20 °C	−30 °C	−40 °C
0.5	16	8	7
1.0	25	15	12
1.5	34	22	16
2.0	44	27	20
2.5	60	35	23

6. Conclusions

The point of this study is to consider the coupled heat transfer between air and soil. According to the theory of unsteady-state conjugate heat transfer, a complete mathematical model was established to analyze the sensitivity analysis of the calculation parameter to the freezing effect of Gongbei Tunnel.

(1) A mathematical model of unsteady-state conjugate heat transfer based on the FSPR method is established, containing three governing equations in the fluid domain (continuity equation, motion equation, and energy equation), differential equation of heat conduction in the solid domain, and continuity condition of conjugate heat transfer interface. The numerical simulation results verify the rationality of the model.

(2) Through the sensitivity analysis of the factors affecting the freezing effect, it is concluded that the brine temperature has a greater impact on the freezing effect, followed by the soil layer, where the formation temperature has the least impact. For muddy silty clay, if the brine temperature is −20 °C, it takes 44 days to meet the design requirements of 2 m. If the brine temperature is −30 °C, 27 days is enough. When the formation temperature is 20 °C, it takes 20 days for medium gravel sand to reach the thickness of freezing curtain, and 32 days for muddy silty clay. Essentially, the difference between soil layers lies in equivalent specific heat and thermal conductivity. When the temperature is −3~−5 °C, equivalent specific heat of medium gravel sand is only 1810.9 J/(kg·K), much smaller than other soil layers, and some can reach up to 9454.42 J/(kg·K). Compared to other soil layers, the freezing effect of the medium gravel sand is relatively better.

(3) In the study, it was found that the initial brine temperature flowing into the freezing pipe would also affect the freezing effect. In engineering practice, the initial temperature in the freezing tube should be controlled reasonably to achieve the best freezing effect without wasting resources and reducing frost heave.

The study provides valuable and operable guidance for practical engineering. However, the model in this study also has some limitations that can be further explored. For example, when establishing the strong coupling general governing equation of unsteady-state conjugate heat transfer, the assumption that the flow state of air is regarded as laminar flow and the assumption that air is approximately regarded as ideal gas still need to be further verified.

Author Contributions: Conceptualization, X.H. and S.S.; methodology, S.D. and S.S.; software, Y.H. and H.L.; validation, Y.H. and H.L.; formal analysis, S.S., F.Z.; resources, X.H.; data curation, Z.H.; writing—original draft preparation, S.S., H.L. and S.D.; writing—review and editing, H.L., S.D. and F.Z.; supervision, S.D. and F.Z. All authors have read and agreed to the published version of the manuscript.

Funding: This research was funded by the National Natural Science Foundation of China (NSFC) (Grant No. 52008208 and 51778287), Natural Science Foundation of Jiangsu Province (Grant No. BK20200707), The Natural Science Foundation of the Jiangsu Higher Education Institutions of China (Grant No. 20KJB560029), China Postdoctoral Science Foundation (Grant No. 2020M671670), Key Laboratory of Soft Soils and Geoenvironmental Engineering (Zhejiang University), Ministry of Education (Grant No. 2020P04), Opening Project of Tunnel and Underground Engineering Research Center of Jiangsu Province (Grant No. 2021SDJJ06).

Institutional Review Board Statement: Not applicable.

Informed Consent Statement: Not applicable.

Data Availability Statement: Not applicable.

Acknowledgments: The funding above is gratefully acknowledged. The author also wishes to thank AL AMIN SARDER at Nanjing Tech University for the language work.

Conflicts of Interest: The authors declare no conflict of interest.

References

1. Hu, X.D.; Deng, S.J.; Hui, R. In situ test study on freezing scheme of freeze-sealing pipe roof applied to the Gongbei tunnel in the Hong Kong-Zhuhai-Macau Bridge. *Appl. Sci.* **2017**, *7*, 27. [CrossRef]
2. Li, J.; Li, Z.H.; Hu, X.D. Analysis on Water Sealing Effect of Freezing-sealing Pipe Roof Method. *Chin. J. Undergr. Space Eng.* **2015**, *11*, 751–758.
3. Hu, X.D.; Fang, T. Numerical simulation of temperature field at the active freeze period in tunnel construction using freeze-sealing pipe roof method. In *Tunneling and Underground Construction, Proceedings of the Geo-Shanghai 2014 International Conference, Shanghai, China, 26–28 May 2014*; Ding, W.Q., Li, X.J., Eds.; American Society of Civil Engineers (ASCE): Houston, TX, USA, 2014; Volume 30, pp. 24–33.
4. Hu, X.D.; She, S.Y. Study of freezing scheme in freeze-sealing pipe roof method based on numerical simulation of temperature field. In *ICPTT 2012: Better Pipeline Infrastructure for a Better Life, Proceedings of the International Conference on Pipelines and Trenchless Technology, Wuhan, China, 19–22 October 2012*; Baosong, M., Mohammad, N., Eds.; American Society of Civil Engineers (ASCE): Houston, TX, USA, 2012; Volume 33, pp. 1008–1020.
5. Duan, Y.; Rong, C.X.; Cheng, H. Model test of pipe curtain freezing temperature field with different pipe jacking combinations. *Glacier Frozen Soil* **2020**, *42*, 479–490.
6. Hu, X.D.; Hong, Z.Q. Analytical solution of steady-state temperature field of special pipe layout form of pipe curtain freezing. *Chin. J. Highw.* **2018**, *31*, 113–121.
7. Hu, X.D.; Deng, S.J.; Wang, Y. Analytical solution to steady-state temperature field with typical freezing tube layout employed in freeze-sealing pipe roof method. *Tunn. Undergr. Space Technol.* **2018**, *79*, 336–345. [CrossRef]
8. Hong, Z.Q.; Hu, X.D.; Fang, T. Analytical solution to steady-state temperature field of Freeze-Sealing Pipe Roof applied to Gongbei tunnel considering operation of limiting tubes. *Tunn. Undergr. Space Technol.* **2020**, *105*, 103571. [CrossRef]
9. Jahangeer, S.; Ramis, M.K.; Jilani, G. Conjugate heat transfer analysis of a heat generating vertical plate. *Int. J. Heat Mass Transf.* **2007**, *50*, 85–93. [CrossRef]
10. Long, W.; Rong, C.X.; Duan, Y. Numerical calculation of temperature field by pipe curtain freezing method in Gongbei tunnel. *Coal Geol. Explor.* **2020**, *48*, 160–168.
11. Afzal, A.; Ansari, Z.; Ramis, M.K. Parallelization of Numerical Conjugate Heat Transfer Analysis in Parallel Plate Channel Using OpenMP. *Arab. J. Sci. Eng.* **2020**, *45*, 8981–8997. [CrossRef]
12. He, L.T.; Oldfieldg, L.M. Unsteady Conjugate Heat Transfer Modeling. *J. Turbomach.* **2011**, *133*, 31022. [CrossRef]
13. Afzal, A.; Samee, A.D.M.; Razak, R.K.A.; Ramis, M.K. Steady and Transient State Analyses on Conjugate Laminar Forced Convection Heat Transfer. *Arch. Comput. Methods Eng.* **2019**, *27*, 135–170. [CrossRef]

14. Samee, A.D.M.; Afzal, A.; Ramis, M.K. Optimal spacing in heat generating parallel plate channel: A conjugate approach. *Int. J. Therm. Sci.* **2019**, *136*, 267–277. [CrossRef]
15. Maffulli, R.; Marinescu, G.; He, L. On the Validity of Scaling Transient Conjugate Heat Transfer Characteristics. *J. Eng. Gas Turbines Power* **2020**, *142*, 31021. [CrossRef]
16. Nouri, M.; Hamila, R.; Perre, P. A double distribution lattice Boltzmann scheme for unsteady Conjugate Heat Transfer: The DD-CHTLB method. *Int. J. Heat Mass Transf.* **2019**, *137*, 609–614. [CrossRef]
17. Voigt, S.; Noll, B.; Aigner, M. Development of a model for unsteady conjugate heat transfer simulations. *Prog. Comput. Fluid Dyn. Int. J.* **2019**, *19*, 69–79. [CrossRef]
18. Makhija, D.S.; Beran, P.S. Concurrent shape and topology optimization for unsteady conjugate heat transfer. *Struct. Multidiscip. Optim.* **2020**, *62*, 1275–1297. [CrossRef]
19. China Communications Second Highway Survey Design and Research Institute. *Special Technical Research Report on Gongbei Tunnel of Zhuhai Connection Project of Hongkong-Zhuhai-Macao Bridge*; China Communications Second Highway Survey Design and Research Institute Co. Ltd.: Wuhan, China, 2012.
20. China Communications Second Highway Survey Design and Research Institute. *Special Technical Research Report on Gongbei Tunnel of Zhuhai Connection Project of Hongkong-Zhuhai-Macao Bridge Abridged Edition (2013 April)*; China Communications Second Highway Survey Design and Research Institute Co. Ltd.: Wuhan, China, 2013.
21. Wang, J. The Application of Computer Simulation of Flow Field in the Design of Heat Treatment Furnace. Ph.D. Thesis, Shanghai Jiao Tong University, Shanghai, China, 2008.
22. Zhang, L.F. Computational Study on Heat Transfer of Internally Cooled Turbine Blades Using Thermal-Fluid Coupling Method. Master's Thesis, Northwestern Polytechnical University, Xi'an, China, 2006.
23. Zhou, G.; Yan, Z.; Xu, S. *Fluid Mechanics*; Beijing Higher Education Press: Beijing, China, 2000.
24. Jenkins, R.; Ray, X.; Meyers, R.A. *Encyclopedia of Physical Science and Technology*, 3rd ed.; Academic Press: Pittsburgh, PA, USA, 2003; pp. 887–902.
25. Pijush, K.K.; Cohen, I.M.; David, R. *Fluid Mechanics*, 6th ed.; Academic Press: Pittsburgh, PA, USA, 1981; pp. 109–193.
26. Shi, T.Z. Narrow del operator and symbolic operation. *J. Henan Financ. Univ.* **2007**, *4*, 16–19.
27. Wang, Q.; Chen, B.Y.; Miao, X.P. Research on Numerical Method of Three-Dimensional Unsteady Coupled Heat Transfer Problem. In *The Fourth National Symposium on New Technologies of Refrigeration and Air Conditioning, Proceedings of the Conference of China Refrigeration Society, Nanjing, China, 7–9 April 2006*; China Academic Journal Electronic Publishing House: Beijing, China, 2006.
28. Guo, L.K. *Numerical Calculation of Heat Transfer*; Anhui Science and Technology Press: Hefei, China, 1987.
29. Jiang, P.X.; Ke, D.Y.; Ren, Z.P. PHOENICS Solve the Coupled Problem of Unsteady Conduction Convection and Radiation Heat Transfer. *J. Tsinghua Univ.* **1999**, *4*, 113–117.
30. Sugawara, M.; Komatsu, Y.; Beer, H. Melting and freezing around a horizontal cylinder placed in a square cavity. *Heat Mass Transf.* **2008**, *45*, 83–92. [CrossRef]
31. Guo, L.K.; Kong, X.Q.; Chen, S.N. *Computational Heat Transfer*; University of Science and Technology of China Press: Hefei, China, 1988.
32. Pimentel, E.; Papakonstantinou, S.; Anagnostou, G. Numerical interpretation of temperature distributions from three ground freezing applications in urban tunnelling. *Tunn. Undergr. Space Technol.* **2012**, *28*, 57–69. [CrossRef]
33. Wang, T.; Zhang, F.; Furtney, J. A review of methods, applications and limitations for incorporating fluid flow in the discrete element method. *J. Rock Mech. Geotech. Eng.* **2022**, *14*, 1005–1024. [CrossRef]
34. Niu, J.; Jiang, L.; Luo, X.; Deng, Z. Structured Method of Dynamic Fully Coupled Rheological Model for Seepage Field and Stress Field in Concrete Dam. In *IOP Conference Series Earth and Environmental Science, Proceedings of the 7th International Conference on Environmental Science and Civil Engineering, Seoul, Kerea, 11–13 April 2017*; IOP Publishing Ltd.: London, UK, 2021.
35. Institute of Rock and Soil Mechanics Chinese Academy of Sciences. *Experimental Report on the Physical and Mechanical Parameters of Artificially Frozen Soil at the North Port of the Zhuhai Connecting Side Arch of the Hong Kong-Zhuhai-Macao Bridge*; Institute of Rock and Soil Mechanics Chinese Academy of Sciences: Wuhan, China, 2012.
36. Wang, A.; Xie, Z.; Feng, X.; Tian, X.; Qin, P. A soil water and heat transfer model including changes in soil frost and thaw fronts. *Sci. Chin.* **2014**, *57*, 1325–1339. [CrossRef]
37. Zhang, M.; Zhang, J.; Lai, Y. Numerical analysis for critical height of railway embankment in permafrost regions of Qinghai–Tibetan plateau. *Cold Reg. Sci. Technol.* **2005**, *41*, 111–120. [CrossRef]

buildings

MDPI

Article

Resistance Characteristic Parameters Estimation of Hydraulic Model in Heating Networks Based on Real-Time Operation Data

Peng Luo [1,2], Hao Wang [1,2], Yongxin Liu [1,2,*], Qingting Du [1,3] and Zhengshuai Zhang [1,2]

1 School of Architecture, Harbin Institute of Technology, Harbin 150001, China; luopeng_hit@yeah.net (P.L.); wang_hao_hit@126.com (H.W.); aces1982@sina.com (Q.D.); zhangmahit@163.com (Z.Z.)
2 Key Laboratory of Cold Region Urban and Rural Human Settlement Environment Science and Technology, Ministry of Industry and Information Technology, Harbin 150001, China
3 Shenzhen Institute of Building Research Co., Ltd., Shenzhen 518049, China
* Correspondence: liuyongxin@hit.edu.cn; Tel.: +86-451-86286338

Abstract: Heating systems are essential municipal infrastructure in winter, especially in severe cold cities of China. The safety and efficiency of heating systems greatly affect building energy efficiency and indoor thermal comfort. Heating networks (HNs), playing the role of transportation, are the key parts of heating systems. In HNs, hydraulic models could be affected by the accuracy of resistance characteristic parameters, which are expressed by pipe friction parameters (PFPs) in this paper. As the uniqueness of the estimation results of PFPs has not been discussed in previous studies, this paper builds an estimation method of PFPs by dividing two types of pipes, substituting variables and establishing a split-step linearization method. Combining with the theory of matrix equations, the decision conditions and solution methods for obtaining the unique estimation results of PFPs are determined. Theoretical analysis and case study results show that estimation values of PFPs can be obtained by utilizing measured data under multiple hydraulic conditions. In the example of DN and the simple actual HN, the average estimation deviation of PFPs is 1.42% and 1.86%, which are accurate enough for actual engineering. Estimation results of PFPs obtained by this method guarantee the accuracy of analysis and regulation in heating systems and improve social energy utilization efficiency.

Keywords: energy efficiency; hydraulic model; pipe friction parameters; civil engineering

Citation: Luo, P.; Wang, H.; Liu, Y.; Du, Q.; Zhang, Z. Resistance Characteristic Parameters Estimation of Hydraulic Model in Heating Networks Based on Real-Time Operation Data. *Buildings* **2022**, *12*, 743. https://doi.org/10.3390/buildings12060743

Academic Editor: Weixin Ren

Received: 11 May 2022
Accepted: 28 May 2022
Published: 30 May 2022

1. Introduction

With rapid urbanization and residents' requirements for thermal comfort, the total heating floor area in Northern China reached 10.8 billion m^2 in 2019. Approximately 5% of the total energy consumption of the country was utilized for space heating [1]. Reducing the energy consumption of heating systems is the key to achieving carbon peak and carbon neutrality in China. Heating networks (HNs) are the carrier of heat medium (hot water) in heating systems, and their major function is to transport and distribute heat medium on demand [2]. An HN needs to make the flow and thermal energy meet the requirements at the same time, that is, to achieve hydraulic balance and thermal balance. Otherwise, uneven heating, low energy efficiency or a high possibility of equipment failure appear easily. In heating systems, hydraulic balance is an essential precondition for thermal balance, and the hydraulic balance depends on the hydraulic model. Thus, obtaining the hydraulic model of HN is the key to realizing the hydraulic and thermal balance.

The hydraulic model has two essential factors, which are the topology and resistance characteristics of an HN. The design documents can provide the topology but not resistance characteristics of an HN. There are three reasons listed as follows:

1. Due to the complexity of an HN, it is impossible to ensure that the design and actual resistance characteristics of the HN are completely consistent in detail;
2. Due to the initial adjustment of an HN, the resistance characteristics of the HN are uncertain when it is just put into use;
3. Due to corrosion or blockage of pipelines (especially the long service ones), the resistance characteristics may be far away from the initial state.

It is recognized as an effective method in that the resistance characteristics are determined by the real-time measured values of operation parameters (such as node heads and outflows). In actual engineering, pipe friction parameters (PFPs) are commonly used to express the resistance characteristics. If the estimation values of PFPs can be obtained, a real-time hydraulic model of the HN can be constructed. Wang et al. [3] concluded that the pump power consumption can be reduced to 85% by using the identified PFPs in the hydraulic model. Moreover, the analysis of hydraulic and thermal conditions based on estimation values of PFPs will greatly improve the energy efficiency of the HN, which will be helpful in achieving carbon peak and carbon neutrality.

In actual HNs, the nodes are usually equipped with pressure sensors. Real-time measured values of node heads can be obtained. HNs, being different from water supply networks, are closed networks and composed of supply networks and return networks. (In this paper, supply networks are taken as an example to analyze, and the analysis process of return networks is the same as that of supply networks.) In the supply networks, the flows of heat sources and heat users can be regarded as node outflows, which can be obtained usually. Based on these conditions, the measured conditions for real-time estimation of PFPs can be determined.

At present, the research on characteristic parameters estimation mainly focuses on water supply networks. Scholars have studied estimation and calibration problems based on least-squares methods of water supply networks for several decades. Nash and Karney [4] calibrated hydraulic models based on the least-squares method, and an objective function was expressed as the difference between measured and calculated values. Reddy et al. [5] combined a least-squares method with Gauss–Newton method. This calibration method was verified in a small-scale water supply network. Savic and Walters [6] researched a simulation and calibration problem of water supply networks. The research result shows that the accuracy of a calculation result by utilizing measured data over a period of time is better than only relying on measured data under one hydraulic condition. Meirelles et al. [7] considered operational flows as a basis of PFPs calibration. Meirelles's method relieves adverse effects on PFPs identification results obtained by using inadequate data. Shamloo and Haghighi [8] added a genetic algorithm to solve an optimization model by utilizing the sequential quadratic programming method. Better results can be obtained. However, the time consumption of these calculation processes is enormous, especially for PFPs estimation under multiple hydraulic conditions. Optimization methods based on least-squares methods have been widely applied in engineering. The objective is to search for an acceptable result for hydraulic calculation equations. However, ensuring the acceptable solution (PFPs values) is the same as the actual resistance characteristic cannot be guaranteed.

The mainstream of PFPs identification methods is based on least-squares principles. However, analytic methods also play an important role in PFPs identification studies. Lansey and Basnet [9] proposed a method based on the gradient method and non-linear programming technology to estimate unknown parameters of water supply networks. Calculation processes contain parameter estimation, calibration assessment and data collection design. Jun et al. [10] chose measured data under multiple hydraulic conditions and expressed results of non-linear equations. This method can be applied to inadequate data. By using the Taylor series approach, Datta and Sridharan [11] proposed a method to estimate unknown parameters by utilizing measured values of node heads and pipe flows. Kapelan et al. [12] combined Levenberg–Marquard algorithm with genetic algorithms. Relatively reliable results could be obtained with this method. For a water supply network, Liu et al. [13,14] proposed an identification method of PFPs based on Moore–Penrose pseudo-inverse solution. Identification results are accurate enough for engineering when the number of measured sites is relatively small. Further, in order to meet the current demands of digital twin technology and intelligent heating technology, it is necessary to develop a method to obtain real-time PFPs values [15].

Being different from water supply networks, an HN needs to be decomposed into two sub-HNs, which increases calculation quantity. More importantly, in order to meet the requirements of heating demand, the flows of HNs are usually constant and huge. Therefore, accurate estimation of PFPs values is more important for HNs. Wang et al. [3] developed an identification method to obtain hydraulic resistance of a branch HN. However, 500 operating conditions need to be provided if relatively small errors of the identified hydraulic resistances are expected. Liu et al. [16] proposed a method to determine the PFPs of a branch HN uniquely by using measured data of heads and outflows in all the heating substations. However, it is difficult to apply this method directly to real-time PFPs analysis of an HN, including loops. Tol [17] and Zheng et al. [18] applied Newton iteration method in PFPs estimation of a branch HN. In the studies mentioned above, estimation methods have been proposed for branch HNs rather than loop HNs, which are more common in engineering. Equations described in hydraulic models are non-linear ones, also creating a barrier for the HN study.

By utilizing measured data under multiple hydraulic conditions, this paper suggests a new expression and decision condition to find unique estimation results of PFPs. Following the four steps mentioned below, this study presents two examples of HNs, illustrating the ability to obtain PFPs unique estimation values by utilizing measured data under multiple hydraulic conditions. First, build a PFPs calculation equation by using mass and energy conservation equations of an HN, and eliminate dependent variables by the relationship of pipe flows according to the types of the pipes. Second, describe energy conservation equations corresponding to tree pipes by relationships of operational data between different hydraulic conditions. Third, transform matrix equations of a tree into a linear form. Finally, analyze theoretically the probability of obtaining unique estimation values of PFPs, and express the results of PFPs estimation by a solution of the corresponding matrix equation.

2. Methods

Hydraulic calculation equations (mass and energy conservation equations) can be established by using operational data or PFPs of an HN. In the design calculation of HNs, PFPs are viewed as known variables. In the estimation process of PFPs, some measured data such as node heads and outflows are viewed as known variables. Pipes in an HN can be divided into a spanning tree (tree) and a corresponding cotree (cotree). PFPs calculation equations of an HN are first described by partitioned matrices and then transformed into linear algebraic equations. According to theories of a linear algebraic equation and matrix analysis, in order to obtain a unique solution, the number of equations being more than or equal to the number of unknown variables should be ensured. However, in actual engineering, the number of measurement sites is limited. In most cases, the number of unknown variables is more than the number of (independent) equations in PFPs calculation equations. This paper researches the PFPs estimation problem of an HN by increasing the number of hydraulic conditions and establishes a theoretical framework for estimating the PFPs values and obtaining unique values.

2.1. Basic Equations

Two matrix functions are defined before further study. Considering a matrix $\boldsymbol{M}$ of dimension $m \times n$, $abs\ (\boldsymbol{M})$ is a matrix of dimension $m \times n$. Every element of the matrix $abs\ (\boldsymbol{M})$ is the absolute value of the element of the matrix $\boldsymbol{M}$ at the corresponding position. Considering a vector $\boldsymbol{X}$ of dimension $n \times 1$, a matrix transform function $\boldsymbol{D} = \boldsymbol{D}\ (\boldsymbol{X})$ is defined. $\boldsymbol{D}$ is a diagonal matrix of dimension $n \times n$. The ith diagonal element of $\boldsymbol{D}$ satisfies this relationship $\boldsymbol{D}\ (i, i) = \boldsymbol{X}\ (i)$, $(i = 1, 2, \ldots, n)$.

Considering an HN with $n + 1$ nodes and b pipes, hydraulic calculation equations can be expressed in the following matrix form.

$$\boldsymbol{AG} = \boldsymbol{Q}, \tag{1}$$

$$abs\left(\boldsymbol{A}^T\boldsymbol{P}\right) = abs(\Delta\boldsymbol{H}) = D^2(\boldsymbol{G}) \cdot \boldsymbol{S}, \tag{2}$$

where $\boldsymbol{A}$ is a basic incidence matrix, which contains information about which pipes are connected to a particular node. The value of every element in matrix $\boldsymbol{A}$ can be defined as -1 or 1 according to the pipe flow being towards or away from the node; else, if the pipe flow does not connect with the node, the corresponding element is defined as 0. $\boldsymbol{G}$ is a pipe flows vector, $\boldsymbol{Q}$ is a node outflows vector, $\boldsymbol{P}$ is a node heads vector, $\Delta\boldsymbol{H}$ is a head losses vector and $\boldsymbol{S}$ is a PFPs vector.

Generally, some operational data (node heads and pipe flows) can be obtained in the PFPs estimation study. The main aim of this study is to search for unique values of PFPs by using operational data. If all measured data of node heads and outflows under a single hydraulic condition are available, there will be $b + n$ equations and $2b$ unknown variables in Equations (1) and (2). In an HN, b is greater than n. Thus, the number of unknown variables is greater than the number of algebraic equations. Equations (1) and (2) are actually underdetermined. Thus, it is not able to find unique results of PFPs in the situation. To obtain more algebraic equations, operational data of more hydraulic conditions are provided. However, these algebraic equations are non-linear and cannot be solved directly. Thus, a linearization method is needed to transform the non-linear equations into linear ones during studying PFPs estimation.

Analyzing the feasibility of obtaining unique values of PFPs, the paper establishes a split-step linearization method for PFPs estimation by using measured data under multiple hydraulic conditions. Since measured data of node heads are widely used in the detection and calibration of hydraulic networks [19,20], considering the most unfavorable estimation conditions, the estimation processes assume there are no pipe flow sensors. (If there are pipe flow sensors, the theory mentioned in this paper can be applied, obtaining better results.)

Since there are unknown variables $\boldsymbol{S}$ needing to be identified in the energy conservation equation (Equation (2)), the mass conservation equation (Equation (1)) can be used to eliminate dependent unknown variables. Flows in tree pipes, satisfying the mass conservation equation, can be written as [21].

$$\boldsymbol{G}_t = -\boldsymbol{A}_t^{-1}\boldsymbol{A}_l\boldsymbol{G}_l + \boldsymbol{A}_t^{-1}\boldsymbol{Q}, \tag{3}$$

where $\boldsymbol{A}_t$ is composed corresponding to tree pipes, $\boldsymbol{A}_l$ is composed corresponding to cotree pipes, satisfying $\boldsymbol{A} = [\boldsymbol{A}_t, \boldsymbol{A}_l]$; $\boldsymbol{G}_t$ is composed by tree pipe flows, $\boldsymbol{G}_l$ is composed by cotree pipe flows, satisfying $\boldsymbol{G} = [\boldsymbol{G}_t, \boldsymbol{G}_l]$.

The energy conservation equation can be divided into two types. Head losses of tree and cotree pipes can be written as follows, respectively.

$$abs\left(\boldsymbol{A}_t^T\boldsymbol{P}\right) = abs(\Delta\boldsymbol{H}_t) = D^2(\boldsymbol{G}_t) \cdot \boldsymbol{S}_t, \tag{4}$$

$$abs\left(\boldsymbol{A}_l^T\boldsymbol{P}\right) = abs(\Delta\boldsymbol{H}_l) = D^2(\boldsymbol{G}_l) \cdot \boldsymbol{S}_l, \tag{5}$$

where $\boldsymbol{S}_t$ is composed of tree PFPs, $\boldsymbol{S}_l$ is composed of cotree PFPs, satisfying $\boldsymbol{S} = [\boldsymbol{S}_t, \boldsymbol{S}_l]$; $\Delta\boldsymbol{H}_t$ is composed of head losses of tree pipes, $\Delta\boldsymbol{H}_l$ is composed of head losses of cotree pipes, satisfying $\Delta\boldsymbol{H} = [\Delta\boldsymbol{H}_t, \Delta\boldsymbol{H}_l]$. Then, PFPs can be solved when head losses and pipe flows are available.

2.2. Expression of PFPs Estimation

A variable representing a relationship of head losses between different hydraulic conditions is introduced here. That is a critical process of solving PFPs calculation equations. Head losses relationship of cotree pipes between multiple hydraulic conditions can be written once the measured value of every node head is available. The first hydraulic condition satisfying the condition $\Delta h_l^{(1)}(i) \neq 0$ is chosen as a reference hydraulic condition. A reference hydraulic condition can be selected randomly and does not impact the estimation study. Then, a ratio of head losses in cotree pipe i between the kth and the first hydraulic conditions can be written as follows:

$$\Delta h_{lP}^{(k)}(i) = \Delta h_l^{(k)}(i)/\Delta h_l^{(1)}(i),\ (i = 1,\ 2, \ldots,\ b-n; k = 1,\ 2, \ldots,\ m) \tag{6}$$

where $\Delta h_l(i)$ is the head loss in cotree pipe i and $\Delta h_{lP}(i)$ is a ratio of head losses in cotree pipe i under the kth and the first hydraulic conditions. The superscript k is an order number of hydraulic conditions.

Variables in energy equations contain both node heads and pipe flows. To eliminate dependent variables in equations, a relationship between node heads and pipe flows under multiple hydraulic conditions needs to be introduced. Considering pipe flows are directional variables, a new vector variable $\Delta \boldsymbol{H}_{lG}^{(k)}$ of dimension $(b-n) \times 1$ is defined to express relationships of cotree pipe flows under different hydraulic conditions. A ratio of cotree pipe flow i between the kth and the first conditions (the ith element of $\Delta \boldsymbol{H}_{lG}^{(k)}$) is:

$$\Delta h_{lG}^{(k)}(i) = \begin{cases} abs\left(\Delta h_{lP}^{(k)}(i)\right)^{1/2}, when\ \Delta h_l^{(k)}(i) \cdot \Delta h_l^{(1)}(i) \geq 0,\ \Delta h_l^{(1)}(i) \neq 0 & (i = 1,\ 2, \ldots,\ b-n; \\ -abs\left(\Delta h_{lP}^{(k)}(i)\right)^{1/2}, when\ \Delta h_l^{(k)}(i) \cdot \Delta h_l^{(1)}(i) < 0 & k = 1,\ 2, \ldots,\ m) \end{cases} \tag{7}$$

By using Equation (7), a ratio of cotree pipe flow i is written by the ratio of head losses in the corresponding pipe between the kth and the first hydraulic conditions. Combining with cotree pipe flows under the first hydraulic condition, those under the kth hydraulic condition can be expressed as follows:

$$\boldsymbol{G}_l^{(k)} = D\left(\Delta \boldsymbol{H}_{lG}^{(k)}\right) \cdot \boldsymbol{G}_l^{(1)},\ (k = 1,\ 2, \ldots,\ m) \tag{8}$$

As dependent variables, pipe flows of a cotree do not appear except for the first hydraulic condition. Under the kth hydraulic condition, Equation (4) can be written, in view of Equation (8), as follows:

$$abs\left(\boldsymbol{A}_t^T \boldsymbol{P}^{(k)}\right) = abs\left(\Delta \boldsymbol{H}_t^{(k)}\right) = D^2\left(-\boldsymbol{A}_t^{-1} \boldsymbol{A}_l D\left(\Delta \boldsymbol{H}_{lG}^{(k)}\right) \cdot \boldsymbol{G}_l^{(1)} + \boldsymbol{A}_t^{-1} \boldsymbol{Q}^{(k)}\right) \cdot \boldsymbol{S}_t\ (k = 1,\ 2, \ldots,\ m) \tag{9}$$

After eliminating cotree pipe flows under other hydraulic conditions except the first one, Equation (4) can be substituted by Equation (9). Considering Equation (9) expresses a non-linear equation, a new matrix variable is defined to transform Equation (9) into a linear equation. A vector $\Delta \boldsymbol{H}_{tG}^{(k)}$ of dimension $n \times 1$ is introduced here. An element of this vector is $\Delta h_{tG}^{(k)}$, whose absolute value equals head loss in a tree pipe to the half power under the kth hydraulic condition. Considering pipe flows are directional variables, the ith element of $\Delta \boldsymbol{H}_{tG}^{(k)}$ is:

$$\Delta h_{tG}^{(k)}(i) = \begin{cases} \left(\Delta h_t^{(k)}(i)\right)^{1/2}, when\ \Delta h_t^{(k)}(i) \geq 0 \\ -\left(-\Delta h_t^{(k)}(i)\right)^{1/2}, when\ \Delta h_t^{(k)}(i) < 0 \end{cases},\ (I = 1,\ 2, \ldots,\ n;\ k = 1,\ 2, \ldots,\ m) \tag{10}$$

where $\Delta h_t(i)$ is the head loss in tree pipe i. Equation (10) is preparation for transforming PFPs calculation equations into linear ones. A new vector $\boldsymbol{S}_{tp}$ is defined to express a vector related to a cotree PFPs. $\boldsymbol{S}_{tp}$ has the same dimension as the vector $\boldsymbol{S}_t$. The relationship of elements can be written as $s_{tp} = s_t^{-1/2}$, meaning s_{tp} equals one divided by s_t to the half power. The linear calculation matrix equation of PFPs is written as:

$$D\left(\Delta \boldsymbol{H}_{tG}^{(k)}\right) \cdot \boldsymbol{S}_{tp} + \boldsymbol{A}_t^{-1} \boldsymbol{A}_l \cdot D\left(\Delta \boldsymbol{H}_{lG}^{(k)}\right) \cdot \boldsymbol{G}_l^{(1)} = \boldsymbol{A}_t^{-1} \boldsymbol{Q}^{(k)},\ (k = 1,\ 2, \ldots,\ m) \tag{11}$$

In Equation (11), coefficient matrices can be represented by node heads and outflows. Unknown variables include cotree pipe flows and PFPs of a tree. The numbers of the two kinds of unknown variables are $b-n$ and n, respectively. Equation (11) can substitute Equation (9) since it is a linear matrix equation and does not contain dependent unknowns.

Even though measured data of node heads and outflows under one hydraulic condition can be obtained, there are n equations and b independent unknown variables in Equation (11). As the number of pipes is greater than the number of nodes, Equation (11) expresses an equation with the number of unknown variables being more than that of equations. In this situation, accurate, unique solutions are unavailable. Thus, operational data of multiple hydraulic conditions are needed to get more accurate results for PFPs.

2.3. Linear Expression of PFPs

If operational data of m hydraulic conditions are available, an integrated matrix equation of PFPs calculation can be written as follows.

$$\begin{bmatrix} D\left(\Delta \boldsymbol{H}_{tG}^{(1)}\right) & \boldsymbol{A}_t^{-1}\boldsymbol{A}_l \cdot D\left(\Delta \boldsymbol{H}_{lG}^{(1)}\right) \\ D\left(\Delta \boldsymbol{H}_{tG}^{(2)}\right) & \boldsymbol{A}_t^{-1}\boldsymbol{A}_l \cdot D\left(\Delta \boldsymbol{H}_{lG}^{(2)}\right) \\ \cdot & \cdot \\ \cdot & \cdot \\ D\left(\Delta \boldsymbol{H}_{tG}^{(m)}\right) & \boldsymbol{A}_t^{-1}\boldsymbol{A}_l \cdot D\left(\Delta \boldsymbol{H}_{lG}^{(m)}\right) \end{bmatrix} \cdot \begin{bmatrix} \boldsymbol{S}_{tp} \\ \boldsymbol{G}_l^{(1)} \end{bmatrix} = \begin{bmatrix} \boldsymbol{A}_t^{-1}\boldsymbol{Q}^{(1)} \\ \boldsymbol{A}_t^{-1}\boldsymbol{Q}^{(2)} \\ \cdot \\ \cdot \\ \boldsymbol{A}_t^{-1}\boldsymbol{Q}^{(m)} \end{bmatrix} \quad (12)$$

There are $m \times n$ equations and b unknown variables. In Equation (12), if the number of independent equations can reach b, there must be unique solutions. If the number of independent equations is less than b, unique solutions cannot be reached. When the number of independent equations can reach b, unknown variables, including tree PFPs and cotree pipe flows under the first hydraulic condition, can be obtained.

Next, pursuing the values of PFPs can be transferred to solve Equation (12), which is a linear matrix equation. Depending upon the results, there will be unique estimation values of PFPs. To study Equation (12), the theory of matrix analysis is employed. There are two cases according to the number of independent equations in Equation (12). If the number of independent equations equals the number of unknown variables, a unique solution can be calculated directly; if not, a Moore–Penrose pseudo-inverse is introduced to express the solution of Equation (12) [13]. When the number of independent equations equals the number of unknown variables, the special Moore–Penrose pseudo-inverse solution can also express a unique solution of Equation (12) [22]. Estimation results of PFPs can be calculated through the unique solution.

Since Equation (12) is a linear matrix equation, whether there is a unique solution depends on the relationship between the numbers of algebraic equations and unknown variables. When there are m hydraulic conditions, Equation (12) contains $m \times n$ algebraic equations and b unknown variables. In those unknown variables, there are $b-n$ cotree pipe flows and n tree PFPs. A necessary condition to obtain a unique solution is the number of algebraic equations greater than or equal to the number of unknown variables. However, it cannot be guaranteed that the number of independent equations can reach b. If hydraulic conditions of an HN are enough to express b independent equations, unique results of PFPs may be obtained; if not, accurate solutions cannot be obtained.

As mentioned above, if hydraulic conditions are sufficient to obtain a unique solution to Equation (12), the PFPs of a tree can be communicated directly. Cotree PFPs can be expressed by cotree pipe flows under the first hydraulic condition. Inversion processes of variables are written as follows:

$$s_t(i) = \left(s_{tp}(i)\right)^{-2} \ (i = 1,\ 2, \ldots,\ n), \quad (13a)$$

$$\boldsymbol{S}_l = D^{-2}\left(\boldsymbol{G}_l^{(1)}\right) \cdot abs\left(\Delta \boldsymbol{H}_l^{(1)}\right) = D^{-2}\left(\boldsymbol{G}_l^{(1)}\right) \cdot abs\left(\boldsymbol{A}_l^T \boldsymbol{P}^{(1)}\right), \quad (13b)$$

According to the theory of linear equations, the number of independent equations being b is a sufficient and necessary condition of the existence of unique solutions. The number of independent equations in Equation (12) increases when adding hydraulic conditions, increasing the probability of adding the number of independent equations. If the number of hydraulic conditions can guarantee the number of independent equations can reach b, there is no need to keep on adding hydraulic conditions. At the beginning of PFPs estimation, the criteria mentioned here can be used to determine whether accurate, unique estimation values of PFPs can be obtained, bringing the advantage of not needing additional hydraulic conditions when there is no error with measured data.

With errors of measured data being considered, if measurement sensors are sufficient, more accurate estimation results of PFPs will be obtained by increasing hydraulic conditions. If not, increasing hydraulic conditions has no contribution to the accuracy of estimation results of PFPs. There is a possibility of a large difference between estimation results and real results of PFPs if measurement sensors are insufficient. The main processes of this method are shown as in Figure 1.

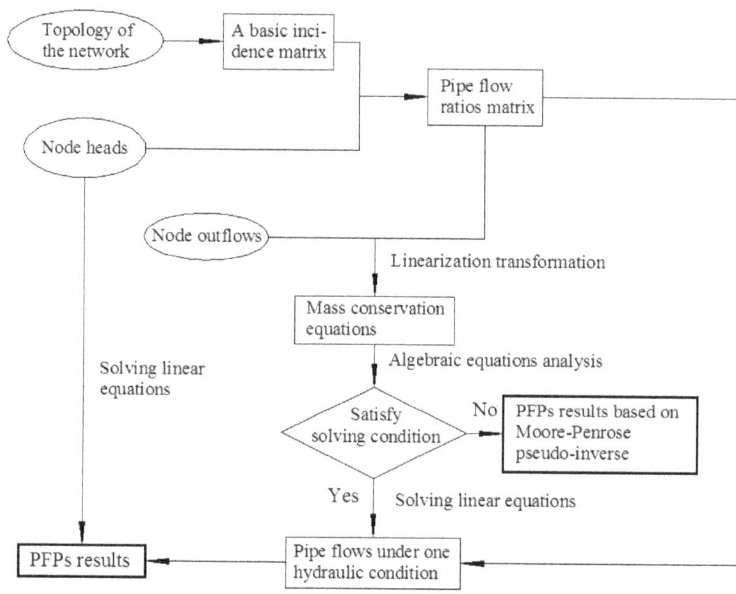

Figure 1. The main processes of the PFPs estimation method.

3. Case Study

To present application processes and demonstrate the effectiveness of the proposed method, two cases of HNs are introduced in this paper.

3.1. An Example HN

The estimation method is verified by an example HN. In the example, PFPs values can be calculated by utilizing measured data of node heads and outflows under two hydraulic conditions. The topology of the HN is shown in Figure 2. This example HN contains twelve pipes and nine nodes, wherein nine nodes represent three heat sources and six heat users. In this HN, setting values of PFPs in pipe 1, 3, 5, 8, 9 or 10 is 3.6×10^{-4} h^2/m^5, in pipe 2 or 4 is 6.0×10^{-4} h^2/m^5, and in pipe 6, 7, 11 or 12 is 7.2×10^{-3} h^2/m^5. By using the values of PFPs, node heads and outflows under two hydraulic conditions used in the next sections (estimation processes) are computed by hydraulic calculation equations. In this example, node heads and outflows of different hydraulic conditions are assumed as "known values", and PFPs and pipe flows are considered as unknown variables. The known values under two hydraulic conditions are shown in Figure 2.

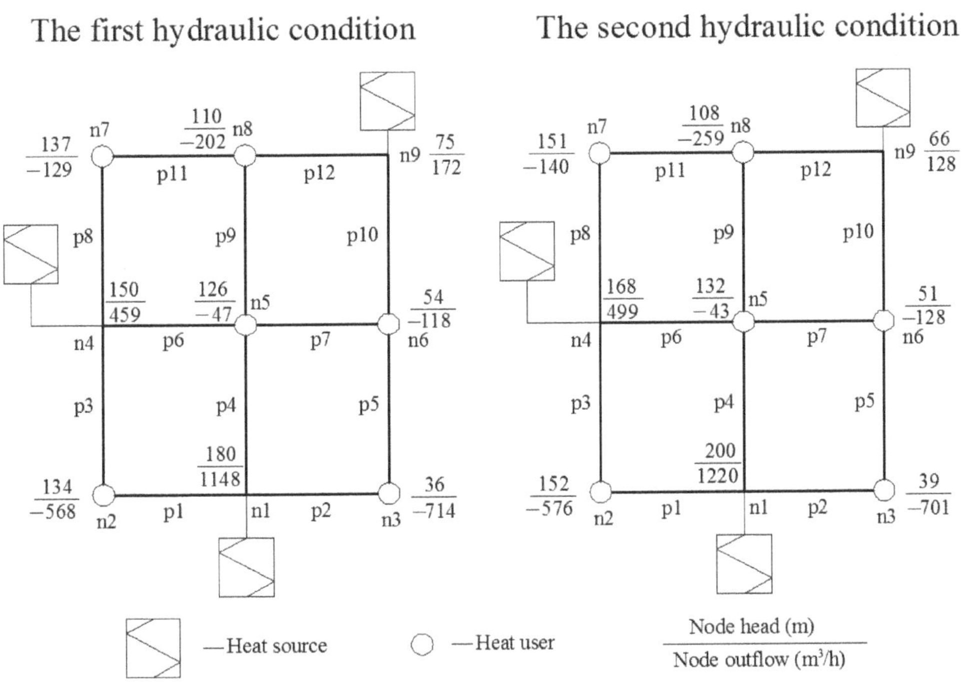

Figure 2. Schematic diagram of an example HN and known variables under two hydraulic conditions.

Pipes 6, 7, 11 and 12 in this HN are selected as cotree pipes. A linear equation is obtained by submitting "known values" in Equation (12). The example HN contains 12 pipes and nine nodes. According to Equation (12), when there are two hydraulic conditions available for PFPs estimation of the HN shown in Figure 2, there are 16 equations and 12 unknown variables, including eight variables related to PFPs of a tree and pipe flows of a cotree under the first hydraulic condition, illustrating a unique solution of Equation (12) exists. Further, in Equation (12), the number of independent equations can reach 12. According to the theory of linear equations, there are unique solutions to Equation (12). Those unknown variables can be solved by calculation tools included in Scilab 6.1.0. Then, the PFPs of the entire HN are calculated by Equation (13). The estimation results are presented in Table 1.

By comparing with setting values of PFPs, the accuracy of PFPs estimation results can be evaluated. Estimation values of PFPs are found very close to setting values, except for the cotree pipes 6, 7, 11 and 12. The main reason can be concluded as follows. In the inversion process of variables, PFPs values of a tree and a cotree are calculated out by Equation (13). Since there is a one-to-one correspondence between PFPs values of a tree and variables representing the values in Equation (13a), other operational data will not be used in the inversion process. This process may give a small error. PFPs values of a cotree need to be calculated by Equation (13b), in which pipe flows and node heads need to be calculated through the inversion process. Not as direct as calculation of tree PFPs, this process brings about limited calculation errors, leading to calculation errors accumulating in PFPs values of a cotree. The maximum estimation deviation of every PFP is less than 5%, and the average estimation deviation is 1.42%, which is accurate enough for engineering.

Table 1. The estimation results of the example HN.

Pipe No.	Setting Values of PFPs (h^2/m^5) $\times 10^{-4}$	Estimation Values of PFPs (h^2/m^5) $\times 10^{-4}$	Estimation Deviation of PFPs (%)
1	3.6	3.58	0.56
2	6.0	6.0	0.00
3	3.6	3.63	0.83
4	6.0	6.04	0.67
5	3.6	3.60	0.00
6	72	68.4	5.00
7	72	73.4	1.94
8	3.6	3.63	0.83
9	3.6	3.55	1.39
10	3.6	3.57	0.83
11	72	73.6	2.22
12	72	70.0	2.78

3.2. A simple Actual HN

To further test the effectiveness of the proposed method, a simple actual HN, located in Harbin City, China, is employed here. This HN represents a primary network that connects heat sources and heat users (heating substations). The simple actual HN is shown in Figure 3.

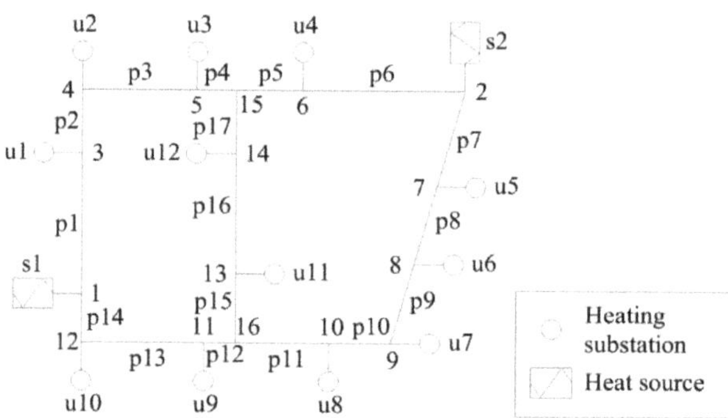

Figure 3. Schematic diagram of a simple actual HN.

As shown in Figure 3, this simple actual HN contains 17 pipes and 16 nodes (including a reference node, s1). In this HN, there are 14 nodes representing 2 heat sources and 12 heating substations, and pipe 10 and pipe 15 are selected as cotree pipes. All node elevations are the same at 12 m. In this heating system, there are two circulating pumps operating in parallel in both heat sources. Performance parameters of circulating pumps are shown in Table 2.

Table 2. Performance parameters of circulating pumps.

Heat Source	Pump Series	Flow (m^3/h)	Head (m)	Revolution (r/min)	Power (kW)
s1	KQSN/L350-M6	756	148	1480	710
s2	KQSN/L300-N4	563	96	1480	250

Setting values of all PFPs in the simple actual HN are shown in Table 3.

In this paper, multiple hydraulic conditions can be achieved by adjusting the valve in every heating substation. However, the valves of pipes in both supply and return pipelines are not adjusted, which can keep PFPs unchanged under different hydraulic conditions. Based on previous study results, when there are measured data of two hydraulic conditions available for PFPs estimation of the simple actual HN shown in Figure 3, unique estimation results of PFPs could be obtained. In this estimation process, node heads and outflows under these two hydraulic conditions are considered "known values", which are listed in Table 4.

Table 3. Details of pipes in the simple actual HN.

Pipe No.	From Node	To Node	Length (m)	Diameter (mm)	PFP (h^2/m^5) × 10^{-5}
1	1	3	500	300	3.01
2	3	4	400	300	2.41
3	4	5	800	250	12.5
4	5	15	400	250	9.4
5	6	15	400	250	9.4
6	2	6	1000	300	4.81
7	2	7	400	350	1.07
8	7	8	300	300	1.8
9	8	9	300	250	4.7
10	10	9	500	200	25.3
11	16	10	500	250	7.83
12	11	16	500	200	25.3
13	12	11	700	250	11
14	1	12	700	300	4.21
15	13	16	700	200	35.4
16	14	13	400	250	6.27
17	15	14	400	300	2.41

Table 4. "Known values" for the simple actual HN.

Node No.	Node Outflows (m^3/h)		Node Heads (m)	
	The First HC [1]	The Second HC	The First HC	The Second HC
1	1055	700	200	180
2	900	552.36	173.10	170.09
3	112	56	193.01	177.36
4	46	32.2	189.71	175.97
5	68	61.2	176.54	170.54
6	211	84.4	170.42	168.96
7	256	163.84	168.38	168.39
8	167	56.78	165.38	167.39
9	172	92.88	162.65	165.89
10	321	250.38	161.44	164.04
11	79	26.86	177.23	170.44
12	287	246.82	186.18	173.14
13	100	79	167.14	166.59
14	136	102	168.46	167.40
15	0	0	170.37	168.52
16	0	0	166.42	166.16

[1] HC—hydraulic condition.

Details of the PFPs estimation processes are almost the same as those in Section 3.1. The estimation processes are not repeated here. The estimation results are presented in Figure 4.

As shown in Figure 4, estimation results of PFPs are close to those setting values shown in Table 3. As a cotree pipe, the relative estimation deviation of pipe 15 (grey mark in Figure 4) is 11.40% (the largest one). As the estimation result of pipe 15 cannot be reached by direct calculation, the relative estimation deviation of pipe 15 is larger than those of the rest of the pipes. This result seems not particularly ideal. If more accurate results are expected, this estimation method theoretically should employ operational data of more hydraulic conditions. In addition, pipe 10 (grey mark in Figure 4) is another cotree pipe with a relatively larger estimation deviation, whose relative estimation deviation is 3.05%.

(Relative estimation deviations of tree pipes are smaller than those of cotree pipes.) The average estimation deviation of every PFP is 1.86%, which illustrates that the estimation results of PFPs are accurate enough for actual engineering. The overall accuracy of those estimation results of the simple actual HN is close to those of the example HN introduced in Section 3.1. (As a matter of fact, relative deviations of PFPs estimation results could be reduced further by adding operation data of more hydraulic conditions or choosing different tree and cotree pipes division plans.)

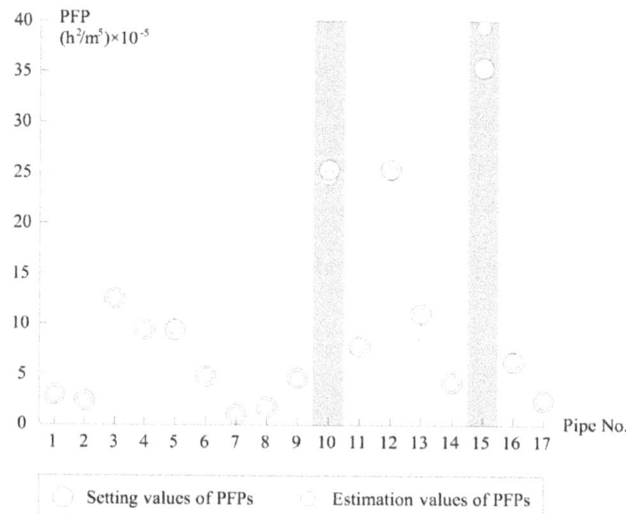

Figure 4. Estimation results of PFPs in the simple actual HN.

4. Discussion

The two HNs mentioned above show the unique values of PFPs cannot be obtained unless measured data of node heads and outflows under multiple hydraulic conditions are available. When hydraulic conditions are sufficient, the number of independent equations may reach the number of unknown variables. In the estimation processes, using the least number of hydraulic conditions to identify unique values of PFPs can relieve the difficulty of providing more hydraulic conditions. During the estimation processes, estimation results of tree PFPs calculated directly may be more accurate. Both estimation results of tree and cotree are accurate enough in engineering.

However, there are also two major limitations in this paper:

1. In actual engineering, PFP estimation is inevitably affected by measurement errors. The estimation principle herein is similar to that of the publication [16], and the unique estimation results can be obtained theoretically. According to the research results of Liu et al. [16], the influence of measurement errors on this series of methods is also limited. Estimation results can satisfy the requirements of engineering applications. Thus, error analysis was not included here.
2. In this paper, a large number of measurement sites are needed for obtaining operational data on HNs. Currently, only high monitoring levels or newly built heating systems could meet the above requirement. In order to achieve PFPs estimation of heating systems with lower monitoring levels or experiencing temporary failure in individual sensors, PFP estimation based on deep neural networks is worthy of being explored in the future.

5. Conclusions

In this paper, PFPs, representing the resistance of an HN, are introduced. Then, a framework is proposed to identify PFPs values by measured data of nodes heads, but not pipe flows. The estimation processes can be achieved by solving matrix equation, which makes a theoretical basis for obtaining accurate resistance models for actual HNs.

In the estimation processes, mass and energy conservation equations were divided by a tree and a cotree. By eliminating dependent unknown variables and expressing relationships between pipe flows and head losses under different hydraulic conditions, a non-linear PFPs calculation equation was developed. This equation is then transformed into linear algebraic equations, providing the probability of further analysis. From studying these linear algebraic equations under multiple hydraulic conditions, the numbers of independent equations and unknown variables are determined, which can be used as a tool to judge a PFPs estimation problem. Estimation values of PFPs can be accomplished by rapidly solving the matrix equations. Utilizing operational data under multiple hydraulic conditions is highly advantageous to PFPs estimation because the number of independent algebraic equations in the PFPs calculation matrix equation increases with the addition of hydraulic conditions. If hydraulic conditions are sufficient, unique estimation results of PFPs can be achieved, and no more hydraulic conditions need to be provided.

This method can find unique results of PFPs directly by solving matrix equations under the least number of hydraulic conditions (usually from two to three hydraulic conditions), being essentially different from previous estimation or calibration methods. The PFPs estimation method mentioned in this paper improves the overall regulation and control performance of heating systems and energy efficiency in buildings.

Author Contributions: Conceptualization, Y.L.; formal analysis, Y.L.; methodology, Q.D. and Y.L.; supervision, P.L.; investigation, Z.Z.; writing—original draft preparation, Q.D. and H.W.; writing—review and editing, H.W. and Y.L.; funding acquisition, P.L. and Y.L. All authors have read and agreed to the published version of the manuscript.

Funding: This research was funded by the National Natural Science Foundation of China, grant number 52078156; and the Financial Assistance under Heilongjiang Postdoctoral Fund, grant number LBH-Z15083.

Institutional Review Board Statement: Not applicable.

Informed Consent Statement: Not applicable.

Conflicts of Interest: The authors declare no conflict of interest.

References

1. Zhang, Y.; Zheng, W.; Fang, H.; Xia, J. Clean heating in Northern China: Regional investigations and roadmap studies for urban area towards 2050. *J. Clean. Prod.* **2022**, *334*, 130233. [CrossRef]
2. Shen, Y.; Chen, J.; Fu, Q.; Wu, H.; Wang, Y.; Lu, Y. Detection of district heating pipe network leakage fault using UCB arm selection method. *Buildings* **2021**, *11*, 275. [CrossRef]
3. Wang, N.; You, S.; Wang, Y. Hydraulic resistance identification and optimal pressure control of district heating network. *Energy Build.* **2018**, *170*, 83–94. [CrossRef]
4. Nash, G.A.; Karney, B.W. Efficient inverse transient analysis in series pipe systems. *J. Hydraul. Eng.* **1999**, *125*, 761–764. [CrossRef]
5. Reddy, P.V.; Sridharan, K.; Rao, P.V. WLS method for parameter estimation in water distribution networks. *J. Water Resour. Plan. Manag.* **1996**, *122*, 157–164. [CrossRef]
6. Savic, D.A.; Walters, G.A. *Genetic Algorithm Techniques for Calibrating Network Models*; University of Exeter: Devon, UK, 1995; p. 44.
7. Meirelles, G.; Manzi, D.; Brentan, B.; Goulart, T.; Luvizotto, E., Jr. Calibration model for water distribution network using pressures estimated by artificial neural networks. *Water Resour. Manage.* **2017**, *31*, 4339–4351. [CrossRef]
8. Shamloo, H.; Haghighi, A. Optimum leak detection and calibration of pipe networks by inverse transient analysis. *J. Hydraul. Res.* **2010**, *48*, 371–376. [CrossRef]
9. Lansey, K.; Basnet, C. Parameter estimation for water distribution networks. *J. Water Resour. Plan. Manag.* **1991**, *117*, 126–144. [CrossRef]
10. Jun, S.; Arbesser-Rastburg, G.; Fuchs-Hanusch, D.; Lansey, K. Response surfaces for water distribution system pipe roughness calibration. *J. Water Resour. Plan. Manag.* **2022**, *148*, 04021105. [CrossRef]
11. Datta, R.S.; Sridharan, K. Parameter estimation in water-distribution systems by least squares. *J. Water Resour. Plan. Manag.* **1994**, *120*, 405–422. [CrossRef]
12. Kapelan, Z.S.; Savic, D.A.; Walters, G.A. A hybrid inverse transient model for leakage detection and roughness calibration in pipe networks. *J. Hydraul. Eng.* **2003**, *41*, 481–492. [CrossRef]

13. Liu, Y.; Zou, P.; He, Z. Pipe friction parameters identification method based on Moore-Penrose Pseudo-inverse solution. *J. Hydraul. Eng.* **2012**, *138*, 100–104.
14. Liu, Y.; Du, Q.; Luo, P.; Zou, P.; He, Z. Pipe resistance coefficients identification of water networks considering solvable conditions. *Can. J. Civ. Eng.* **2021**, *48*, 1223–1230. [CrossRef]
15. Gu, J.; Wang, J.; Qi, C.; Yu, X.; Sunden, B. Analysis of a hybrid control scheme in the district heating system with distributed variable speed pumps. *Sustain. Cities Soc.* **2019**, *48*, 101591. [CrossRef]
16. Liu, Y.; Wang, P.; Luo, P. Pipe hydraulic resistances identification of district heating networks based on matrix analysis. *Energies* **2020**, *13*, 3007. [CrossRef]
17. Tol, H.I. Development of a physical hydraulic modeling tool for district heating systems. *Energy Build.* **2021**, *253*, 111512. [CrossRef]
18. Zheng, X.; Sun, Q.; Wang, Y.; Zheng, L.; Gao, X.; You, S.; Zhang, H.; Shi, K. Thermo-hydraulic coupled simulation and analysis of a real large scale complex district heating network in Tianjin. *Energy* **2021**, *236*, 121389. [CrossRef]
19. Tanyimboh, T.; Burd, R.; Burrows, R.; Tabesh, M. Modelling and reliability analysis of water distribution systems. *Water Sci. Technol.* **1999**, *39*, 249–255. [CrossRef]
20. Covelli, C.; Cozzolino, L.; Cimorelli, L.; Della Morte, R.; Pianese, D. A model to simulate leakage through joints in water distribution systems. *Water Sci. Tech.-Water Sup.* **2015**, *15*, 852–863. [CrossRef]
21. Rahal, H. A cotree flows formulation for steady state in water distribution networks. *Adv. Eng. Softw.* **1995**, *22*, 169–178. [CrossRef]
22. Chen, Y.; Feng, J. Efficient method for Moore-Penrose inverse problems involving symmetric structures based on Group Theory. *J. Comput. Civil Eng.* **2014**, *28*, 182–190. [CrossRef]

MDPI
St. Alban-Anlage 66
4052 Basel
Switzerland
www.mdpi.com

Buildings Editorial Office
E-mail: buildings@mdpi.com
www.mdpi.com/journal/buildings

www.ingramcontent.com/pod-product-compliance
Lightning Source LLC
Chambersburg PA
CBHW040839250926
51850CB00007BA/1359